THE DRAGONFLIES
OF EUROPE

THE DRAGONFLIES
OF
EUROPE

(SECOND EDITION)

R. R. Askew BSc, DPhil, FRES

with colour illustrations
and text figures by the author

HARLEY
BOOKS

Harley Books (B. H. & A. Harley Ltd),
Martins, Great Horkesley,
Colchester, Essex CO6 4AH, England

Text set in Linotron 202 Plantin by
Saxon Printing Ltd, Derby, and
Rowland Phototypesetting Ltd,
Bury St Edmunds, Suffolk.
Page make-up by Rowland Phototypesetting Ltd

Colour originated by Adroit Photo Litho Ltd,
Birmingham, Warwickshire, and
Hilo Colour Printers Ltd, Colchester, Essex

Printed and bound by Midas Printing Ltd, China,
in association with Compass Press Ltd, London

The Dragonflies of Europe
Revised edition, 2004
© R. R. Askew

British Library Cataloguing-in-Publication Data available

ISBN 0 946589 75 5

Contents

Foreword

by the Past-President of the British Dragonfly Society and the
Worldwide Dragonfly Association

The popularity and profile of dragonflies continues to increase, due in no small measure to the excellence of reference books, such as this one, that are available to facilitate their study. In keeping with the popularity of dragonflies, both as objects of arresting beauty and as indicators of habitat richness, increasing numbers of research reports have been published documenting dragonflies' behaviour, ecology and taxonomy. The vigour of this field of enquiry is manifest in the existence of two scholarly journals devoted to odonatology: *Odonatologica* and the *International Journal of Odonatology*, as well as journals and newsletters of national societies.

It is therefore timely that a standard reference such as this book should appear in an updated second edition, encouraging its users to feel that they can be abreast of recent developments. Thus, in this second edition we find mention of several recent milestones in the history of British and European odonatology: the first arrival in Europe of a dragonfly that had successfully crossed the Atlantic Ocean, apparently unaided except by wind; the arrival in Britain from continental Europe of *Erythromma viridulum*; the noteworthy demonstration that *Ischnura hastata* maintains parthenogenetic populations on the Azores; and the discovery of an undescribed species of *Somatochlora* in southern Bulgaria.

The second edition of this fine book appears against a vigorous backcloth of odonatology that is steadily adding to our knowledge in the spheres of science and nature conservation. I confidently predict that it will continue to contribute importantly to the progress of odonatology and I welcome this opportunity to wish it continued success.

Crean Mill, St Buryan, Cornwall   Philip S. Corbet
May 2003

Preface to First Edition

In writing this book my aim is to simplify the identification of European dragonflies and thereby to promote interest in an intriguing order of insects. This is the order Odonata which includes two distinct groups, the dragonflies proper and the damselflies; the term dragonfly used in a general sense encompasses both groups.

Dragonflies are intrinsically attractive insects. They are fine subjects for photography and are large enough to observe individually in the field, often for long periods, as many birdwatchers have discovered. There have been several interesting studies of dragonfly behaviour which have contributed substantially to ecological, ethological and neurophysiological research.

Of very limited economic and medical importance, and completely harmless to man, dragonflies are an irreplaceable adornment to our world. Sadly, after enjoying a long and successful history, there are indications that the dragonflies of today, especially those in the more developed parts of the world, are facing increasing threats to their survival. Three species have disappeared from England this century, two since 1950, and the areas occupied by several others have contracted in the face of loss of habitat due to pollution from industry, application of herbicides, and run-off of fertilizers, and to water-level management, drainage and clearance of emergent and waterside vegetation. There is a similar story from the Netherlands where nine species have not been observed since 1950 (Geijskes & van Tol, 1983), from Switzerland where 45 of the 73 breeding species are adjudged local, endangered or extinct (Wildermuth, 1981) and from West Germany where only 22 of the 71 breeding species are thought to have a reasonably secure status (Schmidt, Eb., 1977). Most at risk are species in northern Europe confined to moorlands or rivers. Encouragingly, a scheme to study and identify the precise biotopes of endangered dragonfly species as indicators of the extent of deterioration of aquatic ecosystems began in 1984 under the auspices of the Council of Europe. A greater awareness and understanding of dragonflies is an essential prerequisite for their conservation. It is hoped that this book will encourage hitherto casual observers of dragonflies to take a closer look and become more deeply acquainted with a truly fascinating group of insects.

The first part of this book is devoted to outlining the general biology and morphology of dragonflies, and this is followed by an account of all species that are known to occur in Europe. The verbal descriptions, illustrations and identification keys should ensure the correct naming of any species found; several are so distinctive that, with practice, reference to the colour plates alone should be sufficient for identification.

This is the first book written originally in English to be devoted to the entire European fauna, the only other works covering this ground being by Sélys Longchamps, *Monographie des Libellulidées d'Europe*, and Sélys Longchamps and Hagen, *Revue des Odonates ou Libellules d'Europe*, which were published in 1840 and 1850 respectively and are now, although of great interest, naturally dated and almost unobtainable. Another very useful book in French, *Guide des Libellules d'Europe et d'Afrique du Nord*, by Aguilar, Dommanget and Préchac, was published in 1985 and an English version appeared in 1986. Schmidt, Er. (1929) deals with the central European fauna in the section on Odonata in *Die Tierwelt Mitteleuropas* and Robert (1958) also concentrates upon this fauna in *Les Libellules (Odonates)*. Robert's book includes a wealth of biological observations and is very beautifully illustrated, but it is not intended to be an identification guide. Five other valuable books cover sections of the European fauna: Aguesse (1968), *Les Odonates de l'Europe Occidentale, du Nord de l'Afrique et des Iles Atlantiques*; Conci and Nielsen (1956), *Fauna d'Italia, Odonata*; Geijskes and van Tol (1983), *De libellen van*

7

Nederland (Odonata); Gibbons (1986), *Dragonflies and Damselflies of Britain and Northern Europe*; and Hammond (1977), *The Dragonflies of Great Britain and Ireland*. Books by Jurzitza (1978), *Unsere Libellen*, and by McGeeney (1986), *A Complete Guide to British Dragonflies*, contain splendid colour photographs of the living insects. Other works on national or regional dragonfly faunas will be found in the list of references at the end of the volume and in the introduction to the distribution maps. Keys for the identification of final-stage larvae are to be found in Gardner (1954b), Aguesse (1968), Hammond (1977) and Franke (1979), and are included here, in spite of imperfections arising from our inadequate knowledge, as a final chapter. Dragonfly biology is discussed by Tillyard (1917), *The Biology of Dragonflies (Odonata or Paraneuroptera)*, and later by Corbet (1962), *A Biology of Dragonflies*. These both give very detailed accounts of the subject, and another excellent book is *Dragonflies* by Corbet, Longfield and Moore (1960).

A remarkable resurgence of interest in dragonflies in the last few decades culminated in the foundation of the Societas Internationalis Odonatologica in 1971. S.I.O. has done much towards promoting both the conservation and scientific study of Odonata, in particular through publication of its journals *Odonatologica* (since 1972) (including *Odonatological Abstracts*, which provides very comprehensive coverage of dragonfly literature) and *Notulae Odonatologicae* (since 1978), and the organization of a biennial symposium. The last four symposia, the sixth to ninth, were convened in Switzerland (1981), Canada (1983), France (1985) and India (1988). The address of S.I.O. is S.I.O. Central Office, P.O. Box 256, 3720 AG Bilthoven, The Netherlands. In addition, several countries have national odonatological societies which are affiliated to S.I.O.

This book has taken about nine years to complete and during this time I have been helped in many ways by many people. I am much indebted to those odonatologists who have generously provided me with copies of their papers, to R. H. R. Abbott, M. J. Cooper, W. Ghayeb, M. A. Kirby, A. Quayle, N. E. Stork and D. W. Yalden who have given me specimens, and to R. Andrew, A. Sandhall and D. Sumner who have loaned photographs of living dragonflies used in the preparation of the illustrations. L. Lockey has given much assistance in producing the photographs of dragonfly wings. Ready access to K. J. Morton's fine collection of Odonata in the Royal Museum of Scotland, Edinburgh, has been invaluable, and for this I am most grateful to Mark Shaw and Rodger Waterston. Jean and Carole de Conti and Mike and Mary Copper have kindly allowed me to collect on their land in France, and Georgina Bryan has translated from Dutch into English for me. I would like to thank Philip Corbet, Bastiaan Kiauta, Mike Parr, Eberhardt Schmidt, and my colleagues at the University of Manchester for their encouragement and interest. Suggestions by Rodger Waterston and Philip Corbet for improving the text are much appreciated, and I am very grateful also to Basil and Annette Harley for their hard work and consideration in preparing the book for publication. Finally, my deepest gratitude to my family: to my wife Letitia for her constant support, encouragement and help, to David whose interest and enthusiasm stimulated the project in the first place and who has been of great assistance in the collection of specimens, to John who has so willingly helped with translation, and to my daughter Margaret for her help in the field and in literature searching.

Hale, Altrincham, Cheshire   R. R. ASKEW

December, 1987

Preface to Second Edition

During the fifteen years since publication of *The Dragonflies of Europe*, there has been a quite remarkable increase of awareness and interest in Odonata, and knowledge of the European species has correspondingly expanded. It was not possible to incorporate post-1988 findings in the body of this edition, but instead I have summarized some of the important recent advances in European odonatology, concentrating first upon species newly discovered in Europe and how they may be recognized, and then upon apparent changes in distribution of the European Odonata. These are contained in the Supplement which is printed on pp. 213–222 with additional references on pp. 234–238. New information on habitat requirements of larvae and adults, and on identification characters of larvae and exuviae, is not included. The now considerable volume of work devoted to dragonflies known in Europe only as fossils is also omitted.

Since 1988 there have also been several changes to the political map of Europe. East and West Germany no longer exist as separate nations, whereas the former Czechoslovakia, Yugoslavia and USSR have become divided into a number of independent countries. National boundaries as currently defined are recognized in the Supplement to this second edition.

In preparing this revision I have received much help from Javier Blasco-Zumeta, Arturo Compte-Sart, Basil and Annette Harley, Lesley Lockey, Mark Shaw and my wife, Letitia. Milen Marinov very kindly provided details and photographs of *Somatochlora borisi*, and Martin Schorr made several very helpful suggestions. The colour photograph used on the front cover of this edition, illustrating this new species, has been generously loaned by Burkhard Grebe. The digital photographs used in the preparation of the new Plate 30 were skilfully taken by Ian Miller. The excellent abstracting service of the Societas Internationalis Odonatologica provides *Odonatological Abstracts* published in the journal *Odonatologica*, and these have been invaluable in writing this supplement.

Beeston, Tarporley, Cheshire
December, 2003

R. R. ASKEW

I. INTRODUCTION

Dragonflies are remarkable insects in many ways. So distinctive in appearance that they cannot be mistaken for any other type of insect, and quite unique in many aspects of their biology, they stand isolated in the present world fauna. They are survivors; originating some three hundred million years ago when the plants we now burn as coal provided them with perches, they have persisted through the ages with relatively little change in their basic structure. With about 4875 known species (which may be compared to 4000 mammals), dragonflies have withstood the test of time. Davies & Tobin (1984, 1985) list 2362 described species of Zygoptera and 2513 described species of Anisoptera, estimating a final total somewhat in excess of 5000. A unique feature of the group is that they are generalized predators both as larvae and adults, to a large extent opportunistic in the prey they take, and are so well adapted to this ecological role that no other group of animals has threatened to out-compete and displace them. Of course, their long and continuing history has demanded specialization and the dragonfly of today represents an amalgam of specialized and archaic characters. An efficient basic body design that has proved sufficiently adaptable to ensure survival during environmental change, and the adoption of a highly individual mode of life, are the keys to dragonfly success.

The larval dragonfly is aquatic, feeding on water animals of suitable size. It is thought, however, that the aquatic existence is secondary and that the larvae of ancestral dragonflies were terrestrial, or at least semiterrestrial, perhaps living amongst damp moss and other ground vegetation in the humid conditions of the late Carboniferous period. A few species living today, mainly on oceanic islands of the southern hemisphere, have terrestrial larvae, but these have evolved from species with aquatic larvae and are not primitively terrestrial.

The adult dragonfly during evolution has lived in only one medium. It has always been an aerial creature, preying on flying insects or on those crawling over plants. The major requisite of such an existence is skill in flight coupled with a highly developed visual sense.

Wings of dragonflies are very fully provided with veins which form a network of small cells, a primitive feature, and the wings are entirely membranous lacking both the thickening and hardening seen in grasshoppers, bugs, beetles and other less aerial insects, and also the vestiture of small hairs or scales found on the wings of more advanced insects. Of all the living types of winged insect, only dragonflies and mayflies represent the basic palaeopterous state. That is, they are unable to fold their wings flat along the dorsal surface of the body when at rest. This prevents dragonflies from crawling into small spaces and thereby enjoying an attribute that is important to the success of neopterous insects. It is possible that dragonflies and mayflies evolved wings independently both of each other and of other winged insects, but a more widely held view is that winged (pterygote) insects are monophyletic and that the Neoptera evolved from a palaeopterous ancestor. Sister-group relationships between Ephemeroptera (mayflies) and Odonata plus Neoptera, or alternatively, between Palaeoptera and Neoptera (Riek & Kukalova-Peck, 1984) have been suggested.

All four wings of dragonflies are of fairly similar size, especially so in damselflies in which fore- and hindwings are almost indistinguishable in shape and vein arrangement. This is another primitive feature, dissimilarity between the two pairs of wings occurring in most insects. A further specialization which dragonflies do not have is a coupling device between fore- and hindwings causing them to act as a single functional unit and so preventing the hindwings from beating in air made turbulent by the forewings. Muscles which bring about the down-beat of the wings are directly connected to the wing bases in dragonflies so that the two pairs of wings can beat independently. In *Calopteryx*, the hindwings lead in the downstroke, the forewings lead in the upstroke (Rudolph, 1976). This is not the case in advanced insects in which there is no direct connection between flight muscles and wings and a very rapid rate of synchronized wing beating is effected by distortion of the elastic thorax. Dragonflies cannot achieve a high rate of wing beating, but beating rate and flight speed are not simply related. Large dragonflies may have a beat rate of from 20 to 40 per second and achieve speeds of about 25 to 30 kilometres (16 to 19 miles) per hour whilst, at the other extreme, some midges reach over 1000 wing beats per second but fly at only about 3 kilometres per hour. Dragonflies, despite their basic wing structure and flight muscle system, rival in speed the fastest flying insects such as hawk-moths and horse-flies, and they outclass these in manoeuvrability. The damselflies are much slower than the larger dragonflies but they can hover and turn with precision amongst the herbage in which they often fly. Of all insect orders with the possible exception of the Diptera, the Odonata are the most volatile. Correlated with rapid and versatile flight is the dragonfly's keen vision. Very large compound eyes give a panoramic field of view in which movement can be detected at a range of several metres. By studying the reactions of dragonflies to models, it has been shown that they have also a highly developed ability to discriminate form and colour. The dragonfly's world is a visual one; the other senses are relatively poorly developed and the antennae, the main olfactory organs, are very small.

A small oriental libellulid has a wingspan of little more than two centimetres, but most modern dragonflies are medium-sized to large insects. The largest of them, however, attains only about one-quarter of the wing expanse of some of the fossil forms of the upper Carboniferous and Permian periods. *Meganeura monyi* Brongniart, from the Commentry beds in the Basses Alpes of south-east France laid down at the end of the Carboniferous period 300 million years ago, had an estimated wingspan of 670mm, about the same as that of a kestrel. *Erasipteron bolsoveri* Whalley, 1979, described from a fossil found a thousand metres beneath the surface in Bolsover Colliery, Derbyshire, England in 1978, had a wingspan of 220mm, and *Tupus diluculum* Whalley, 1980, from the same coal-mine, spanned over 500mm. These were among the very first animals to evolve the power of flight, preceding pterodactyls by more than 100 million years and birds by 150 million years. They had the body proportions and wing shape of a large modern dragonfly, with spiny legs and powerful mandibles indicative of a predatory life style, but the wings were without a pterostigma, arculus or node, the prothorax was better developed, and the tarsus of the legs, in specimens in which this has been preserved, had four instead of three segments. These and other differences place the Carboniferous fossil dragonflies in the order Protodonata (Meganisoptera*), now known only from fossils but almost certainly the ancestral group which gave rise to living representatives of the order Odonata. Odonata probably appeared in the lower Permian when Protodonata were still abundant; fossil Zygoptera are known from lower Permian strata and Anisoptera from the Jurassic.

A considerable amount of speculation on the phylogeny of Odonata has failed to produce a generally accepted evolutionary scheme. All theories published so far, based to a large extent on wing venation, have met with objections, and it is apparent that the real relationships within Odonata will be established only after a complete analysis of more data. As an interim measure, however, it may be helpful to present an arrangement of the extant European families of Odonata in a way that illustrates some of their possible relationships (fig. 1). It will be appreciated that this falls far short of a complete phylogeny from the fact that neither extinct nor non-European taxa are included.

Extant Odonata are allocated to three suborders, the Zygoptera (damselflies), the Anisoptera (dragonflies proper) and the Anisozygoptera. The Zygoptera seem to be the most generalized of dragonflies, many of their characters having departed little from the postulated primitive state,

---

*Meganisoptera Martynov, 1932, was erected for *Meganeura* in place of Protodonata Brongniart, 1893, because the latter, as originally defined, included Palaeodictyoptera as well as Meganisoptera.

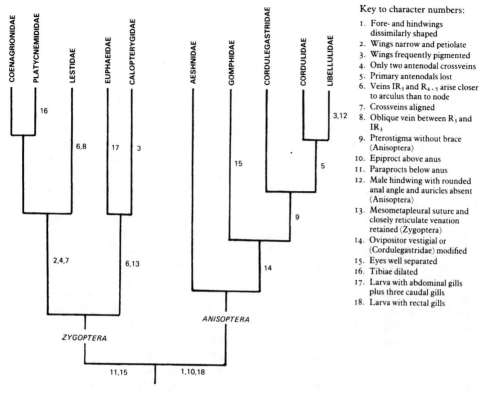

Key to character numbers:

1. Fore- and hindwings dissimilarly shaped
2. Wings narrow and petiolate
3. Wings frequently pigmented
4. Only two antenodal crossveins
5. Primary antenodals lost
6. Veins IR_3 and $R_{4,5}$ arise closer to arculus than to node
7. Crossveins aligned
8. Oblique vein between R_3 and IR_3
9. Pterostigma without brace (Anisoptera)
10. Epiproct above anus
11. Paraprocts below anus
12. Male hindwing with rounded anal angle and auricles absent (Anisoptera)
13. Mesometapleural suture and closely reticulate venation retained (Zygoptera)
14. Ovipositor vestigial or (Cordulegastridae) modified
15. Eyes well separated
16. Tibiae dilated
17. Larva with abdominal gills plus three caudal gills
18. Larva with rectal gills

Figure 1 Diagram to show possible relationships between European families of Odonata

and they probably gave rise to both Anisozygoptera (a relict suborder dating back to at least the Triassic, once widespread but now surviving only in Japan and the Himalayas) and Anisoptera. There are two main types of Zygoptera, the broad-winged damselflies (e.g. *Calopteryx*) and the narrow-winged damselflies (e.g. *Coenagrion*). The broad-winged damselflies are the more generalized. Among the European families of Anisoptera, the Aeshnidae probably had the earliest origin (although a claim could be made for the Gomphidae), and the Libellulidae are the most recent. Consideration of families in this book commences therefore with Calopterygidae and concludes with Libellulidae.

II. LIFE HISTORY

Dragonflies have an incomplete metamorphosis; their life cycle includes egg, larval and adult stages but no pupa. Nearly all dragonflies have aquatic larvae (sometimes called nymphs or naiads) and the female lays her eggs either freely in water or inserted in vegetation growing in or beside water. The latter type of oviposition (endophytic) may be more primitive and is practised by nearly all Zygoptera, Anisozygoptera and, among Anisoptera, by Aeshnidae (Pl. A, fig. 1). A fully-functional, serrated ovipositor enables these species to cut a small slit in plant tissue into which an egg is laid (fig. 2a). Some species select a particular type of plant; *Coenagrion armatum* in Britain was associated with frogbit (*Hydrocharis*), *C. pulchellum* often oviposits in the undersurface of leaves of water-lily (*Nymphaea*; *Nuphar*), and *Aeshna viridis* prefers water-soldier (*Stratiotes*). Eggs may be arranged regularly, in rows in stems or in concentric circles beneath leaves (figs. 2b, c). Sometimes, as in the case of *Lestes viridis* laying its eggs in the bark of willow (*Salix*) branches overhanging water, the injury to the plant caused by oviposition stimulates the formation of a gall. Instances are on record (Kennedy, 1915) of plants being killed after exposure to heavy oviposition. Most endophytic species oviposit in vegetation just below the water surface or in floating plants or debris. Females of some Zygoptera, however, may descend completely beneath the water surface to lay eggs in the submerged parts of plants, although total immersion appears to be largely facultative. *Erythromma najas* has been observed ovipositing at a depth of over half a metre and *Enallagma cyathigerum*, exceptionally, may remain submerged for an hour (Robert, 1958). During immersion these insects respire using a bubble of air trapped between wings and body and encompassing the spiracles; more oxygen is available to the insect than that contained in the original bubble due to diffusion from the surrounding water.

Although avoiding having their eggs displaced to unfavourable locations, endophytic dragonflies attract the attentions of egg parasites. Eggs of Odonata lodged in plant tissue are sometimes destroyed by parasitic Hymenoptera, Zygoptera and Aeshnidae being attacked by species of Mymaridae, Trichogrammatidae, Eulophidae (*Tetrastichus*) and Encyrtidae.

Exophytically ovipositing species lay more spherical eggs than the rather elongated eggs of endophytic species (figs. 2d, e). They lack a specialized ovipositor, normally dipping the tip of the abdomen beneath the water surface whilst in flight so that the eggs are washed off and sink to the bottom (Fraser, 1952); less commonly eggs may be dropped whilst flying low over water, thrust into mud bordering streams (*Cordulegaster*, in which the ovipositor is secondarily developed as a dibble), or dropped on damp ground (some *Sympetrum*). As a rule, exophytic species lay many more eggs, totalling perhaps two thousand or more (Corbet, 1962), than endophytic species, and most use stagnant or very slowly-flowing water so that their free eggs are not washed away. Alternatively, eggs may be laid in gelatinous masses which adhere to submerged plants (fig. 2f) or, in many Gomphidae, the gelatinous material may be restricted to an adhesive disc at the posterior pole of the egg. Inhabitants of very fast-flowing water may even have a long filament at the posterior end of the egg which anchors it by entangling with vegetation.

Dragonflies complete their development in all manner of water, from mountain torrents to temporary ponds, and even, in some subtropical species, saline lagoons. Several European species are tolerant of brackish water and the internal concentration of sodium and chlorine ions can be maintained in almost pure water by active absorbtion through the rectal epithelium (Komnick, 1978). A few dragonflies have terrestrial larvae; a species of *Megalagrion* inhabits

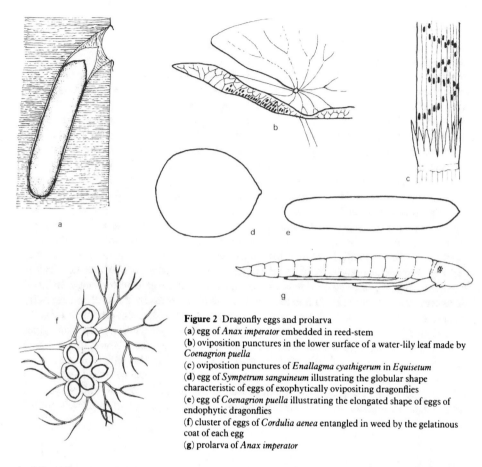

Figure 2 Dragonfly eggs and prolarva
(**a**) egg of *Anax imperator* embedded in reed-stem
(**b**) oviposition punctures in the lower surface of a water-lily leaf made by *Coenagrion puella*
(**c**) oviposition punctures of *Enallagma cyathigerum* in *Equisetum*
(**d**) egg of *Sympetrum sanguineum* illustrating the globular shape characteristic of eggs of exophytically ovipositing dragonflies
(**e**) egg of *Coenagrion puella* illustrating the elongated shape of eggs of endophytic dragonflies
(**f**) cluster of eggs of *Cordulia aenea* entangled in weed by the gelatinous coat of each egg
(**g**) prolarva of *Anax imperator*

leaf-litter in upland forest in Hawaii, and another Zygoptera with a terrestrial larva is known from New Caledonia. Watson (1982) discovered a corduliid larva in rain-forest litter in Queensland. Another interesting larval habitat, again from the warmer parts of the world, is the temporary accumulations of water at the bases of leaves of bromeliads, palms and other plants (called phytotelmata), or in rot-holes in trees, or in broken bamboos. Corbet (1983) records fourteen species from bromeliad leaf axils. Nearly all dragonflies occupying such larval habitats are Zygoptera, some of which (e.g. *Megaloprepus*) have exceptionally long abdomens. In a Bolivian *Mecistogaster* (Pseudostigmatidae), the abdomen is flicked by the hovering female to 'throw' eggs at a small, exposed area of water (Machado & Martinez, 1982).

Eggs hatch in from two (most Zygoptera) to four (most Anisoptera) weeks after being laid, unless they enter diapause. Diapause is a state of suspended development due to physiological changes and it may delay hatching for several months. The British species of *Aeshna* and *Lestes* overwinter as diapausing eggs, diapause being the device whereby the life cycle of the insect is synchronized with its seasonally changing environment and the winter, with temperatures too low for satisfactory feeding and growth, is spent in the resistant egg stage. Development is resumed after diapause has been broken by a period of cold weather and when the ambient temperature has risen sufficiently in spring to permit the metabolic processes involved in growth. Winter may be passed in stages other than the egg but, at least in temperate climates, never as a young larva which is particularly intolerant of cold. Several species overwinter as large larvae, and *Sympecma*, which is exceptional, does so as an adult.

Larval biology

Dragonfly eggs hatch after the completion of embryological development. For those laid above water level, wetting may provide a stimulus to hatching (Corbet, 1962). The young larva bursts open its egg with pulsating expansions of its head. This first-instar larva, the prolarva (fig. 2g), is of very short duration lasting for just a few minutes. Its function is to escape from the egg shell and other investing material, and to reach water. When eggs have been laid in vegetation at the water's edge, this can involve a journey of several centimetres. The appendages of the prolarva are not functional but it is able to move by flexing its body and sometimes, when it does this vigorously, by jumping. Moulting or ecdysis entails a lengthwise splitting of the integument down the middle of the head and top of the thorax, and a transverse split between the eyes. The second-instar larva crawls out of the prolarval integument and rests whilst its system of breathing tubes (tracheae) fill with gas. The respiratory system is closed from the atmosphere, and in the prolarva it is filled with fluid. This fluid is withdrawn from the tracheae in the second stage and replaced by a secretion of gas which soon approximates to the composition of air as a result of gaseous exchange by diffusion between the tracheae and surrounding water.

Oxygen used by dragonfly larvae in respiration is obtained from the water in which they live; it diffuses through the body wall and into the tracheae. Specialized regions, where the cuticle is thin over a relatively large area, have been developed to facilitate gaseous exchange. In larval Anisoptera, the wall of the rectum is thrown into numerous inwardly directed and regularly arranged folds, the rectal gills, each of which is richly supplied with tracheal branches (fig. 3). Water is pumped in and out of the rectum through the anus, and exchange of respiratory gases takes place across the rectal wall between water and tracheae. Zygoptera larvae, in contrast to Anisoptera, have external gills (fig. 4). These are three large, leaf-shaped appendages at the posterior end of the abdomen that contain many tracheal branches and have thin walls across which gaseous exchange occurs. Zygoptera larvae can survive in reasonably well-oxygenated water without their gills, but it is estimated that these supply about half of the larva's normal

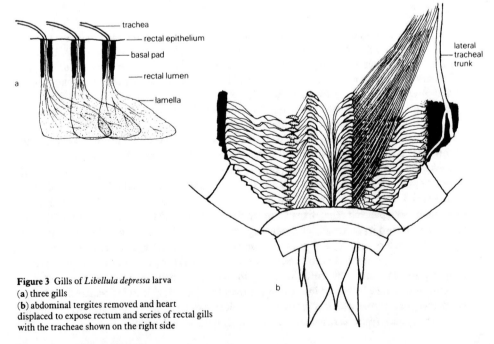

Figure 3 Gills of *Libellula depressa* larva
(a) three gills
(b) abdominal tergites removed and heart displaced to expose rectum and series of rectal gills with the tracheae shown on the right side

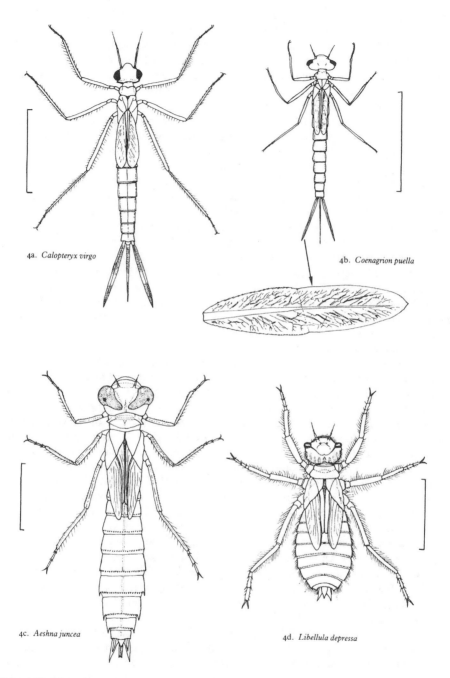

4a. *Calopteryx virgo*

4b. *Coenagrion puella*

4c. *Aeshna juncea*

4d. *Libellula depressa*

Figure 4 Final-instar larvae of Zygoptera (above) and
Anisoptera (below), with an enlarged lateral anal gill of
Coenagrion puella. Scale lines represent 1cm

oxygen requirement, the remainder being obtained through the general body surface and rectal wall. In Euphaeidae, there are paired gills on most of the anterior abdominal segments, in addition to the three caudal gills, and these may be an adaptation to compensate for low oxygen tension when the small streams in which the larva lives partially dry up and become reduced to isolated stagnant pools. Dragonfly larvae will move to the surface or crawl right out of water deficient in oxygen, and most species of Anisoptera are able to survive in air for several days provided that their environment is humid.

Dragonfly larvae, like the adult insects, are carnivores, feeding on a wide range of aquatic organisms. They are not specialist feeders, taking anything of suitable size that is available such as the larvae of midges and mosquitoes, small worms, fish fry and tadpoles. Important stimuli in prey recognition are size and movement; shape and colour are not (Pritchard, 1965). Entomostraca (water fleas, copepods) are the most numerous prey items of *Ischnura elegans* but chironomid larvae provide the greatest biomass. There is no size-selection of prey within the range of animals that can be dealt with and large larvae take both large and small prey whilst small larvae feed only on small prey (Thompson, 1978). Being generalist predators, different species of larval Odonata inhabiting the same water are likely to be potential competitors. It has been shown by Johannsson (1978) that interspecific competition for food between zygopterans is reduced primarily by spatial separation of coexisting species and that this is more important than temporal separation through life cycle displacement. A different conclusion was reached by Carchini & Nicolai (1984) who studied larval diet in coexisting populations of *Lestes barbarus* and *L. virens* and found that diet varied with larval size. Since development of *L. barbarus* is in advance of *L. virens*, available food resources are thought to be partitioned between the two species, enabling their coexistence.

Dragonfly larvae tend not to seek out prey but rather to wait, immobile and concealed in vegetation (Zygoptera, *Anax*, *Sympetrum*, *Leucorrhinia*) or sediment (*Cordulegaster*, Gomphidae, *Libellula*, *Orthetrum*) until a suitable food item comes within range of a forward rush and strike with the mouthparts. Prey is seized by a remarkable and uniquely odonatan adaptation of the lower lip or labium (fig. 5). This is more or less spoon-shaped and armed with a pair of movable hooks which are borne on the labial palps. Contraction of dorsoventral muscles in the abdomen increases internal body pressure and simultaneous relaxation of the labial flexor muscles allows the labium (or mask) to be shot forwards very rapidly (Tanaka & Hisada, 1980). Pritchard (1965) calculates that full extension requires only 15–20 milliseconds. Prey is grasped by the hooks, which are operated by their own muscles, and muscular contraction retracts the labium after an attack to bring the food between the powerful mandibles.

Small larvae detect their prey mainly with their antennae which are supplied with sense organs that are stimulated by water movement. In older larvae, vision plays an increasingly important role in prey detection and the number of units (ommatidia) making up the eyes increases during larval life. An *Aeshna* eye has 170 ommatidia in the first-instar larva and nearly 8000 in the last-instar larva; new ommatidia are added along the front edge of the eye at each moult and they displace older ommatidia posteriorly so that an ommatidium starts by looking forwards and ends by looking backwards (Sherk, 1978a).

The manner in which the labium is used to capture prey requires that the larva must be able to judge distance. A hunting larva will turn its head towards its prey so that the visual axes of corresponding ommatidia in the two eyes intersect on the prey. Pritchard (1965) suggests that as an *Aeshna* larva approaches its prey, the ommatidial angles of the parts of the eyes on which the image falls become smaller until they are at their limit of 1°12′. The prey is then at a distance from the head equal to the extended length of the labium. It is essential for the hunting larva that its prey moves. Larvae do not respond to static prey. The compound eye is an organ that is very sensitive to movement and the flicker threshold (the maximum number of light impressions that can be perceived separately) of a dragonfly larva is about 60 per second (45/s in

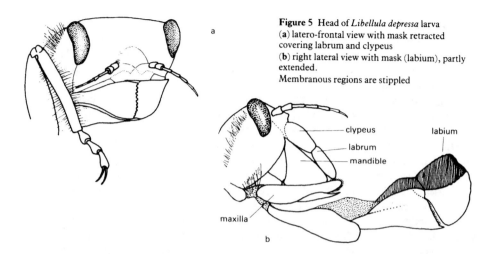

Figure 5 Head of *Libellula depressa* larva
(a) latero-frontal view with mask retracted covering labrum and clypeus
(b) right lateral view with mask (labium), partly extended.
Membranous regions are stippled

man) (Crozier *et al.*, 1937). Larvae of *Cordulegaster* can detect prey at a range of 10–15cm provided that it is moving at an angular velocity of 10°/s or more (Weber & Caillere, 1978).

The dragonfly larva is itself subject to predation, particularly from fish but also from larger dragonfly larvae. The labium, so well adapted for catching food, is not an effective organ of defence. Stabbing movements with the spiny end of the abdomen may be directed at an attacker by anisopteran larvae, but immobility and concealment are the principal protective strategies. Immobility is so important to zygopteran larvae in avoiding predation that they will remain clinging to weed, even when starving, rather than move off in search of prey. Concealment is enhanced in many species by mud particles adhering to the body hairs, and the brown or greenish coloration of the body matches the background. In addition, dragonfly larvae, especially those living amongst vegetation, have a limited capacity to change their colour to blend with their surroundings. Should a larva be detected, further protective devices may be employed. The most spectacular of these is the jet-propelled escape response of Anisoptera larvae; water is rapidly evacuated from the rectum, shooting the larva forwards. Larvae of some species feign death when disturbed, lying completely immobile for several minutes to avoid triggering a movement-stimulated attack by a predator. Zygoptera larvae will shed legs seized by a predator, and most dragonfly larvae have the ability over a few instars to regenerate legs that have been lost.

Larval development

During larval life the dragonfly must grow to a particular size, and the rate at which this is achieved is dependent upon temperature and food supply. A species that spends only a year in the larval stage in southern Europe may require up to three years for its development in higher latitudes. There may even be variation in the larval duration of a species within a single pond. Larval life of the larger Anisoptera in northern Europe may occasionally require five years but this would be exceptional. In Britain, larval life of Zygoptera usually lasts one or two years (sometimes three in *Pyrrhosoma* (Macan, 1974)), whilst most Anisoptera require two or three years, but variation within species and between allied species makes generalization difficult. In the genus *Aeshna* in Britain, for example, *A. juncea* and *A. grandis* take three or more years, *A. cyanea* usually requires two years, and *A. mixta* always develops in one year. In southern Europe, *Sympetrum fonscolombei* and *Crocothemis erythraea* are known to pass through two generations in some summers, and several Zygoptera are similarly bivoltine. *Lestes* has an exceptionally rapid larval development, an adaptation to life in temporary ponds, and in *L.*

sponsa this is completed, even in Britain, in about ten weeks. *L. dryas* in southern France has a larval life occupying only forty-five days. In species developing over two or more years, cohorts of different age-group larvae are present together. The number of larval instars of Odonata is variable but always large, between ten and fifteen (Tillyard, 1917). *Aeshna cyanea* and *Ischnura elegans* both normally have twelve larval instars.

In temperate regions a number of species spend a winter in diapause as final-instar larvae. Such species (*Calopteryx virgo, Pyrrhosoma nymphula, Gomphus vulgatissimus, Anax imperator, Libellula* species) characteristically emerge as adults synchronously the following spring. Sometimes larval diapause is facultative in that not all individuals pass through a diapause state. This is so in British populations of *Anax imperator*, a species thoroughly investigated by Corbet (1957b). In a few advanced individuals that become final-instar larvae before the beginning of June in their second year, there is no diapause and adults emerge later in the same year. Usually, however, the final larval stage is entered after the beginning of June, diapause intervenes, and adults do not emerge until their third year. *A. imperator* larvae are sensitive to increasing day-length and as long as this exceeds two minutes a day they enter a non-diapause path of development. The rate of development after diapause is mainly temperature controlled. Species not diapausing as final-instar larvae are termed 'summer species' (in distinction to the 'spring species') and these (e.g. the larger *Aeshna* species, *Crocothemis erythraea*) tend to have a much more protracted emergence from spring to late summer.

As the final-instar larva approaches the time of adult emergence, it begins to change in appearance and behaviour. The eyes darken and expand, the wing sheaths swell, the thorax becomes more arched and, eventually, the larval labium regresses. The larva frequents shallower water, moving towards the shore or climbing submerged vegetation. Its respiratory rate increases and it partially exposes itself to the air to obtain the extra oxygen required as metamorphosis approaches its climax. Feeding activity stops. These changes, like all insect metamorphoses, are hormone-regulated, the result of a delicate balance between the amount of juvenile hormone (which maintains the larval state) and moulting hormone (which initiates ecdysis) in the blood. Moulting hormone is produced by the ventral gland in the thorax, juvenile hormone by the corpora allata in the brain.

Adult emergence

The metamorphosing larva eventually leaves the water, climbing on to the shore or ascending reeds, tree trunks, or anything else that is available. It is seeking a support to which it can cling securely by its tarsal claws. Larvae may occasionally move ten or more metres from the water before they find a suitable support, but most adults emerge close to water. Different emergence positions on their supports are adopted by larvae of different dragonfly families. Libellulids, corduliids and many gomphids use almost horizontal supports, so that their heads are scarcely raised above their bodies, whereas aeshnids adopt situations in which they can hang at an angle of 90° to 180° to the horizontal with their ventral surfaces uppermost. Zygoptera usually select more or less vertical surfaces to which they cling with their heads upwards. The emergence posture is often similar to the resting posture of the adult insect.

The adult emerges usually between one and three hours after the larva has left the water. A T-shaped slit appears in the back of the larval integument and the adult insect removes all of its body except the abdomen from the investing larval shell (fig. 6). It then rests awhile, in most Anisoptera with the front part of the body dangling and supported only by the still-anchored abdomen. Emergence is completed when the resting support or front of the exuvia is grasped by the adult legs and the abdomen pulled free of the exuvia (Pl. A, fig. 2). Wings are inflated by pressure of body fluid generated by abdominal muscular contraction, a process normally completed in less than thirty minutes. The expanded wings of the adult gradually harden as the insect dries out and loses body-weight preparatory to its maiden flight.

Adult dragonflies may emerge during the night in warm conditions, but in temperate regions emergence is often delayed until early morning when the air temperature has risen sufficiently (Corbet, 1962). Butterfly emergence tends to follow a similar pattern which is, partly at any rate, an adaptation to minimize mortality from diurnal predators, particularly birds. The time of adult emergence is a vulnerable period in a dragonfly's life; in addition to predation, mortality may follow damage sustained in competition with conspecifics for emergence supports.

Figure 6 Stages in the emergence of an adult aeshnid dragonfly

III. THE ADULT DRAGONFLY

The newly-emerged adult dragonfly is sexually immature with a soft cuticle and the bright adult colours not yet formed. Such an insect is said to be teneral and as soon as its wings are sufficiently hard it usually flies away from the water in which it developed as a larva. This maiden flight is often at dawn, provided that the temperature is high enough, and Corbet (1957b) describes a mass maiden flight of *Anax imperator* being preceded by the insects whirring their wings to raise their internal temperature, all individuals eventually flying off before the first birds began to feed. The maiden flight may take the dragonflies only a few metres from their breeding sites, as in some Zygoptera, or it may be extended into a prolonged migration, as in some Anisoptera.

An important function of the maiden flight is to take the immature dragonfly out of the sphere of activity of sexually active males so that it may feed and develop without harassment. Its maturity is signalled by its return to water and this may occur from only two or three days (in *Calopteryx*) to as much as two or three weeks (in some Anisoptera, *Lestes*) after emergence. *Sympecma* does not become reproductively mature until spring of the year following its emergence. The duration of the maturation period is variable, but it tends to be longer in Anisoptera than in Zygoptera and slightly longer in females than in males.

Territoriality

Males, because of their more rapid maturation, usually arrive at breeding sites before their females and each day most male Anisoptera fly from roosting site to breeding area, generally in the forenoon, to establish territories which they defend against intruding males. Males often make several visits a day to the breeding sites, interspersing bouts of territorial behaviour with periods of feeding. A territory-holding dragonfly will approach an intruding dragonfly, either of its own species or sometimes another, usually by flying up to it from below. If it is a conspecific female, the territory-holder will attempt to mate, but otherwise it may strike the intruder with its dorsal surface and the pair will skirmish in flight, each attempting to get below the other so that they spiral downwards. Such skirmishes between large dragonflies are accompanied by a distinctly audible clashing of wings, and the contestants often sustain wing damage. They are usually of short duration and result in one individual, generally the intruder, leaving the area.

The area of defended territory varies approximately with the size of the dragonfly species so that a small pond may be defended by a single male *Anax* or *Aeshna* but its margins may support dozens of Zygoptera. Territory size is correlated both with the distance over which a dragonfly can see movement (Moore, 1953) and the density of males; as male density rises the territories of a species become more closely packed and reduced in area until a certain minimum size is reached after which male density remains constant, the highest steady density (Moore, 1964). The highest steady density is characteristic of a species and is greatest for the smallest species. Moore, working in Dorset in southern England, expressed the highest steady density as the number of males observed over a one-hundred metre stretch of pond margin. The following values were obtained:

Aeshna juncea 1	*Orthetrum coerulescens* 15	*Pyrrhosoma nymphula* 44
A. cyanea 1	*Sympetrum striolatum* 16	*Enallagma cyathigerum* 60
Anax imperator 2	*S. danae* 17	*Lestes sponsa* 65
Cordulia aenea 2	*Libellula quadrimaculata* 28	*Coenagrion puella* 115
Libellula depressa 8	*Ischnura elegans* 40	*Ceriagrion tenellum* 147

To reduce interspecific conflict, the territories of different species may be zoned, both vertically and horizontally. Aeshnids and corduliids hold territories over open water, the former at a height of two metres or so and the latter just above its surface and usually close to the margin. Libellulids often occupy beds of reeds whilst gomphids select areas of sparsely vegetated ground such as shingle spits adjacent to the breeding sites. Most Zygoptera keep to marginal vegetation, but male *Erythromma najas* frequently perch singly on water-lily leaves.

It is implied above that the behaviour of male Odonata at the breeding site is territorial in the sense that a single male defends a certain fixed area for a period of time. This requires qualification. The 'territories' of Zygoptera (except *Calopteryx*) are apparently very transient and difficult to define because males are moving almost continually from place to place. Similarly, in some Anisoptera, males are not territorial in the strictest sense. They are aggressive towards conspecific males but are not attached to a constant site. Such is the case in *Onychogomphus forcipatus* and *Cordulegaster boltoni* (Kaiser, 1974c, 1982). The following papers may be referred to for more detailed discussions of territoriality in Odonata: Moore (1952a, 1964), Jacobs (1955), Kormondy (1961), Johnson (1964), Miller & Miller (1981) and Parr (1983a).

A clear example of territoriality has been studied by Zahner (1960) in *Calopteryx splendens*. Males defend territories about two to three square metres in area which encompass river bank, emergent vegetation and open water. Males court females within their territories. After copulation a couple disengage but the female is guided by the male to an oviposition site, again within the territory. The male stays perched close to his mate whilst she is ovipositing, often submerged, and drives off other males with a flicking of his colourful wings. Male *C. splendens* sometimes maintain the same territory for several consecutive days.

Male dragonflies generally interact by fighting, but in *Calopteryx* (Heymer, 1972) and *Platycnemis* (Heymer, 1966; Aguesse, 1968) physical conflict is replaced by a ritualized aggressive display which, because it is unlikely to cause injury, may be thought of as more efficient. In *Platycnemis*, competing males face each other in flight and display their enlarged and conspicuous tibiae. Aggressive displays and fighting are initiated by the sight of a potential rival, significant visual features being colour of wings (*Calopteryx*) or body (*Orthetrum albistylum*), shape of abdomen (*Cordulia aenea*), and movement of wings and abdomen (*Leucorrhinia*) (Frantsevich & Mokrushov, 1984).

Mating behaviour

It is by sight that a male dragonfly recognizes a mate as well as a rival. Buchholtz (1951, 1955), using models, demonstrated that male *Calopteryx* recognize females of their own species by the colour and transparency of their wings, as well as by their wing and body size. *C. splendens* exists in Europe as a number of geographical races, the females of which differ in the strength of their wing pigmentation and therefore in their wing transparency. Males of each race are most attracted by wings that allow a particular percentage of light to pass through, the percentage that is characteristic of females of their own race. Flight movements by the female may be the visual cue that males of some species use for recognition. This is the case in *Platycnemis pennipes* (Buchholtz, 1956), the males of which respond to a horizontally zig-zagging model provided that the thorax of the model is patterned like that of a real female. The red colour of *Pyrrhosoma nymphula* and the subapical abdominal blue spot of *Ischnura elegans* are signals that release either a sexual or an aggressive response in the males of these species. In *Leucorrhinia dubia* and *L. rubicunda* the flight patterns of the sexes, not their coloration, provide the important specific and sexual recognition signals (Pajunen, 1964a), and likewise in *Aeshna cyanea* it is the female's flight movement at the breeding site that attracts the male (Kaiser, 1974a). That dragonflies are able visually to discriminate between similar species was shown by Loibl (1958) who found that male *Ischnura elegans* responded to female *I. pumilio* with only about one-quarter of the frequency of their response to females of their own species.

Many damselfly species have polymorphic females, some morphs resembling the male in coloration (andromorphs) and others being more or less dissimilar to the male in appearance (heteromorphs). This situation poses species-recognition problems for the male and it is usually explained as a balanced polymorphism. Andromorph females have the advantage of enhanced attractiveness to males but they may suffer from excessive male interference whilst attempting to lay their eggs (Paulson, 1974); alternatively, heteromorphs suffer from more interspecific mating attempts than do andromorphs, but they may benefit from reduced predation because of their more cryptic coloration (Johnson, 1975).

Male dragonflies are usually attracted by flying females but in many species they will also pounce upon settled or ovipositing females. In *Libellula* and *Leucorrhinia*, males usually seize ovipositing females, but in Aeshnidae the down-curved abdomen of an ovipositing female is a deterrent to males and unresponsive females will, in flight, indicate their refusal to mate by a stereotyped down-turn of the abdomen. In Zygoptera, unreceptive females raise the tips of their abdomens and partially open their wings.

Calopteryx and *Platycnemis*, but few others, engage in a courtship display. Here there must be reciprocal recognition between the sexes. When a female *Calopteryx* flies into a male territory, the perched male half opens his wings and raises the tip of his abdomen to display the ventral surfaces of the apical segments which are differently coloured in different species. A receptive female, on seeing this signal, will settle, and the male will flutter around her in a characteristic fashion, always facing her, before attempting to copulate (Buchholtz, 1951). Courtship is reported in very few Anisoptera but Martens (1984) describes a male *Trithemis arteriosa* flying in a figure-of-eight pattern below a hovering female before trying to grasp her.

The manner in which copulation is effected by dragonflies is quite unique and fits nicely their evolutionary position as one of the very first groups of flying insects. The first insects were wingless and they gave rise to the spring-tails (Collembola) and bristle-tails (Thysanura) of today. In neither of these groups is there union between male and female genitalia. Instead, males deposit their sperm, which is enclosed in protein packets called spermatophores, on the ground and the females pick these up in their vulvae. There is no union between the primary genitalia in dragonflies either, and in this respect dragonflies differ from all other winged insects except a few Heteroptera (bugs) which practise hypodermal impregnation. Male dragonflies, instead of depositing spermatophores on the ground, place their sperm in secondary or accessory genitalia on the ventral surface of their second abdominal segment, and it is from this structure that the female is impregnated during copulation. Spermatophores are not produced by modern Odonata, having presumably been lost at some stage in their evolution.

The male grasps the female's thorax with his legs and curls his abdomen forwards so that his posterior abdominal appendages can take hold of the female, either by the prothorax (Zygoptera) (fig. 7d) or by the head (most Anisoptera) (fig. 8e) or by both (Aeshnidae). Denting of female eyes as a result of pressure from the male epiproct is not uncommon in Gomphidae and Aeshnidae (Dunkle, 1979). The parts of the female held by the male's appendages, as well as the appendages themselves, are often specifically modified, possibly to ensure a firm hold and to serve as another reproductive isolating mechanism discouraging union between individuals of different species. Such mechanical isolation is especially important in those species whose females do not differ very much in appearance from those of allied species, as for example in *Lestes* (Loibl, 1958). In *Enallagma* and *Ischnura*, where also the visual recognition system is limited, there is a tactile recognition system whereby females recognize conspecific males by the mechanical stimulation they receive on their mesostigmal plates from the male superior appendages (Robertson & Paterson, 1982).

Once the abdominal appendages have grasped the female, the male releases his leg hold on her thorax, straightens his abdomen, and the pair fly for a time 'in tandem' (figs 7g,9a).

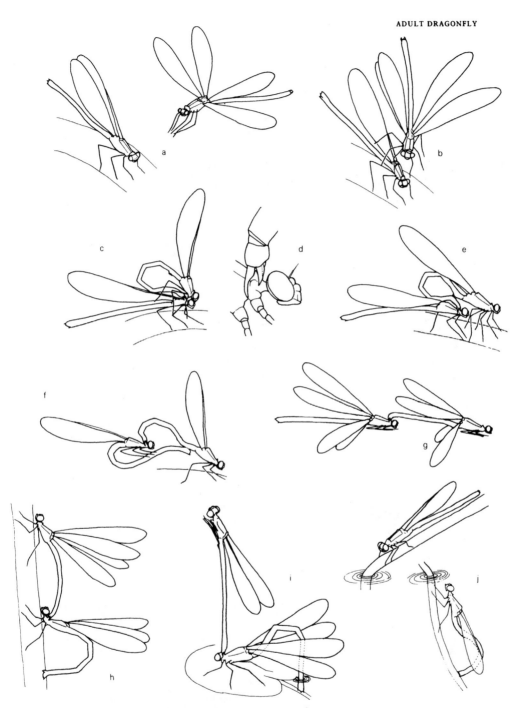

Figure 7 Reproductive and oviposition behaviour in Zygoptera
(a) male approaching settled female (with courtship flight in Calopterygidae)
(b) male flying on to thorax of female
(c) male seizing female pronotum in terminal abdominal appendages
(d) male's grasp on the female's pronotum
(e) male priming accessory genitalia whilst in tandem
(f) copulation
(g) flying in tandem
(h) lestid ovipositing in aerial stem
(i) coenagrionid ovipositing in submerged stem from floating vegetation
(j) coenagrionid ovipositing in submerged stem whilst totally immersed, the male on guard

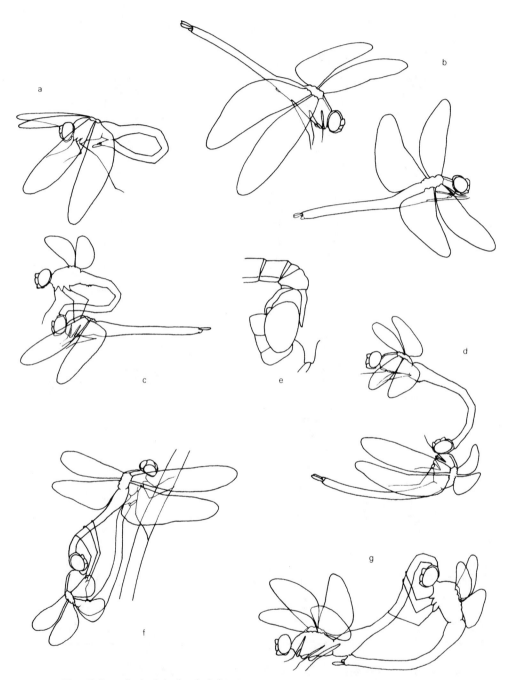

Figure 8 Reproductive behaviour in Anisoptera:
(a) male priming accessory genitalia in flight (see text)
(b) male flying to female as she visits oviposition site
(c) (d) male seizing flying female
(e) *Cordulia aenea* male's grasp on the female's head
(f) copulation on a perch
(g) copulation in flight

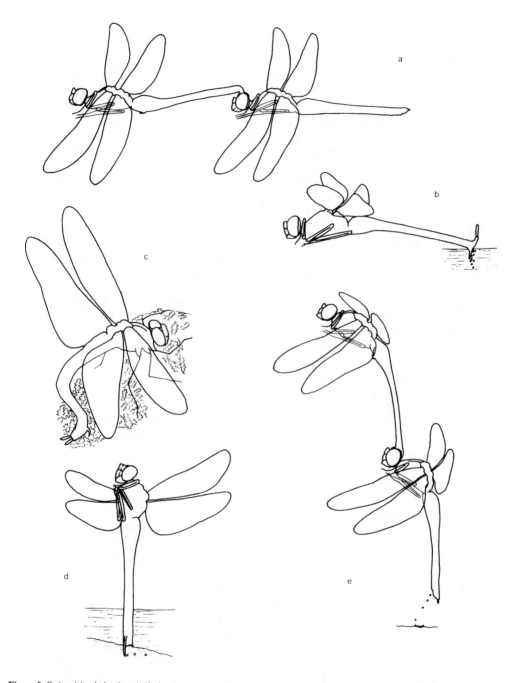

Figure 9 Oviposition behaviour in Anisoptera:
(a) pair of libellulids flying in tandem
(b) female *Somatochlora metallica* ovipositing unaccompanied at water surface
(c) *Aeshna cyanea* ovipositing into moss
(d) *Cordulegaster boltoni* ovipositing into mud
(e) female *Sympetrum sanguineum* in tandem, 'throwing' eggs on to wet ground

Typically, the male primes his accessory genitalia with sperm soon after tandem formation by curving the apex of his abdomen beneath its base (at the same time carrying the female forwards) and bringing the posterior genital opening into contact with the secondary genitalia (fig. 7e). Sperm translocation (intra-male sperm transfer) has also been observed prior to female seizure while the male is alone (fig. 8a) but such behaviour may be anomalous (Utzeri, 1985). In *Enallagma* and *Ischnura* sperm translocation is preceded by the female briefly touching her genitalia against those of her mate (Robertson & Tennessen, 1984). Copulation occurs whilst the pair are settled (Zygoptera) (Pl.A, fig. 3) or in flight (most Anisoptera). It is effected by the female curling her abdomen forwards to establish contact between her primary genitalia and the male's accessory genitalia. This is the 'wheel position' (figs 7f,8g) and when it has been assumed the male's intromittent organ of the accessory genitalia can implant spermatozoa in the female's vulva. Copulation may be encouraged by leg movements of the male, and the small lobes (auricles or oreillets) on the second abdominal segment in males of several Anisoptera have been said to direct the female genitalia into a copulatory position (Fraser, 1943).

Visual and tactile stimuli figure very prominently in dragonfly reproduction; olfactory and auditory signals, so important in many insect groups, do not seem to be employed by dragonflies at all. The visual response by the male to the female, and sometimes *vice versa*, the mechanical coupling in tandem, and specific behavioural patterns, are all conducive to the union of two animals of the same species. Nevertheless, mistakes are not infrequent. Male Odonata may attempt to form tandems with other males, and Moore (1952a) found that males took up the tandem position with tethered males just as frequently as with tethered females. He concludes that the aerial clashes between males are probably based upon sexual rather than aggressive territorial behaviour. Bisexual but heterospecific tandems involving, for example, *Anax* and *Aeshna*, two *Leucorrhinia* or two *Sympetrum* species, *Enallagma* and *Coenagrion*, or even a combination of two families with *Anax* and *Cordulegaster*, have all been observed (Jurzitza, 1966; Bick & Bick, 1981; Heidemann, 1982). Paulson (1974), working in the United States, showed that in four species of *Enallagma* the males were unable visually to distinguish their own females, and it was not until attempted tandem formation between heterospecifics that the mating attempt usually broke down due to the inability of the male abdominal appendages to secure a firm grasp of the female prothorax. In one pairing, however, individuals of different species were regularly successful in establishing tandems. In these instances of heterospecific tandems, the only remaining barrier to cross-insemination is the act of copulation, and in this the onus passes from the male to the female.

In most species of Anisoptera the male releases the female after copulation, but in several species of *Sympetrum*, *Aeshna affinis* and usually *Anax parthenope* the tandem position continues whilst the female lays eggs (fig. 9e). Most Zygoptera also oviposit in tandem, although probably not obligatorily and never in *Calopteryx* and *Ischnura* (Schmidt, Eb., 1965, and fig. 7j). In many coenagrionids the male 'stands' on the tip of his abdomen, supported only by his grip on the prothorax of his mate, whilst the female immerses her abdomen to lay eggs in plants beneath the water surface (fig. 7i). In *Cercion lindeni*, the female may submerge completely during oviposition, causing the male to release his grasp, but he generally awaits her re-emergence from the water, perching on a nearby plant, until the tandem position can once more be assumed. In other Zygoptera, both members of the tandem may be submerged whilst the female is ovipositing. Couples of *Lestes sponsa* may remain underwater for about thirty minutes. Variations in oviposition behaviour and the extent of male involvement are shown in Table 1.

The tandem coupling of dragonflies permits both partners to fly actively, an advantage over the tail to tail union of, for example, butterflies, in which one partner must remain passive if the pair take flight. In those libellulids that oviposit in tandem, the couple fly low over the water surface and the male, by lowering his abdomen, dips the end of his partner's abdomen in the water so that her eggs are washed off. Even among Odonata that separate after copula-

Table 1 Categories of oviposition behaviour in European taxa of Odonata (after Schmidt, Eb., 1965)

	Oviposition whilst standing on substrate: ENDOPHYTIC	Oviposition in flight: EXOPHYTIC
Oviposition in tandem	Lestidae Platycnemididae Coenagrionidae usually *Anax parthenope* usually, sometimes female alone *Aeshna affinis* usually, sometimes female alone	*Sympetrum* usually, sometimes female alone
Oviposition with pair separated, male guarding	*Calopteryx*, in male territory *Cercion lindeni*	some *Libellula* some *Orthetrum* *Leucorrhinia* *Brachythemis*
Oviposition unaccompanied	*Anax imperator* *Aeshna* (except *A. affinis*) Other Aeshnidae *Ischnura*	Gomphidae Cordulegastridae Corduliidae

tion, such as *Libellula fulva*, *Orthetrum brunneum*, *O. coerulescens*, *Leucorrhinia* and *Calopteryx*, the male keeps guard and drives off other males whilst the female oviposits. In *Calopteryx maculata*, an American species, guarded mates enjoy about twelve to fifteen minutes of undisturbed oviposition compared with unguarded females which average only one or two minutes of oviposition before they are interrupted by males (Waage, 1979b). In species in which females oviposit alone and unguarded, egg-laying generally takes place after the peak of male sexual activity and in hidden situations.

The male's prolonged association with the female reduces the possibility of her being inseminated by another male. Waage (1979a,b) showed that male *Calopteryx maculata* can poke out sperm deposited in the female at an earlier mating, using the intromittent organ or penis as a probe. Between 88 and 100 per cent of sperm stored by a female from a previous mating may be removed by a male before he transfers his own sperm. Sperm do not fertilize the eggs until these are about to be deposited. Sperm displacement has been demonstrated in a number of Zygoptera following Waage's initial discovery, and in some Libellulidae the barbed flagellum on the penis is used to scoop sperm from the spermathecae. Such behaviour puts selection pressure on a male to ensure that another male does not gain access to a female that he has inseminated. Maintaining the tandem position (contact guarding), and also non-contact guarding, are effective means of achieving this end and maximizing a male's reproductive success when opportunities for multiple mating are restricted (Alcock, 1979).

Because he produces millions of spermatozoa, a male's best strategy in the absence of sperm displacement would be to inseminate as many females as possible, spending as short a time as possible with each. Peter Miller has shown copulation in *Crocothemis erythraea* to be brief (10–15 seconds) and females are inseminated by a succession of males. There is no evidence of sperm displacement in this species and females gradually accumulate sperm throughout each visit to the breeding site. *Orthetrum cancellatum* females, in contrast, maintain a fairly constant sperm volume during oviposition visits to ponds, males displacing their predecessor's sperm. Female *O. cancellatum* are sometimes mated away from water by young males and copulation is then more protracted, lasting up to fifteen minutes. By examining females at increasing intervals after the beginning of copulation, Miller has shown that most time is spent in removing already deposited sperm and it is only right at the end of a copulation that the male deposits his own sperm in the female, these two phases of copulation being distinguishable in coenagrionid damselflies by rapid thrusting movements of the male's anterior abdominal segments during sperm removal, and a slower rocking movement during sperm deposition. When sperm displacement is practised by a species, males usually adopt the tactic of mate-guarding. By so doing, they limit the success of rivals but they also reduce the amount of time they have available for mating other females. Some male libellulids, however, may mate with a second female whilst the one they are guarding is ovipositing, and thereby risk being usurped. Male *Ischnura elegans*, in contrast, maintain contact with a single female for an average of over five hours (Miller, 1987), releasing her only late in the afternoon when male sexual activity has ceased.

Feeding

Immature adult dragonflies spend much of their time in feeding, usually commencing on the first or second day after emergence (Corbet, 1962), and sexually mature insects also spend a part of each day in feeding activity. Like their larvae, adult dragonflies are generalized predators and they will feed on any other insect that is not too large and can be crushed between the powerful mandibles. Anisoptera prey mainly on Diptera, especially mosquitoes and midges, and also on mayflies, caddis flies, and occasionally on Lepidoptera, Hymenoptera and other Odonata. These insects are generally caught and also eaten in flight, unless they are too unwieldy to carry. *Aeshna grandis* has been observed taking small frogs from the ground (Geest, 1905) and a hummingbird was seen to be seized, killed and carried off by the American *Anax junius* (Hofslund, 1977). *Orthetrum chrysostigma* is said to catch tsetse flies on the backs of Africans. Zygoptera, being smaller and slower in flight than Anisoptera, tend to feed more often on settled insects, and aphids, which are picked off vegetation, are a major item in the diet of many temperate species. *Pyrrhosoma nymphula*, which is a relatively large damselfly, can cope with insects as large as *Sialis* and *Chrysopa* (Killington, 1925), and American *Megaloprepus* pick spiders off webs as they hunt in forest glades.

Feeding dragonflies are not aggressive towards each other, even when sexually mature, and large numbers sometimes congregate where prey is swarming.

The dragonfly thorax is constructed with the segments sloping obliquely forwards ventrally (figs 12,13, p.42); this shifts the leg insertions far forwards and the legs are directed almost horizontally beneath the head rather than being in the vertical plane. The legs are seldom used for walking and instead are adapted for catching and holding prey, their toothed tarsal claws and long bristles retaining food items just below the mouthparts.

Potential prey is perceived visually, and the large compound eyes of a dragonfly are particularly well able to detect movement. Large species can see movement at a range of twenty metres or more. Anisoptera have only a small binocular zone in their visual field and presumably do not use binocular vision to discover their distance from flying prey, but in Zygoptera the binocular zone is much larger and binocular vision probably enables damselflies to detect prey against a complex background of plant stems and foliage.

Adult Odonata may be divided into 'fliers' and 'perchers' (Corbet, 1962). Fliers hawk for their prey, seldom settling, and they also patrol their territories in flight. Perchers, on the other hand, survey their territories and often their feeding grounds from some vantage point, such as the top of a reed or a dead twig, which allows them an uninterrupted view over a wide area. Aeshnidae and Corduliidae are typical fliers; Zygoptera, Libellulidae and Gomphidae may usually be described as perchers (Pl. A, fig. 6).

Temperature relations

Insects do not maintain a constant body temperature and the ambient temperature is a critical factor in their lives. When the temperature is low, muscle metabolism is retarded and an insect may be unable to fly. Some large insects like dragonflies, however, are able to generate considerable internal heat by whirring their wings, and this they have to do before flying in cool conditions. The internal thoracic temperature at take-off of fliers after wing-whirring varies with their mass but is between 27° and 39°C (Vogt & Heinrich, 1983). Fliers are less dependent on ambient temperature than are perchers, small perchers being unable to fly until their thoracic temperature has climbed, without wing-whirring, to at least 19°C. Perchers, however, derive some freedom from air temperature control by varying their perching positions in relation to the sun. Internal air sacs and hairs on the thorax serve as insulation and reduce heat loss. It is no accident that the hairiest dragonflies, such as *Brachytron* and *Cordulia*, are among the very first dragonflies to appear as adults in spring in northern latitudes.

Whilst low temperatures are probably the main climatic hazard in temperate regions, high temperatures in the warmer parts of the world also cause difficulties. A flying dragonfly runs the risk of overheating if it is unable to dissipate metabolic heat when ambient temperatures are high. For this reason, the ratio of fliers to perchers decreases in hot regions, and the fliers tend to be crepuscular, feeding only at dawn and dusk and retiring to shaded forest during the heat of the day. Even in Britain, Moore (1953) found that *Anax imperator* stopped flying for a period around noon. A number of day-active fliers in tropical countries have hindwings with enlarged anal areas enabling them to glide and so reduce the amount of heat generated in the thorax (Corbet, 1962; May, 1978). Some species of Aeshnidae are known to possess an impressive capacity for heat transfer from thorax to abdomen (May, 1976; Heinrich & Casey, 1978). Occlusion of the heart in the abdomen eliminates this ability and it is concluded that the endogenous thoracic heat is transferred by the blood to the long, slender abdomen from which it can be dissipated. Such heat-shunting does not occur during the pre-flight warm-up. The dimensions of the fossil Protodonata indicate that they too must have needed to get rid of endogenous thoracic heat (May, 1982).

Most perchers are diurnal. The duration of their perching periods is long when temperatures are low, short when temperatures are intermediate, and long again when temperatures are high (Heinrich & Casey, 1978). On a perch, their posture varies with temperature. As it becomes hotter their wings droop to reflect the sun's rays from the sides of the thorax, and they may point the abdomen towards the sun so as to reduce the area exposed to direct rays. This necessitates raising the abdomen almost vertically ('obelisk' position) in the tropics at noon. As the air cools after a sunny day, perchers often adopt perches closer to the ground surface, or in unshaded places, or they may select light-coloured rocks for a resting site so as to receive heat from the substrate. *Sympetrum* is so named for its habit of sometimes resting on stones.

Longevity and natural enemies

Odonata are relatively long-lived as adults. *Pyrrhosoma nymphula* can live 46 days (Corbet, 1952), *Anax imperator* 60 days (Corbet, 1957b), male *Enallagma cyathigerum* 39 days (Parr, 1976), male *Ischnura elegans* 42 days and female *I. elegans* 50 days (Parr, 1973). The record for longevity must be held by adult *Sympecma* which emerge in late summer, hibernate, and do not begin egg-laying until the following spring. Most adult dragonflies, however, die before old age. Population studies on sexually mature insects at water have provided the following estimates of average life expectancy: *P. nymphula* 7 days (Corbet, 1952), *E. cyathigerum* 12 days (Parr, 1976), *A. imperator* 14 days (Corbet, 1957b). Mortality of sexually mature dragonflies appears to be very largely independent of their age, death striking young and old insects with equal frequency. Predators, and damage sustained in conflict with others of their species, kill most adult dragonflies, although they have other hazards to contend with as well. Parasitic gregarine Protozoa may damage the mid-gut epithelium and expose it to bacterial infection. Larvae are not infected by gregarines and the degree of adult infection increases with age. Adult dragonflies are often found with red mite larvae attached to thorax, leg articulations or wing veins (Pl.A, fig. 4). These parasitic mites (*Arrenurus* spp.) destroy the epidermis at their feeding sites (Abro, 1982), and heavy infestations weaken the host. Species of *Lestes*, Coenagrionidae and *Sympetrum* seem to be particularly prone to attack.

Birds are probably the main predators of Odonata. The larger Anisoptera, capable of speeds up to about thirty kilometres per hour and with considerable manoeuvrability, can evade most avian predators, although bee-eaters and small falcons such as hobbies take their toll. Four species of Zygoptera and nineteen species of Anisoptera are recorded in the prey of the European bee-eater (Fry, 1984). Zygoptera, and freshly emerged Anisoptera, fall prey to sparrows, reed buntings, wagtails, warblers and no doubt other birds, and many become entangled in spiders' webs. Wasps and robber flies sometimes feed on Zygoptera, and small Odonata are very liable to be devoured by larger species. Dragonflies may be killed even by insectivorous plants. I have seen both *Leucorrhinia dubia* and *Enallagma cyathigerum* entrapped on the leaves of sundew (Pl.A, fig. 5). 'Grappling hook' hairs on stems of the non-insectivorous *Picris echioides* (bristly oxtongue) can snag the wings of *Ischnura elegans* (Whalley, 1986).

A threatened dragonfly will usually resort to flight, but if it is too cold for this, it may make abdominal movements which might intimidate an attacker. Zygoptera will often slip around the stem on which they are resting to interpose it between themselves and an approaching source of danger, thus effectively concealing the body whilst the eyes protrude on either side keeping watch (Askew, 1982).

Dispersal

It is the adult stage of dragonflies that must seek out new breeding sites. This is necessary in all species because no habitat is for ever unchanging; rivers become silted, ponds dry up, and the aquatic and marginal vegetation on which most species depend at some stage of their lives undergoes successional change. Dispersal is most important, however, to those species that breed in the more temporary habitats, and the species which engage in mass migrations are those that breed in ponds rather than in rivers.

All dragonflies leave their breeding sites, although perhaps travelling only a very short distance, on their maiden flights. Some return on reaching sexual maturity but others seek new sites. Fertilized female dragonflies, and males unable to establish territories at fully occupied sites, may also search for new breeding areas. The rapid colonization of newly-formed ponds and artificial lakes bears witness to dragonfly mobility and exploratory capacity. The island of Heligoland is 50 kilometres from the mainland and only one species, *Ischnura elegans*, breeds there continuously, but a total of 35 species has been recorded (Schmidt, Eb., 1980b). Only a proportion of individuals of most species disperse in search of new breeding sites. Sexually mature individuals of some Anisoptera fly once or a number of times a day to the breeding site from the feeding or roosting sites, returning to the same part of the breeding site on several consecutive days.

Long distance mass movements of dragonflies have frequently been reported. The individuals are usually sexually immature, and it seems that such migrations represent an extended maiden flight. Amongst our European dragonflies, *Libellula quadrimaculata*, several species of *Sympetrum* and *Aeshna mixta* have most often been seen engaging in mass flights, sometimes settling in large numbers on ships at sea or flying in swarms, together with other migrating insects, through high mountain passes. Frozen specimens of *Sympetrum* have been found on snow and glacier ice at altitudes of around 3000m in the Alps. The flight direction of migrating dragonflies is orientated northwards in Europe in early summer, and a return southerly movement in autumn has been reported for a few species.

Pantala flavescens is a well-known migratory dragonfly. It has probably the widest distribution of any dragonfly species, being found in North and South America, the West Indies, Africa, India, south-east Asia, Australia and occasionally elsewhere. Like several migratory species, *Pantala* has an enlarged hind wing anal area which facilitates gliding and slope-soaring, and conserves energy on long flights (Gibo, 1981). The flight direction of most migrating dragonflies may be maintained by sun-compass orientation, but Corbet (1962) suggests that *Pantala* might fly with the wind. This would have the effect, over much of its range, of taking it into the inter-tropical convergence zone where rainfall, and hence pools in which to breed, are assured.

1. An ovipositing *Anax imperator* lays its eggs through a small slit cut in a reed-stem (p. 14)

2. A newly-emerged *Ischnura elegans* ♂ rests, after its emergence, above its exuvia while its wings harden (p. 20)

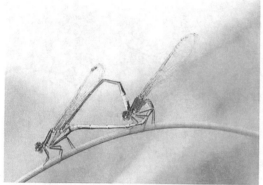

3. Copulating *Ischnura elegans*: the ♀ is of the immature *violacea* form, very rarely seen in copulation (see p. 93)

4. *Sympetrum meridionale* with parasitic mites (*Arrenurus* sp.) attached to its wing-veins. *Sympetrum* species are particularly prone to attack (p. 31)

5. *Leucorrhinia dubia* ♂ caught by sundew (*Drosera rotundifolia*), from which, despite its size, it may not escape

6. *Calopteryx virgo* ♂ on a perch in its territory – a vantage point which allows it an uninterrupted view over a wide area (p. 30)

Plate A: Biology

1. Whixall Moss, Shropshire, England. Species found here include *Lestes sponsa, Libellula quadrimaculata, Sympetrum danae* and *Leucorrhinia dubia*

2. An artificial lake, Woodchester Park, Gloucestershire, England. Habitat for *Pyrrhosoma nymphula, Erythromma najas, Coenagrion puella, Enallagma cyathigerum, Ischnura elegans, Aeshna cyanea, Cordulia aenea* and *Libellula depressa*

3. A man-made lake near Bergerac, Dordogne, France. Species include the three *Platycnemis* spp., *Cercion lindeni, Ceriagrion tenellum, Anax imperator, Gomphus pulchellus, Somatochlora metallica, Libellula depressa, Orthetrum cancellatum, O. albistylum, O. brunneum, Sympetrum striolatum* and *S. sanguineum*

4. The River Lot near Cahors, Dordogne, France. Habitat for *Gomphus vulgatissimus, G. graslini, Onychogomphus forcipatus, O. uncatus* and *Oxygastra curtisi*

5. A dyke in the Camargue, France. Suitable habitat in which to find *Erythromma viridulum, Aeshna mixta, A. isosceles, Anax parthenope, Crocothemis erythraea* and other species

Plate B: Habitats

6. A small lake in the Flumsberg region, St. Gallen, Switzerland. Among species found here are *Aeshna juncea, Somatochlora alpestris* and *S. metallica*. By the lake stand Bastiaan Kiauta and Norman Moore during an excursion organized for participants at the S.I.O. meeting at Chur, August 1981

IV. THE DISTRIBUTION OF EUROPEAN DRAGONFLIES

The European dragonfly fauna is a medley of species originating outside Europe, to the south or to the east, or within Europe. There are, however, only a few species which are confined to Europe and these are all centred in the Iberian Peninsula and south-west France.

Dragonflies are usually associated with warm and sunny conditions, but there are some, especially Anisoptera, that are characteristic elements of the insect fauna flourishing in the brief summers of far northern latitudes. Fifteen species are recorded from the Lake Inari region of Finnish Lapland (Hämäläinen, 1984c). Such species have very broad ranges, extending from Europe across Siberia and often reaching to Japan (Euro-Siberian species), and some are also found in North America (Holarctic species). Iceland, alone among European countries, has no resident Odonata, but this is attributable more to its isolation than to its climate.

Somatochlora sahlbergi is the sole European representative of the tundra dragonfly fauna that is quite rich in species in North America, but somewhat further south, in the boreal coniferous forest or taiga zone, species such as *Aeshna caerulea, A. subarctica, Somatochlora arctica* and *S. alpestris* make their appearance. These are widespread in Scandinavia and northern Russia, but further south in Europe they are restricted to mountainous regions, and in some cases are well established only in the Alps outside their northern ranges. Such species are described as having a boreo-alpine distribution. The broad zone of mixed and deciduous temperate forest accommodates the species of *Leucorrhinia, Aeshna viridis*, and the northern species of *Coenagrion* such as *armatum, lunulatum* and *johanssoni*. This zone has been severely man-modified and its typical dragonflies tend now to be very localized in their distribution. In the south of Europe, in the Mediterranean region, the characteristic elements include *Lestes viridis, Coenagrion mercuriale*, several Gomphidae, *Aeshna affinis, A. mixta, Anax parthenope, Crocothemis erythraea* and many others. Several Mediterranean species extend to lowland areas quite far north, and likewise the western Mediterranean region has been invaded by northwards-thrusting Ethiopian species (e.g. *Orthetrum chrysostigma, Brachythemis leucosticta, Diplacodes lefebvrei, Trithemis annulata*). All of these Anisoptera probably crossed the Sahara from more southerly areas of Africa during wetter periods in the Pleistocene. A number of species of west Mediterranean origin (e.g. *Calopteryx haemorrhoidalis, Coenagrion caerulescens, Ischnura graellsi, Ceriagrion tenellum, Boyeria irene, Gomphus simillimus*) are found in both Europe and North Africa. This can probably be explained in terms of a land-bridge connection between the two continents during the Wurm glaciation in the Pleistocene. A westwards movement of Asian dragonflies has also made a contribution to the European Mediterranean fauna, bringing *Lindenia tetraphylla* and *Orthetrum ramburi* to south-east Europe, *Orthetrum albistylum* almost to the Atlantic, and *Sympetrum decoloratum* to Spain by way of North Africa. These African and eastern components of the dragonfly fauna of southern Europe give it a west-east differentiation which is not apparent in the fauna of northern Europe (Schmidt, Eb., 1978a). This differentiation is accentuated by the presence in south-west Europe of a few endemic forms such as *Macromia splendens, Gomphus graslini, Platycnemis acutipennis* and *P. latipes*.

Some species, such as *Coenagrion hylas*, are extremely localized in Europe, but several have very broad ranges which encompass many of the zones and regions mentioned above. Such ubiquitous species can be identified from the distribution maps and include *Lestes sponsa, Platycnemis pennipes, Coenagrion puella, Aeshna cyanea, Libellula depressa* and *Sympetrum striolatum*. All of these breed in lowland pools, a relatively unstable and impermanent habitat

Table 2 Species of Odonata characteristic of different aquatic biotopes
(modified from Schmidt, Eb., 1982)

Water type	Characteristic Odonata
Cool, montane streams	*Cordulegaster* spp. *Calopteryx virgo* *Ophiogomphus cecilia*
Rivers and warmer streams	*Gomphus* spp. *Calopteryx splendens*
Ditches, warm pools with weak flow	*Coenagrion ornatum* *C. mercuriale* *Orthetrum brunneum*
Oligotrophic to mesotrophic bogs and moorland pools	*Somatochlora arctica* *Sympetrum danae* *Leucorrhinia pectoralis* *L. albifrons* *L. caudalis*
Mesotrophic to eutrophic fens and heathland waters	*Coenagrion hastulatum* *Aeshna juncea* *Leucorrhinia dubia*
Eutrophic ponds with floating vegetation	*Erythromma* spp. *Anax imperator*
Eutrophic ponds, sometimes temporary, with reed-beds	most Coenagrionidae *Lestes* spp. *Aeshna mixta* *Sympetrum* spp.
Eutrophic ponds with bare banks and open water	*Libellula depressa* *Orthetrum* spp.
Lakes	*Epitheca bimaculata*

which demands of their odonate fauna a capacity for exploratory flight and colonization. Other wide-ranging species that inhabit such a biotope, for example *Sympetrum flaveolum*, *S. fonscolombei* and *Libellula quadrimaculata*, are strongly migratory and may sometimes reach areas that are unsuitable for permanent colonization. This is true also for *Hemianax ephippiger* and *Pantala flavescens*. Localized and more permanent habitats, such as upland lakes and river sources, provide an environment for dragonflies for which distant dispersal is not advantageous, being unlikely to result in the discovery of new breeding habitats. It is in such waters, where physical conditions such as oxygen content and temperature remain fairly constant, that the 'less advanced' species tend to persist; lower down a river system are those species adapted to tolerate more varying conditions and having a broader geographical range (Table 2). Very generally, the families Corduliidae, Cordulegastridae and Gomphidae are best represented towards river sources whilst Coenagrionidae and Libellulidae dominate waters on lowland plains.

Another trend in distribution concerns the ratio of Zygoptera to Anisoptera. The proportion of Zygoptera species tends to fall in moving from warmer to cooler latitudes. Odonata almost certainly evolved in tropical or subtropical conditions and the Zygoptera, the more ancient suborder, includes fewer genera than the Anisoptera that have adapted to the more varying conditions of a temperate environment.

Coverage of this book

Odonata having a broad distribution frequently develop local forms and it is often difficult to decide upon the taxonomic status of these. In this book I have adopted a rather parsimonious approach and consider *Calopteryx splendens xanthostoma* and *Somatochlora metallica meridionalis*, regarded by several authors as good species, to be of no more than subspecific status. On the other hand, the following are treated here, with varying degrees of confidence, as specifically distinct: *Gomphus schneideri* and *G. simillimus, Cordulegaster heros* and *C. picta, Sympetrum nigrescens* and *S. striolatum*. The total number of recognized species of European Odonata is 114 – 37 Zygoptera and 77 Anisoptera. Also mentioned are some North African species that reach the southern Mediterranean littoral, and some eastern species that extend to the Caucasus or to the islands of the eastern Mediterranean.

The eastern boundaries of Europe, as politically defined, do not correspond with zoogeographical discontinuities, and the fauna of the eastern Mediterranean, particularly of Cyprus and the islands of the eastern Aegean, merges without sharp transition into that of Anatolia and the Middle East.

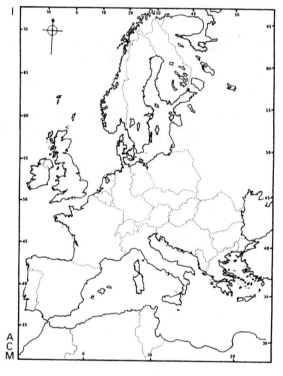

Figure 10 Distribution map showing area covered by this book and the national boundaries marked. Beside the map,

I = Iceland
A = Azores
C = Canary Isles
M = Madeira (including Porto Santo

The map numbers in the text correspond with the species' check list numbers on pp. 49–51

The distribution maps

An attempt has been made to illustrate, by means of a map (fig. 10), the European distribution of each species. The maps have been compiled from published information supplemented by personal observations and data labels on museum specimens. Published records of dragonfly distribution in Europe are scattered through entomological literature and uneven in coverage. It is not claimed that more than a good proportion of these papers have been seen. Detailed species' distribution maps are available for the British Isles, Holland, Belgium and Scandinavia, and numerous papers record the dragonfly fauna of various areas of West Germany, Italy, and to a lesser extent France and East Germany. Iberia, excepting the south of Spain, and much of eastern Europe, particularly the Balkans, are under-recorded areas. Some of the principal and mostly recent publications on distribution are listed below by country or region. I am very aware of the incompleteness of this list, but it should be sufficient to introduce an enquirer to the literature of a region and provide a basis for further search in the reference lists appended to the cited papers. Authors and dates only are given here: the full titles of the works are to be found in the list of references at the end of this volume.

Albania	Fudakowski, 1930a; Bilek, 1966
Austria	Lödl, 1976
Balearic Islands	Compte Sart, 1963
Belgium	Cammaerts, 1979
British Isles	Lucas, 1900; Longfield, 1937; Fraser, 1949c; Hammond, 1977; McGeeney, 1986
Bulgaria	Beshovski, 1964; Scheffler, 1973; Dumont, 1977d
Corsica	Dommanget & Martinez, 1984
Czechoslovakia	Teyrovský, 1977
European Turkey and the eastern Mediterranean islands	Dumont, 1977c; Yazicioglu, 1982
Finland	Valtonen, 1980
France	Aguesse, 1968 (covers the whole of the Mediterranean region); Dommanget, 1981; Francez & Brunhes, 1983
Germany	Ris, 1909; May, 1933
Germany (East)	Königstedt & Schmidt, 1980; Müller, 1980; Klausnitzer et al., 1978
Germany (West)	Lohmann, 1980; many papers by Eberhardt Schmidt
Greece	Cowley, 1940; Buchholz, 1954; Bilek, 1967; Galletti & Pavesi, 1983; Kemp & Kemp, 1985
Hungary	Sándor, 1957; Steinmann, 1984
Italy	Conci & Nielsen, 1956; recent papers by Balestrazzi, Bucciarelli, et al.
Latvia	Cowley, 1937
Malta	Valetta, 1949
Netherlands	Geijskes & van Tol, 1983
North Africa (Algeria)	Martin, 1910; Dumont, 1978; Koch, 1979
North Africa (Canary Islands)	Valle, 1931; Belle, 1982; Baez, 1985
North Africa (Madeira, Azores)	Gardner, 1963
North Africa (Morocco)	Lieftinck, 1966; Dumont, 1973
North Africa (Sahara, north-west)	Nielsen, 1956
North Africa (Tunisia)	Dumont, 1977b; Koch, 1979
Norway	Sømme, 1937; Aagaard & Dolmen, 1977
Romania	Cirdei & Bulimar, 1965; Dumont, 1977d
Russia	Bartenef, 1915, 1919
Sardinia	Nielsen, 1941; Castellani, 1951; Bucciarelli, 1977
Scandinavia	Valle, 1952
Sicily	Capra, 1934; Bucciarelli, 1971, 1977
Spain	Benitez Morera, 1950; Compte Sart, 1965; Ferreras Romero & Puchol Caballero, 1984 (Andalucia)
Sweden	Sahlén, 1985
Switzerland	Bischof, 1971, 1973, 1976; De Marmels, 1979a, 1979b; Dufour, 1982
Yugoslavia	Adamovic, 1967; Dumont, 1977d

The maps are of too small a scale to allow for precise details of distribution. In particular, areas indicated as being occupied by a species may include metropolitan centres, mountain peaks, lowland swamps or other biotopes which are quite unsuitable for the species in question. Old records of localities where the species is now extinct (e.g. *Coenagrion armatum* and *C. scitulum* in England) are shown. Breeding ranges of the species are indicated by solid black areas and seemingly isolated localities or records of vagrant individuals appear as small spots. In defining ranges, areas for which no records have been traced are included only if the area is under-recorded and there is no reason to doubt that the species in question is present. Caution has been exercised and actual ranges may exceed those indicated. In particular, most difficulty was experienced in defining the limits of a species within Russia, the central Balkans and, to a lesser degree, North Africa, and the boundaries here are based upon scanty data. Question marks are placed on areas where a species might well occur but for where no reliable data could be found. Records of species from the Azores, Canary Islands and Madeira (including Porto Santo) are shown by the letters A, C and M respectively at the foot of the map, and for Iceland (only *Hemianax ephippiger*) by an I at the top of the map. To facilitate establishing the co-ordinates of localities and ranges, national boundaries are represented by dotted lines.

V. MORPHOLOGY OF
THE ADULT DRAGONFLY

Some knowledge of the names applied to the various external features of damselflies and dragonflies is needed before the identification keys can be effectively used. In this chapter odonate topography is described, with emphasis placed on those features that have particular taxonomic value.

Head

In Zygoptera, the head in dorsal view (fig. 11a) is spindle-shaped, narrow medially in the antero-posterior axis, and with the almost hemispherical compound eyes well-separated. A group of three simple eyes or *ocelli* is arranged in a triangle, apex to the front, centrally on the dorsal surface or *vertex* of the head. The anterior or median ocellus is slightly larger than the two lateral ocelli. The vertex runs into the *occiput* posteriorly and the *frons* anteriorly. These three sclerites are seldom clearly demarcated. The occiput in most Coenagrionidae has a pair of brightly coloured spots, the *postocular spots*. The *antennae* are inconspicuous organs inserted at the junction of vertex and frons in line with the anterior ocellus and the inner front corner of each eye. Each antenna is composed of seven (occasionally six) segments (antennomeres), the basal two (scape and pedicel) being much stouter than the terminal ones which constitute a filamentous or filiform flagellum. The flagellum starts as a single segment with apical and basal growth centres, but during larval development it divides four times, thrice at the basal centre and once at the apical. In Zygoptera and Aeshnidae the basal divisions precede the apical division, but in Libellulidae and Corduliidae the apical division is advanced and followed by one or two basal divisions (Miyakawa, 1977). The structures forming a roof over the mouthparts project forwards in a semicircle below the frons, the *clypeus* and *labrum* forming respectively a dorsal and anterior cover. The clypeus is divided by a suture into a broad, horizontal *postclypeus* and below it a small *anteclypeus*. The labrum is connected to the ventral edge of the anteclypeus by a suture. The mouthparts are of the biting type and include a pair of powerful *mandibles*, a pair of *maxillae* and, posteriorly, the *labium*. The labium is deeply cleft medially in damselflies, but among dragonflies only Cordulegastridae show this feature which may be a relic of the ancestral paired condition of the labium.

The large, heavy head of Anisoptera (figs 11b, c) is very narrowly connected to the thorax and highly mobile. In turning during flight the rest of the body tilts but the head remains vertical, serving as a gyroscope. Rolling movements of the rest of the body stimulate pads of sensory hairs projecting from the prothorax and contacting the occipital head surface. The anisopteran head is differently shaped from the head of Zygoptera being more or less globular, with the major part of its surface occupied by the very large *compound eyes*. The eyes, except in Gomphidae, are not dorsally separated and are sometimes in contact over a considerable distance. The large aeshnids may have more than 28,000 ommatidial units in each eye (Sherk, 1978b). These are indicated externally by small hexagonal facets. When the eye of a living dragonfly is examined, a black spot or pseudopupil will be seen. This is formed from those ommatidia that are viewed along their optical axes, absorbing all light. In addition, a ring of accessory pseudopupils often encircles the central pseudopupil. The expansion of the anisopteran eye reduces and separates the *vertex* and *occiput*, the latter being a small, triangular plate, the *occipital triangle*, at the back of the dorsal surface of the head. In front of the eyes, the three ocelli are accommodated on the vertex which is mostly an elevated protuberance above the frons. Only in Gomphidae is the vertex broad and relatively flat. The *antennae*, as in Zygoptera,

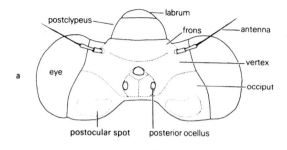

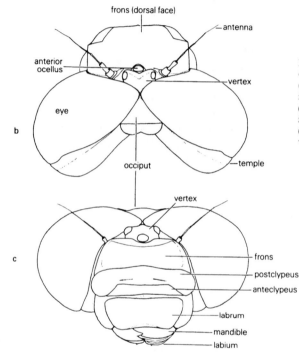

Figure 11 Morphology of the dragonfly head
(**a**) *Coenagrion pulchellum* (Zygoptera) in dorsal view
(**b**) *Cordulegaster boltoni* (Anisoptera) in dorsal view
(**c**) *Cordulegaster boltoni* in frontal view

are very small, and they carry just one type of minute sense organ (coeloconic sensilla) in contrast to the varied arrays of sensillae present on the antennae of less visually-biased insects. The *frons* is divided by a horizontal crest (in Zygoptera seen only in *Nehalennia* and *Ceriagrion*) separating its dorsal and anterior surfaces. The clypeus is composed of *postclypeus* and *anteclypeus* as in Zygoptera, but both are vertically positioned in Anisoptera.

Thorax

This is the locomotor centre of insects. It is composed of three segments, each of which carries a pair of legs, and the posterior two are each provided with a pair of wings. Externally, the surfaces of the thoracic segments are divided by a number of grooves or sutures which mark the positions of internal ridges to which wing and leg muscles are attached. The first segment of the thorax, the *prothorax*, is relatively small and slightly movable, whilst the wing-bearing *mesothorax* and *metathorax* are intimately fused and together comprise the *synthorax* (or pterothorax). In Zygoptera the dorsal surface of the prothorax may be modified for pairing and it provides characters of taxonomic value. The synthorax slopes obliquely so that in lateral view

41

(fig. 12) it appears triangular, its true dorsal surface and the wing bases being inclined posteriorly. The pleurites of the synthorax are produced forwards and upwards to form a false dorsum to the thorax, sloping up from the prothorax to a peak in front of the anterior wing bases. A *median carina* or mid-dorsal keel divides the false dorsum and lying on either side of it in Zygoptera, at the very front of the synthorax, are two small triangular or semicircular plates, the *mesostigmal lamellae*, associated with the mesothoracic spiracles and contacted by the males' superior abdominal appendages during copulation. Behind the median carina a series of more or less parallel sutures can be located: the *humeral* (or *mesopleural, first lateral*) *suture* defines the posterior limit of the *mesepisternum*, the *mesometapleural* (or *intersegmental, second lateral*) *suture* marks the posterior edge of the *mesepimeron* and the lateral division between mesothorax and metathorax, and the *metapleural* (or *metepisternal-epimeral, third lateral*) *suture*

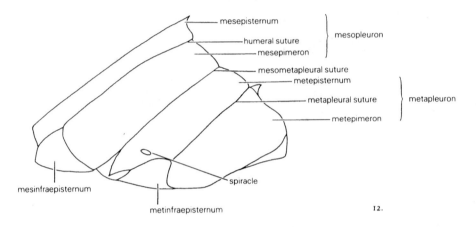

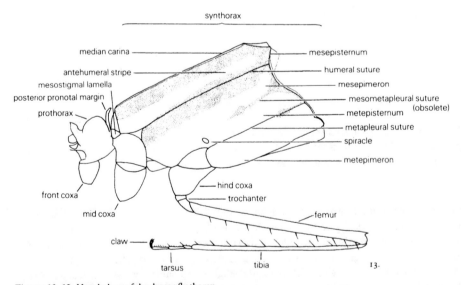

Figures 12, 13 Morphology of the dragonfly thorax
(12) (above) synthorax of *Calopteryx splendens* in left lateral view
(13) (below) thorax and hind leg of *Coenagrion puella* in left lateral view; darkened areas in the meso- and metapleura are stippled

separates the *metepisternum* from the *metepimeron*. This arrangement will be more easily understood by referring to figures 12 and 13. The mesometapleural suture is almost obsolete in all Odonata except Calopterygidae and Euphaeidae. Each mesepisternum may be marked by a pale, longitudinal *antehumeral stripe* approximately midway between dorsal carina and humeral suture, and there is sometimes a pale *humeral stripe* along the humeral suture. A *spiracle* is situated on the metepisternum above the insertion of the hind leg.

The *legs* (fig. 13), because of the slope of the thorax, are directed forwards and better suited to grasping vegetation than to standing on the ground. Only Gomphidae habitually rest on flat surfaces, and in all Odonata walking is an exceptional activity. The basal leg segments (podomeres) are conical *coxae* which articulate with the thorax. Following the coxa in succession are the *trochanter, femur, tibia* and *tarsus*. The trochanter is divided by a constriction into a short basal part and a much longer distal section. The femur and tibia of a leg are about equal in length, except in the ground-resting Gomphidae in which the hind tibia is much shorter than the hind femur. The tarsus is composed of three segments (tarsomeres), the first of which is short and the third is furnished with a pair of toothed claws. The femora and tibiae are equipped with two ventral rows of spines or stout setae which are employed in securing prey, the forwardly-directed legs forming a catching basket for flying insects. The front tibia in both Zygoptera and Anisoptera carries a comb of closely-set, small spines which are paler than the rest and are used to clean the eyes.

Wings

The wings are elongated structures, the two pairs similar in Zygoptera (figs 14,15, p.44) but differing in shape and venation in Anisoptera (figs 16,17, p.45). They are held together along the back of the body, dorsal surfaces apposed, in resting Zygoptera (partially opened in Lestidae), but in Anisoptera they are held more or less horizontally at right angles to the body. The membrane of the wing is bare but small hairs, *microtrichia*, are scattered along some of the veins. At the base of the wing are five longitudinal veins but some of these fork distally and they are connected by a system of crossveins which divides the wing into many small cells. The wings are not flat but pleated for stiffness, with the longitudinal veins marking the summits of convexities (costa, radius + median, anterior median, anal) or the troughs of alternating concavities (subcosta, radial sector, cubital). The longitudinal veins and major crossveins have names and their arrangement (venation) is of great importance in odonate taxonomy at all levels down to the species. The major venational features are best identified by referring to the wing maps of Zygoptera and Anisoptera (figs 14–17). Nomenclature of dragonfly wing veins is controversial and several systems have been proposed, each following a different interpretation of vein homology. Most recent odonatologists apply a system based upon that of Tillyard & Fraser (1940), and this is followed in the present work. It should be noted, however, that the Comstock & Needham (1898) system is still sometimes used, and a new system proposed by Carle (1982) appears to have the advantage of simplicity. Notation for these three systems is as follows:

Tillyard-Fraser and this work	C	Sc	R_1	R_2	R_3	IR_3	R_{4+5}	MA	CuP	A_1	
Comstock-Needham	C	Sc	R	M_1	M_2	Rs	M_3	M_4	Cu_1	Cu_2	A_1	A_2
Carle		CA	CP	RA	RP_1	RP_2	MA	MP	CuA	CuP	A_1	A_2 A_3

The anterior margin of the wing is formed by the *costa* (C) which is broken at a variable distance from its base at a point termed the *node*. The *subnodal crossvein* below the node marks the distal limit of the *subcostal vein* (Sc). The *radius* (R_1) runs from wing base to apex and is connected with the costa in the distal part of the wing by a darkened, sclerotized cell termed the *pterostigma*. The pterostigma functions as an inertial regulator of wing twisting. Between the node and pterostigma, veins R_1 and C are connected by a series of *postnodal crossveins*. *Antenodal*

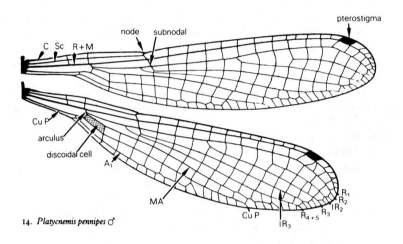

14. *Platycnemis pennipes* ♂

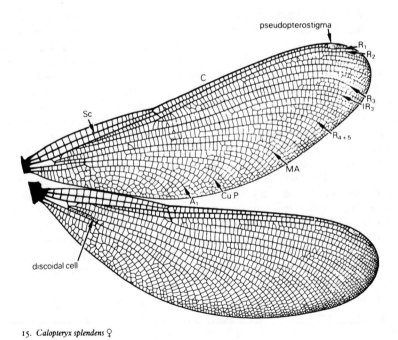

15. *Calopteryx splendens* ♀

Figures 14–17 Right fore- and hindwings of Zygoptera (above) and Anisoptera (right) to illustrate venation. Some key areas are stippled

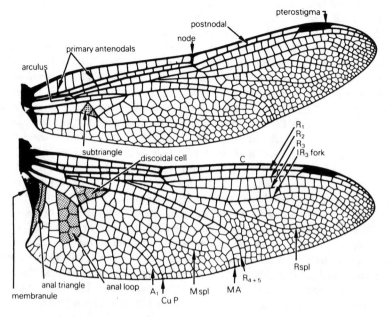

16. *Aeshna mixta* ♂

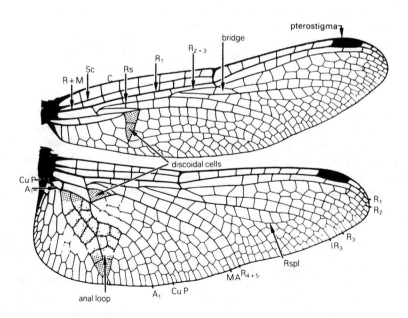

17. *Sympetrum sanguineum* ♂

crossveins link Sc and C basal to the node, and there may or may not be an alignment with crossveins between R_1 and Sc. In Zygoptera (except Calopterygidae and Euphaeidae) there are only two antenodals and these unite with two crossveins between R_1 and Sc. In Anisoptera there are several antenodals but two, equivalent to those in most Zygoptera, are stouter than the rest and are known as the *primary antenodals*. Only in the most specialized Anisoptera (Corduliidae, Libellulidae) are the primary antenodals indistinguishable. The radius and *median veins* are fused ($R+M$) at the base of the wing, but after a short distance the median vein separates sharply from R_1 in an oblique transverse section, the *arculus*, and then continues as the anterior median vein (MA) to the posterior margin of the wing. The posterior branch of M is absent in Odonata. The *radial sector* (Rs) arises from the arculus and divides into three concave branches (R_2, R_3, R_{4+5}) which have intercalated veins between them (IR_2 between R_2 and R_3 in Zygoptera, IR_3 between R_{2+3} and R_{4+5} in Anisoptera). IR_3 forks in many of the larger Anisoptera, and a vein running posterior to it, the *radial supplementary* or radial planate (Rspl), is sometimes developed. Between the separation of R_{2+3} and IR_3, basal to the subnodal vein, is a shallow, triangular cell, the *bridge* vein defining its posterior border, traversed by one or more crossveins. Vein MA, like IR_3, may have in Anisoptera a *supplementary vein* (Mspl) running behind it. Posterior to $R+M$ is the *cubital vein* (CuP, the anterior cubital CuA not being represented in dragonflies) which is gently sinuous in Zygoptera but angled abruptly downwards, more or less below the arculus, in Anisoptera. The area basal to the arculus between veins $R+M$ and CuP is the *median* (or *basal*) *space* which in some Anisoptera is traversed by crossveins. The *discoidal cell* (or *quadrilateral*) in Zygoptera is four-sided and has CuP as its posterior edge (except in Calopterygidae, Euphaeidae), whilst in Anisoptera the discoidal cell (or *triangle*) is triangular with CuP forming its basal edge. In Anisoptera there is another triangular cell, the *subtriangle*, sharing its base with the discoidal cell. Distad to the discoidal cells, between MA and CuP, is the *trigonal space*. Posterior to CuP is the convex *anal vein* (A_1) which has a recurved branch basally forming a loop, the *anal loop*, enclosing a group of cells of different configuration in different families. In many male Anisoptera there is a well-defined triangular area at the base of the hindwing, the *anal triangle*, and the wing margin is often more or less angled. Behind A_1 at the base of the wing in Anisoptera is a small flap of opaque tissue, not enclosed by a vein, which is termed the *membranule*.

Abdomen

The abdomen of Odonata is long, usually cylindrical and narrow, but sometimes in Anisoptera it is dorsoventrally flattened and quite broad. In male Anisoptera it may be narrowed at the third segment, and in some genera it is somewhat expanded distally (club-shaped). Females nearly always have a rather stouter abdomen than their males. In both Zygoptera and Anisoptera there are ten abdominal segments, conventionally numbered S1 to S10, S1 being the basal segment following the thorax. A mid-dorsal keel or carina is often present.

Males have *accessory* (or *secondary*) *genitalia* ventrally on S2+3, sperm being transferred to this apparatus from the gonopore on S9 prior to insemination. The female genitalia are near the apex of the abdomen and receive sperm from the male's secondary genitalia when the female curls her abdomen beneath that of the male to effect a union (figs 7,8). Males of Anisoptera, except *Anax*, *Hemianax* and Libellulidae in the European fauna, possess a pair of ear-shaped processes termed *auricles* (or oreillets) on S2 and it has been suggested (Fraser, 1943) that these function as guides to the female in securing proper positioning of her abdomen during copulation. Presence of auricles is correlated with an angulation of the anal margin of the male hindwing which prevents contact with the auricles during flight. The anisopteran male secondary genitalia (fig. 18) consist of a three-segmented *penis* arising from the anterior margin of sternum S3 and bending back on itself when at rest. In this position it is shielded by a guard or *ligule*. Before copulation, the penis is charged with sperm which are transferred from the genital opening on S9 to a cavity at the penis apex. This is achieved by curling the tip of the abdomen forwards below its ventral surface. Anterior to the ligule are two pairs of *hamules*, and in front of the hamules is a development of sternum S2, the *anterior lamina*. This latter is cleft in Aeshnidae to accommodate the female's ovipositor during copulation. The hamules hold the female abdomen and they vary in shape a great deal; in Aeshnidae the anterior hamules are well

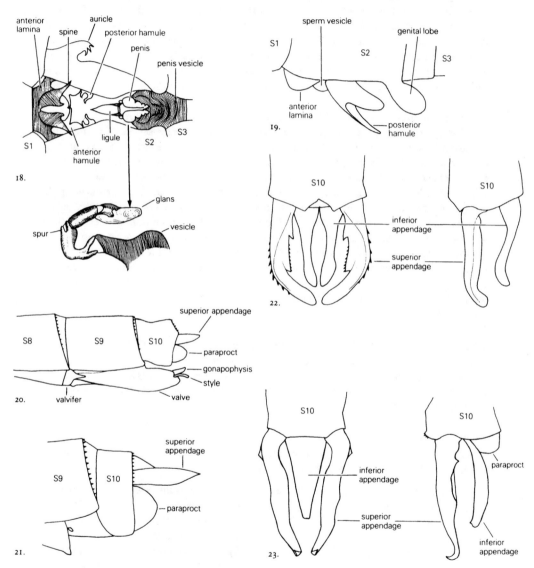

Figures 18–23 Morphology of the terminal abdominal segments and male accessory genitalia. Abdominal segment numbers are indicated S1, S2, S3, etc.
(18) male accessory genitalia of *Aeshna juncea*, ventral view, with the penis (enlarged) in left lateral view
(19) male accessory genitalia of *Sympetrum striolatum* in left lateral view
(20) last three abdominal segments of female *Lestes sponsa* from the left
(21) last two abdominal segments of female *Sympetrum striolatum* from the left
(22) abdominal appendages of male *Lestes sponsa* in dorsal and right lateral views
(23) abdominal appendages of male *Somatochlora metallica* in dorsal and right lateral views

developed, but in Libellulidae (fig. 19) the posterior pair are most prominent and provide important taxonomic characters. The penis too adopts a variety of forms and may be furnished with various processes; it is used, at least in some genera, to remove any sperm deposited in the female at a previous mating. The zygopteran penis is not three-segmented and probably not homologous with the anisopteran organ.

Females have no secondary genitalia and no auricles. The primary genitalia are on S8+9. In Zygoptera and Aeshnidae there is a well-developed *ovipositor* (fig. 20), made up of three pairs of valves, serrated, and used to insert the eggs in slits cut in plant tissue. *Cordulegaster* females also have a conspicuous ovipositor to deposit eggs in the mud of stream beds, but it is unserrated and of much-simplified construction. In other families of Anisoptera (fig. 21), there is no ovipositor and the eggs are usually simply washed into the water as they are extruded from the genital aperture.

The final abdominal segment (S10) carries the *anal appendages*. In Zygoptera (fig. 22), the male has a pair of *superior appendages* (or cerci) and a pair of *inferior appendages*, the latter being situated below the anus. In male Anisoptera (fig. 23), there is again a pair of superior appendages but the inferior appendage (or epiproct) is a single structure (though sometimes biramous) and situated above the anus. Inferior appendages are not homologous in the two sub-orders. They are absent in females. The anal appendages of the male are used to grasp the female during mating and they show considerable diversity of form; they are important aids to both us and the insects in interspecific discrimination. The specific uniqueness of these structures reduces the likelihood of consummation of attempted interspecific couplings. In Anisoptera, the male superior appendages press against the back of the female's head and the inferior appendage engages the occiput or eyes (fig. 8), sometimes dinting the latter. In Zygoptera, the male anal appendages grasp the posterior lobe of the female's prothorax, the inferior appendages on the anterior face, the superiors behind.

Body length

Where body length is quoted in this book, the measure taken is from the front of the head to the apices of the longer anal appendages (apex of the ovipositor in female *Cordulegaster*), the abdomen being held in a straight line with the rest of the body. This measurement is preferred to that of wing or abdomen length, given in several works on dragonflies, because it can be taken more easily and quickly.

Body weight

The live weight of insects is seldom measured, and it may be of value to quote here weights of a selection of Odonata as presented by Greenewalt (1962):

Calopteryx virgo	91mg	*Gomphus vulgatissimus*	638mg
Pyrrhosoma nymphula	38	*Cordulia aenea*	201
Ischnura elegans	20	*Libellula depressa*	245
Aeshna juncea	958	*Orthetrum coerulescens*	248
Anax imperator	1200	*Sympetrum striolatum*	232

VI. THE EUROPEAN SPECIES OF ODONATA

The framework for the following classification of European dragonflies is adapted mainly from Davies' (1981) synopsis of world genera, although families are listed in different order. As remarked on page 12, it is not yet possible to construct an evolutionary tree of dragonfly families, but relationships between European species are indicated by their groupings into higher taxa. In this book, descriptions follow a general progression from what appear to be the more generalized insects to what seem the more specialized. However, all species living today have survived a very long evolutionary history and no extant species can be described as unspecialized. All that can be accurately said is that some characters of a species seem to be further from or closer to a postulated primitive condition.

CHECK LIST

ORDER	**Odonata**	
SUBORDER	**ZYGOPTERA**	
SUPERFAMILY	**Calopterygoidea**	
FAMILY	**CALOPTERYGIDAE**	
	Calopteryx Leach, 1815	
1	*virgo* (Linnaeus, 1758)	
2	*splendens* (Harris, 1782)	
2a	*s. xanthostoma* (Charpentier, 1840)	
3	*haemorrhoidalis* (Vander Linden, 1825)	
FAMILY	**EUPHAEIDAE**	
	Epallage Charpentier, 1840	
4	*fatime* (Charpentier, 1840)	
SUPERFAMILY	**Lestoidea**	
FAMILY	**LESTIDAE**	
SUBFAMILY	LESTINAE	
	Lestes Leach, 1815	
5	*viridis* (Vander Linden, 1825)	
6	*barbarus* (Fabricius, 1798)	
7	*virens* (Charpentier, 1825)	
8	*macrostigma* (Eversmann, 1836)	
9	*sponsa* (Hansemann, 1823)	
10	*dryas* Kirby, 1890	
SUBFAMILY	SYMPECMATINAE	
	Sympecma Burmeister, 1839	
11	*fusca* (Vander Linden, 1820)	
12	*annulata* (Sélys, 1887)	
SUPERFAMILY	**Coenagrionoidea**	
FAMILY	**PLATYCNEMIDIDAE**	
	Platycnemis Burmeister, 1839	
13	*pennipes* (Pallas, 1771)	
14	*latipes* Rambur, 1842	
15	*acutipennis* Sélys, 1841	

FAMILY	**COENAGRIONIDAE**	
SUBFAMILY	COENAGRIONINAE	
	Pyrrhosoma Charpentier, 1840	
16	*nymphula* (Sulzer, 1776)	
	Erythromma Charpentier, 1840	
17	*najas* (Hansemann, 1823)	
18	*viridulum* (Charpentier, 1840)	
	Coenagrion Kirby, 1890	
19	*mercuriale* (Charpentier, 1840)	
20	*scitulum* (Rambur, 1842)	
21	*caerulescens* (Fonscolombe, 1838)	
22	*hastulatum* (Charpentier, 1825)	
23	*lunulatum* (Charpentier, 1840)	
24	*armatum* (Charpentier, 1840)	
25	*ornatum* (Sélys, 1850)	
26	*johanssoni* (Wallengren, 1894)	
27	*hylas* (Trybŏm, 1889)	
28	*puella* (Linnaeus, 1758)	
29	*pulchellum* (Vander Linden, 1825)	
	Cercion Navas, 1907	
30	*lindeni* (Sélys, 1840)	
SUBFAMILY	ISCHNURINAE	
	Enallagma Charpentier, 1840	
31	*cyathigerum* (Charpentier, 1840)	
	Ischnura Charpentier, 1840	
32	*pumilio* (Charpentier, 1825)	
33	*elegans* (Vander Linden, 1820)	
34	*genei* (Rambur, 1842)	
35	*graellsi* (Rambur, 1842)	
SUBFAMILY	NEHALENNIINAE	
	Nehalennia Sélys, 1850	
36	*speciosa* (Charpentier, 1840)	
	Ceriagrion Sélys, 1876	
37	*tenellum* (Villers, 1789)	

SUBORDER **ANISOPTERA**

SUPERFAMILY **Aeshnoidea**

FAMILY **AESHNIDAE**

SUBFAMILY AESHNINAE

TRIBE Aeshnini

Aeshna Fabricius, 1775

38	*caerulea* (Ström, 1783)
39	*juncea* (Linnaeus, 1758)
40	*subarctica* Walker, 1908
41	*crenata* Hagen, 1856
42	*serrata* Hagen, 1856
43	*mixta* Latreille, 1805
44	*affinis* Vander Linden, 1823
45	*cyanea* (Müller, 1764)
46	*viridis* Eversmann, 1836
47	*grandis* (Linnaeus, 1758)
48	*isosceles* (Müller, 1767)

TRIBE Anactini

Anax Leach, 1815

49	*imperator* Leach, 1815
50	*parthenope* (Sélys), 1839

Hemianax Sélys, 1883

51	*ephippiger* (Burmeister, 1839)

SUBFAMILY BRACHYTRONINAE

TRIBE Brachytronini

Brachytron Sélys, 1850

52	*pratense* (Müller, 1764)

Caliaeschna Sélys, 1883

53	*microstigma* (Schneider, 1845)

TRIBE Gomphaeschnini

Boyeria McLachlan, 1896

54	*irene* (Fonscolombe, 1838)

FAMILY **GOMPHIDAE**

SUBFAMILY GOMPHINAE

Gomphus Leach, 1815

55	*flavipes* (Charpentier, 1825)
56	*vulgatissimus* (Linnaeus, 1758)
57	*simillimus* Sélys, 1840
58	*schneideri* Sélys, 1850
59	*pulchellus* Sélys, 1840
60	*graslini* Rambur, 1842

Paragomphus Cowley, 1934

61	*genei* (Sélys, 1841)

Ophiogomphus Sélys, 1854

62	*cecilia* (Fourcroy, 1785)

Onychogomphus Sélys, 1854

63	*forcipatus* (Linnaeus, 1758)
64	*uncatus* (Charpentier, 1840)
65	*costae* Sélys, 1885

SUBFAMILY LINDENIINAE

Lindenia De Haan, 1826

66	*tetraphylla* (Vander Linden, 1825)

SUPERFAMILY **Cordulegastroidea**

FAMILY **CORDULEGASTRIDAE**

Cordulegaster Leach, 1815

67	*boltoni* (Donovan, 1807)
68	*picta* Sélys, 1854
69	*heros* Theischinger, 1979
70	*insignis* Schneider, 1845
71	*bidentata* Sélys, 1843

SUPERFAMILY **Libelluloidea**

FAMILY **CORDULIIDAE**

SUBFAMILY CORDULIINAE

Cordulia Leach, 1815

72	*aenea* (Linnaeus, 1758)

Somatochlora Sélys, 1871

73	*metallica* (Vander Linden, 1825)
73a	*m. meridionalis* Nielsen, 1935
74	*alpestris* (Sélys, 1840)
75	*arctica* (Zetterstedt, 1840)
76	*sahlbergi* Trybŏm, 1889
77	*flavomaculata* (Vander Linden, 1825)

Epitheca Charpentier, 1840

78	*bimaculata* (Charpentier, 1825)

SUBFAMILY GOMPHOMACROMIINAE

Oxygastra Sélys, 1871

79	*curtisi* (Dale, 1834)

SUBFAMILY MACROMIINAE

Macromia Rambur, 1842

80	*splendens* (Pictet, 1843)

FAMILY **LIBELLULIDAE**

SUBFAMILY LIBELLULINAE

Libellula Linnaeus, 1758

81	*quadrimaculata* Linnaeus, 1758
82	*fulva* Müller, 1764
83	*depressa* Linnaeus, 1758

Orthetrum Newman, 1833

84	*trinacria* (Sélys, 1841)
85	*chrysostigma* (Burmeister, 1839)
86	*cancellatum* (Linnaeus, 1758)
87	*albistylum* (Sélys, 1848)
88	*nitidinerve* (Sélys, 1841)
89	*brunneum* (Fonscolombe, 1837)
90	*coerulescens* (Fabricius, 1798)
91	*ramburi* (Sélys, 1848)

SUBFAMILY	SYMPETRINAE
	Diplacodes Kirby, 1889
92	*lefebvrei* (Rambur, 1842)
	Brachythemis Brauer, 1868
93	*leucosticta* (Burmeister, 1839)
	Crocothemis Brauer, 1868
94	*erythraea* (Brullé, 1832)
	Sympetrum Newman, 1833
95	*striolatum* (Charpentier, 1840)
96	*nigrescens* Lucas, 1912
97	*vulgatum* (Linnaeus, 1758)
98	*meridionale* (Sélys, 1841)
99	*decoloratum* (Sélys, 1884)
100	*fonscolombei* (Sélys, 1840)
101	*flaveolum* (Linnaeus, 1758)
102	*sanguineum* (Müller, 1764)
103	*depressiusculum* (Sélys, 1841)
104	*danae* (Sulzer, 1776)
105	*pedemontanum* (Allioni, 1766)

SUBFAMILY	LEUCORRHINIINAE
	Leucorrhinia Brittinger, 1850
106	*caudalis* (Charpentier, 1840)
107	*albifrons* (Burmeister, 1839)
108	*dubia* (Vander Linden, 1825)
109	*rubicunda* (Linnaeus, 1758)
110	*pectoralis* (Charpentier, 1825)

SUBFAMILY	TRITHEMISTINAE
	Trithemis Brauer, 1868
111	*annulata* (Palisot de Beauvois, 1807)

SUBFAMILY	ZYGONICHINAE
	Zygonyx Hagen, 1867
112	*torridus* (Kirby, 1889)

SUBFAMILY	TRAMEINAE
	Pantala Hagen, 1861
113	*flavescens* (Fabricius, 1798)

SUBFAMILY	UROTHEMISTINAE (MACRODIPLACINAE)
	Selysiothemis Ris, 1897
114	*nigra* (Vander Linden, 1825)

These 114 species are recognized as breeding within Europe; that is, as far east as the Ural Mountains and south to the north coast of the Caspian Sea, the Caucasus, the northern coasts of the Black Sea and Sea of Marmara. Some extra-limital species of North Africa, Anatolia, Rhodes and Cyprus are mentioned in the text but excluded from the check list. The status of forms of uncertain specific status is commented upon on page 36.

Keys to the identification of adult insects in European suborders and families follow, preceding the sequential description of each family, genus and species. Keys to genera will be found below the family descriptions and keys to species after the generic descriptions.

For each species, the original designation, type locality, more important synonyms and vernacular names (English, French, German) are given. The species descriptions stress the diagnostic characters of each species and are followed by notes, with bibliographic references, on biology and habitat, flight period and distribution.

NOTE

A further eleven breeding species have been discovered in Europe since 1988, raising the total from 114 to 125. These additional species are listed below, numbered with a suffix according to their position in the Check List and with a page reference to the Supplement. Recently recognized valid names for species previously appearing in the list are also given below after their original species number:

5a	*Lestes parvidens* (Artobelevski, 1929) (p.215)
12	*Sympecma paedisca* (Brauer, 1882) (p.213)
28a	*Coenagrion intermedium* Lohmann, 1990 (p.215)
35a	*Ischnura fountainei* Morton, 1905 (p.216)
50a	*Anax immaculifrons* Rambur, 1842 (p.216)
54a	*Boyeria cretensis* Peters, 1991 (p.216)
70a	*Cordulegaster helladica* (Lohmann, 1993) (p.213)
73a	*Somatochlora meridionalis* Nielsen, 1935 (p.213)
73b	*Somatochlora borisi* Marinov, 2001 (p.216)
85a	*Orthetrum taeniolatum* (Schneider, 1845) (p.217)
91	*Orthetrum anceps* (Schneider, 1845) (p.213)
91a	*Orthetrum sabina* (Drury, 1770) (p.217)
99	*Sympetrum sinaiticum* Dumont, 1977 (p.213)
111a	*Trithemis festiva* (Rambur, 1842) (p.217)

Key to suborders of European Odonata

1   Fore- and hindwings similarly shaped (figs 14,15, p.44; 24–27 p.52), the basal portions adjoining the thorax narrow and, except in Calopterygidae (figs 15,27), parallel-edged (petiolate). Discoidal cell four-sided. Eyes well separated on top of the head (fig. 11, p.41). Females with an ovipositor (fig. 20, p.47), males with two superior and two inferior anal appendages (fig. 22, p.47). Mostly smaller and more slightly-built insects, resting with wings held vertically together over the abdomen (slightly apart in Lestidae). Damselflies .... Zygoptera

–   Forewings narrower than hindwings (figs 16,17, p.45), wings never petiolate. Discoidal cell three-sided. Eyes, except in Gomphidae, touching on top of the head. Females without an ovipositor (except Aeshnidae, Cordulegastridae), males with two superior and one inferior anal appendages (fig. 23, p.47). Mostly larger and more robust insects, resting with wings held horizontally or (some Libellulidae) even drooping, at right angles to the long axis of the body. Dragonflies .... Anisoptera

Key to families of European Zygoptera

1   Wings not petiolate and with twelve or more antenodal cross-veins (figs 15, p.44; 27), wings often pigmented. Mesometa-pleural suture complete (fig. 12, p.42). (Calopterygoidea) .. 2

  –   Wings petiolate and with only two antenodal crossveins (figs 14, p.44; 24–26), hyaline. Mesometapleural suture incomplete (fig. 13, p.42) 3

2(1)   Wings (fig. 27) broad and more or less pigmented, eighteen or more antenodals, pterostigma absent (a white, false pterostigma is present in females only). Abdomen slender ...
.... Calopterygidae (p.54)

  –   Wings narrow, hyaline or pigmented only apically, twelve to fourteen antenodals, pterostigma brown to black, well developed in both sexes. Abdomen stout ....
.... Euphaeidae (p.57)

3(1)   Several wing cells five-sided, veins IR_3 and R_{4+5} originate far basal to the node, pterostigma distinctly longer than broad (fig. 26). Body usually strongly metallic green to bronze. (Lestoidea) .... Lestidae (p.58)

  –   Most wing cells four-sided, veins IR_3 and R_{4+5} originate beneath the node, pterostigma often not or scarcely longer than broad (figs 14, p.44; 24,25). Body at most only weakly metallic. (Coenagrionoidea) 4

4(3)   Discoidal cell almost rectangular (figs 14,25). Tibiae more or less expanded .... Platycnemididae (p.67)

  –   Discoidal cell trapezoidal (fig. 24). Tibiae not expanded ......
.... Coenagrionidae (p.71)

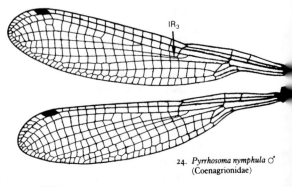

24. *Pyrrhosoma nymphula* ♂
(Coenagrionidae)

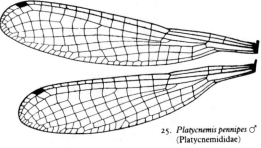

25. *Platycnemis pennipes* ♂
(Platycnemididae)

oblique crossvein

IR_3

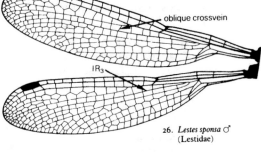

26. *Lestes sponsa* ♂
(Lestidae)

pseudopterostigma

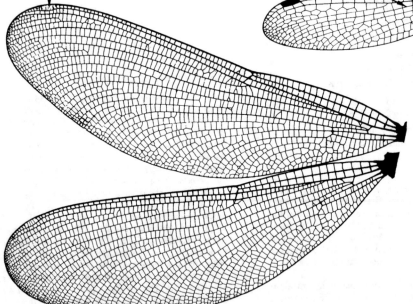

27. *Calopteryx splendens* ♀
(Calopterygidae)

Figures 24–27
Left wings of Zygoptera

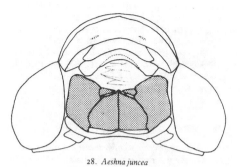

28. *Aeshna juncea*

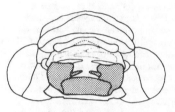

29. *Gomphus pulchellus*

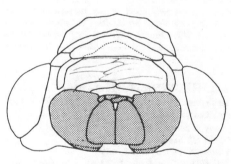

30. *Cordulegaster boltoni*

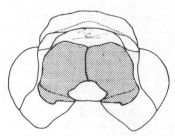

31. *Orthetrum cancellatum*

Figures 28–31 Heads, in ventral view, to show the differing forms of the labium (median and lateral lobes stippled)

Key to families of European Anisoptera

1   Discoidal cells in fore- and hindwings similarly shaped and more or less equidistant from arculus; two primary antenodals prominent, remaining antenodals not aligned over costal and subcostal spaces (fig. 16, p.45). Middle lobe of labium large, lateral lobes widely separated (figs 28–30) 2

  –   Discoidal cells in fore- and hindwings differently shaped, much nearer to arculus in hindwing than in forewing; no primary antenodals distinguishable, all in alignment across costal and subcostal spaces (fig. 17, p.45). Middle lobe of labium small and concealed by the contiguous lateral lobes (fig. 31). (Libelluloidea) 4

2(1) Pterostigma with an oblique crossvein (brace) at its proximal end (figs 16,32). Middle lobe of labium entire or with only a minute apical notch (figs 28,29). (Aeshnoidea) 3

  –   Pterostigma without a proximal brace (fig. 33). Middle lobe of labium deeply cleft apically (and with complete median suture) (fig. 30). (No green or blue pigmentation). (Cordulegastroidea) .... Cordulegastridae (p.136)

3(2) Eyes contiguous dorsally, usually for an appreciable distance. Middle lobe of labium with median suture (fig. 28). Blue pigmentation often present .... Aeshnidae (p.98)

  –   Eyes widely separated dorsally. Middle lobe of labium without a median suture (fig. 29). Body colour blackish and yellow to green, never blue .... Gomphidae (p.120)

4(1) Posterior eye margin with prominence (fig. 34). Males with auricles on S2 and angled anal margin to hindwing. Body metallic (except *Epitheca*). Anal loop of hindwing (figs 256,258,260, p.144) not always foot-shaped .... Corduliidae (p.145)

  –   Posterior margin of eye without a conspicuous prominence. Males without auricles and with hindwing anal angle smoothly rounded. Body seldom (*Trithemis, Zygonyx, Diplacodes*) with metallic areas. Anal loop of hindwing distinctly foot-shaped (fig. 17, p.45) .... Libellulidae (p.156)

32. *Onychogomphus forcipatus*

brace

33. *Cordulegaster boltoni*

Figures 32, 33 (above) Pterostigmata of right wing (stigmal brace of *Onychogomphus forcipatus* indicated)

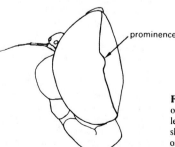
prominence

Figure 34 (left) Head of *Cordulia aenea* in left lateral view to show the prominence on the posterior eye margin

53

Suborder: Zygoptera (Damselflies)

CALOPTERYGIDAE

Synonym: Agriidae

This family is represented in Europe by a single genus, *Calopteryx*. These are the so-called demoiselle flies, large (body length about 45–48mm) and colourful insects with bright metallic reflections from their bodies. They have a fluttering flight, rarely straying far from water, and their broad wings are conspicuously pigmented. Calopterygids are the only Odonata in the European fauna to have a distinct courtship display, their coloration playing an important part in this. The very large number of wing veins and minute cells distinguish Calopterygidae from other families of Zygoptera. In addition, mesometapleural suture is complete (as in Euphaeidae), and there is no true pterostigma in either sex although females have a white pseudopterostigma divided by crossveins.

Water inhabited by *Calopteryx* larvae must be well oxygenated but not too strongly-flowing to prevent growth of water plants amongst which the larvae live. Males establish territories in which the females oviposit alone but guarded by the males. During oviposition, the abdomen is deeply immersed below the water surface to lay eggs in plant stems; often the female becomes completely submerged.

CALOPTERYX Leach, 1815

Agrion Fabricius, 1775

The name to be applied to this genus, whether *Calopteryx* or *Agrion*, has been long and often discussed in entomological literature (see Montgomery, 1954, and under *Coenagrion*). All of Linnaeus' described species of Odonata were allocated to the single genus *Libellula*, and it was Fabricius who first divided the group when he erected the genus *Agrion* and included in it *puella* and *virgo*. No type species was, of course, indicated by Fabricius. Latreille (1810) appears to designate *virgo* as the type of *Agrion*, but Leach (1815) (widely circulated at least five years before this publication date) included *virgo* in his genus *Calepteryx* (emended to *Calopteryx* by Burmeister, 1839), fixing *puella* by elimination as the type of *Agrion*. However, 'type' species designation by these early authors is not explicit. An objective of zoological nomenclature is to provide nominal stability and this is best achieved when based upon common (accepted) usage. Although a few British authors still use *Agrion*, the majority of other European and American writers adopt *Calopteryx* (meaning 'beautiful wings'), and this course is followed here in accord with the preference of most odonatists. As Calvert observes, 'What *Calopteryx* is is known, what *Coenagrion* is is known. *Agrion* is the uncertainty' (Calvert *et al.*, 1949).

There are about thirty known species of *Calopteryx*, found in the Palaearctic region, Africa and North America. In Europe, there are three well-defined, reproductively isolated species, but each of these has a number of named, more or less geographically separated populations which are mostly poorly-differentiated morphologically. These are usually accorded the rank of subspecies. Using electro-

phoresis to determine isoprotein electromorphosis of a range of these forms, it has recently been shown (Maibach, 1985) that, with one exception, there is only very weak differentiation between the forms within each of the three species. The exception concerns *C. splendens xanthostoma* which is discussed below. Maibach (1987) provides a key to the species and subspecies of *Calopteryx* found in western Europe and North Africa.

Key to species of *Calopteryx*

1 Body of male coppery to purplish in colour, not green; female hindwing weakly infuscated in proximal three-quarters but with the apical part brownish, the pigmentation strongest at the junction of the two regions. Tibiae partly pinkish. Apex of male abdomen rose-coloured beneath. Wings of male with brown pigmentation extending right to the base between radius and subcosta, lining the antenodal veins ..... *haemorrhoidalis* (p.56)

– Body of male blue-green; female hindwing uniformly weakly brownish (*virgo*) or almost hyaline (*splendens*). The apical third of the female hindwing of *virgo* may be slightly darker than elsewhere, but there is never a distinctly differentiated brownish tip to the wing as in *haemorrhoidalis*. Tibiae black. Apex of male abdomen yellowish to brick-red beneath. Wings of male basally hyaline 2

2 Male wings broadly hyaline basally with pigmentation extending only slightly proximal to the node. Female wings hyaline or faintly yellowish. Wings relatively narrower with coarser venation than in *virgo* .... *splendens* (p.55)

– Male wings narrowly hyaline basally, the pigmentation extending to at least half-way between base and node. Female wings brownish. Wings broader with very fine venation, particularly in the male .... *virgo* (p.54)

CALOPTERYX VIRGO (Linnaeus)   Plate 1 (1,2)

Libellula virgo Linnaeus, 1758, *Syst. Nat.* (Edn. 10) **1** : 545.
Type locality: Sweden.

   Agrion nicaensis Risso, 1826
   Agrion colchicus Eichwald, 1837
   Agrion vesta Charpentier, 1840

E – Beautiful Demoiselle, **F** – Le Caloptéryx vierge,
D – Blauflügel-Prachtlibelle

DESCRIPTION OF ADULT

The wings of the mature male are extensively dark brown with blue reflections, only the base and sometimes the extreme apex remaining clear. Females and immature males have uniformly light brown wings (occasionally the hindwing of the female may be darker over the apical third). The pseudopterostigma is relatively far from the wing apex, the distance from wing apex to the centre of the pseudopterostigma being greater than one-fifth of the distance between wing apex and node. The wings are broader in *virgo* than in either of the other species and those of the male usually have exceedingly dense venation. The body is metallic blue-green (male) or greenish bronze (female), and the legs are black. The last three sternites of the abdomen of the mature male are reddish. Homeochrome (male-coloured) females are sometimes reported.

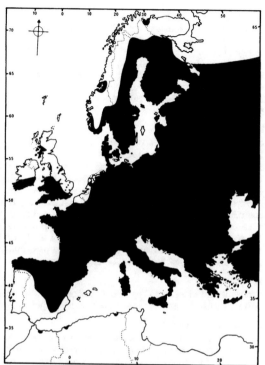

Map I – *Calopteryx virgo*

The male of the northern subspecies, *C. v. virgo*, has forewings clear only up to about the level of the arculus, but the extreme apex also is clear. *C. v. meridionalis* Sélys, 1873, however, has male wings clear from base to half way to node, but pigmentation then extends right to the apex. A third subspecies, *C. v. padana* Conci (in Conci & Nielsen, 1956), described from northern Italy, has much of the base of the male forewing clear as in *meridionalis*, and the apex also is clear, but confusingly there is a form (*schmidti* Conci in Conci & Nielsen, 1956) of this subspecies with entirely pigmented male wings. Maibach (1986) concludes that *padana* and form *schmidti* are synonyms of *C. v. virgo*. *C. v. festiva* (Brullé, 1832), described from Greece (Morea), also has heavily pigmented wings with only the bases narrowly hyaline, but the wings are broader and the body size larger than in the other forms.

BIOLOGY

A conspicuous insect which flies slowly and settles frequently on trees and bushes bordering streams and rivers with clean, rather rapidly-flowing water. An account of male territoriality and aggressive behaviour is given by Pajunen (1966b). Both this species and the next are especially sensitive to waterway management (clearance of vegetation), and this may account for their disappearance from parts of central Europe, particularly much of Switzerland.

FLIGHT PERIOD

Mostly June and July to early August in northern Europe, but in southern Europe it appears in May or even the end of April and continues to the beginning of September.

DISTRIBUTION (map 1)

From the west of Scotland, Ireland and Scandinavia south to North Africa. *C. v. meridionalis* (type locality: Provence) flies in Algeria, Morocco, the Iberian Peninsula, the Mediterranean islands, Italy except the north, the southern half of France (south of a line from Brittany to Jura) and Ticino in Switzerland, to be replaced by the typical subspecies in the northern part of the range excepting parts of northern Italy where *C. v. padana* occurs. *C. v. festiva* is known from southern Yugoslavia (Montenegro), Albania, Greece, south-eastern Bulgaria, Romania (Banat), Turkey (Dumont, 1977c) and the Caucasus. The species (or superspecies) extends across Asia to Japan (*C. v. japonica* Sélys, 1869); Miyakawa (1983) distinguishes *japonica* as a separate species and Kiauta (1968) finds chromosomal differences between European and east Asian forms.

CALOPTERYX SPLENDENS (Harris)
Plate 2 (3–5)

Libellula splendens Harris, 1782, *Expos. English Ins.* : 99. Type locality: England.

    Agrion parthenias Charpentier, 1840
    Calopteryx ludoviciana Sélys, 1840

E – Banded Demoiselle, **F** – Le Calptéryx éclatant, **D** – Gebänderte Prachtlibelle

DESCRIPTION OF ADULT

The wing of this species is less broad than that of *C. virgo* and the basal part as far as the node is clear in the male. Wing veins of the female are metallic green, the membrane is slightly yellowish, and the white pseudopterostigma is nearer the tip, about one-seventh of the distance between wing apex and node, than in female *C. virgo*. Females with male-coloured wings are prevalent in some populations. The body colour is metallic blue-green (male) to green or green-bronze (female), and the legs are black.

Four major forms, differing mainly in the extent of pigmentation on the male forewings, have been recognized in Europe. *C. s. splendens* has a broad band of pigmentation from the node to within four to six millimetres of the wing apex. *C. s. xanthostoma* (Charpentier, 1825) has relatively narrow wings (Plate 2(4)) with pigmentation in both fore- and hindwings extending from the nodes to the wing apices, the yellow colour on lower face and thorax is more extensive, and the apex of the male abdomen is yellowish beneath, not whitish as in *C. s. splendens*. *C. s. caprai* Conci in Conci & Nielsen, 1956, is intermediate between *splendens* and *xanthostoma* with respect to the extent of the pigmented wing-band in the male, and in both sexes venation is denser than in the other forms. Males of *C. s. balcanica* Fudakowski, 1930b, also have an extensive blue wing-band which reaches, or almost reaches, the wing apex and proximally may extend to half-way between wing base and node; the band has a rather jagged proximal edge (Dumont, 1977d).

European populations of *C. splendens* constitute a 'formenkreis' (Schmidt, Eb., 1978a), an assemblage of slightly differing forms each of which is found in a particular region. It is an oversimplification to consider *C. splendens* to include four European subspecies of equal rank. Maibach (1985) has demonstrated by electrophoresis that whilst *C. s. splendens* and *C. s. caprai* are rather weakly differentiated, *C. s. xanthostoma* is quite well-defined. On these and

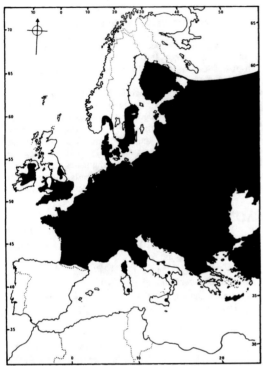

Map 2A – *Calopteryx splendens* excluding *C. s. xanthostoma*

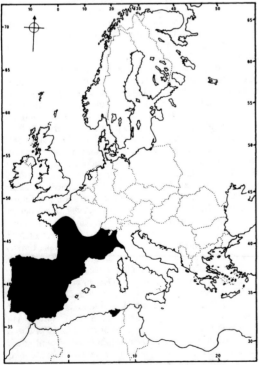

Map 2B – *Calopteryx s. xanthostoma*

morphological grounds, Maibach (1986) proposes that west European populations are referable to three subspecies (*splendens*, *caprai* and *xanthostoma*), at the same time recognizing the higher level of differentiation of *C. s. xanthostoma*. Dumont (1972) and others treat *xanthostoma* as a good species, but the fact that it interbreeds with *C. s. splendens* in their contact zone suggests that it is more appropriately considered a subspecies, or perhaps a semi-species.

BIOLOGY

C. s. xanthostoma, like *C. virgo*, is associated with rather fast-flowing, clear streams and rivers, but *C. s. splendens* and *C. s. caprai* frequent sluggish rivers, canals, or occasionally lakes with more or less muddy beds. Both sexes spend much time at rest on waterside vegetation, preferring reeds and herbage to trees and bushes, in contrast to *C. virgo*. Detached wings of male *C. splendens* are sometimes found in large numbers beside water inhabited by the species, presumably the residue of an avian predator's meal. Adult biology of this and the preceding species are described by Zahner (1959, 1960).

FLIGHT PERIOD

From the end of April to beginning of September, but in northern latitudes mainly from June to early August.

DISTRIBUTION (maps 2A, 2B)

Similar to that of *C. virgo*, extending east to China. *C. s. splendens* occurs from Britain and southern Scandinavia over most of northern Europe, to be replaced (with some overlap) by *C. s. xanthostoma* in southern France (north to the Charente), northern Italy, western and southern Switzerland (Geneva, Lugano), the Iberian Peninsula, and Algeria. *C. s. caprai* is found in parts of northern and central Italy and in the Mediterranean region of southern France (north to the Lot). *C. s. balcanica* and other named forms inhabit the east Mediterranean parts of Europe. *C. s. splendens* has declined considerably in abundance during the last two or three decades.

CALOPTERYX EXUL Sélys, 1853

This is an endemic species in Morocco, Algeria and Tunisia. It is characterized chiefly by narrow, almost clear wings which lack any pigmented bands, and by some yellow coloration on the inner surfaces of the hind femora, but morphologically it resembles *C. splendens*. Other clear-winged species inhabit Turkey and the Middle East.

CALOPTERYX HAEMORRHOIDALIS
(Vander Linden)   Plate 3 (6–8)

Agrion haemorrhoidalis Vander Linden, 1825, *Monograph. Libell. Europ. Spec.* : 34.
Type locality: Lazio, Italy.

F – Le Caloptéryx hémorroïdal, Le Caloptéryx méditerranéen

DESCRIPTION OF ADULT

An easily recognizable species. The body-colour of the male is metallic purplish red and, when mature, the ventral surfaces of the apical abdominal segments are rose-coloured. The wing has a broad, pigmented band. Tibiae in both sexes are pinkish. The body of the female is metallic green to bronze, and the hindwing has a conspicuous brown band towards the apex.

As in the two other European species of *Calopteryx*, there is variation in the extent of male wing pigmentation and this is used to differentiate forms. The typical subspecies has pigmentation reaching almost or quite to the apex of the hindwing and nearly to the apex of the forewing, whereas in *C. haemorrhoidalis occasi* Capra, 1945, the band extends at most to midway between node and wing apex of the forewing and the tip of the hindwing is clear. A form of doubtful status, *papyreti* Zeller in Sélys & Hagen, 1854, resembles *C. h. occasi* in coloration but is very small; it was described from specimens caught on the banks of the River Cyane in Sicily, but has not been rediscovered. Maibach (1986) finds *C. h. occasi* to merge with *C. h. haemorrhoidalis* without sharp morphological or geographical transition and he concludes that *C. haemorrhoidalis* in Europe is not subspecifically divisible.

BIOLOGY

A species of clean, running water.

FLIGHT PERIOD

From the end of April to the beginning of September (Majorca), but mainly June to the beginning of August.

DISTRIBUTION (map 3)

Spain, Portugal, southern France, peninsular Italy, Sicily, Sardinia, Corsica and North Africa. *C. h. occasi* occupies the more northerly part (France, northern Italy) of this range and *C. h. haemorrhoidalis* the southern part.

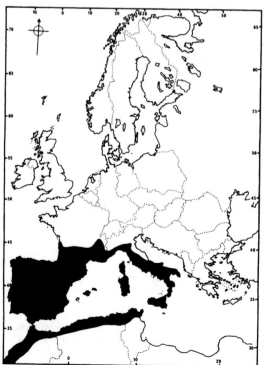

Map 3 – *Calopteryx haemorrhoidalis*

EUPHAEIDAE

Synonyms: Epallaginidae, Epallagidae

A family allied to Calopterygidae but differing from *Calopteryx* in having fewer wing veins, larger wing cells and a pterostigma in both sexes. In some genera (e.g. *Euphaea*, but not *Epallage*) occur structures analogous to the auricles of Anisoptera. The larva is exceptional in having seven lateral pairs of external gills on S2–8, an adaptation to living in temporary waters.

EPALLAGE Charpentier, 1840

The single European species is a large insect, somewhat resembling *Calopteryx* but from which it can easily be distinguished by its venation. There are only twelve to fourteen antenodal veins and the discoidal cell is rectangular and undivided by veins. The posterior margin of the wing joins the anal vein at some distance before its base so that the wings are more petiolated than those of *Calopteryx*. An elongated, brown (yellowish in immatures) pterostigma is present in both sexes and the wings are narrower than those of *Calopteryx*.

EPALLAGE FATIME (Charpentier)   Plate 4 (9,10)
Agrion fatime Charpentier, 1840, *Libell. europ. descr. depict.*: 132.
Type locality: European Turkey.
  Euphaea fatime (Charpentier) Sélys in Sélys & Hagen, 1850

DESCRIPTION OF ADULT

Body dark brown to black, extensively pruinescent in the adult male on head, thorax, abdomen and even legs. There are five yellow rays on each side of the thorax and a yellow median dorsal line. The abdomen has a fine, mid-dorsal yellow line and, on the anterior segments, a yellow line on each side. The head is extensively yellow-marked. All of this patterning is, however, obscured by pruinescence in the mature male. The wings are mainly clear but may be more or less yellow-tinted towards the bases, and the tip of the wing from the apex of the pterostigma is sometimes strongly brown-pigmented. The head is broad, the eyes are large, and the antennae are short with the second segment no longer than the rest. The abdomen is relatively short and robust (hence the name *fatime*), as also are the legs which have only short spines. The male superior appendages are broad and, in dorsal view, apically rounded, the tips being curved downwards and inwards; the inferior appendages are short.

BIOLOGY

Associated with fast-flowing rivers and streams, sometimes those that seasonally dry up. Adults rest on rocks or vegetation.

FLIGHT PERIOD

Specimens examined have been taken from April to July.

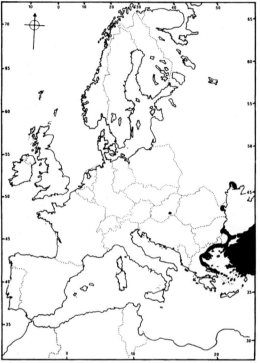

Map 4 – *Epallage fatime*

DISTRIBUTION (map 4)

From north-west India, Afghanistan, Pakistan and Kashmir, Iran, the Arabian Peninsula, Caucasus, Middle East, and the eastern Mediterranean to Rhodes, Cyprus, Greece, Bulgaria and Romania. A recent record for Kiskunság, Hungary (Steinmann, 1986) represents a considerable westward extension of the known range.

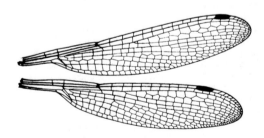

Figure 35 Right fore- and hindwings of *Sympecma fusca*

LESTIDAE

Lestids are very easily recognized damselflies, having many of the cells in their wings (figs 26, p.52; 35) pentagonal and not almost all quadrilateral as they are in platycnemidids and coenagrionids. Further, the pterostigma is at least twice as long as broad, wing veins IR_3 and R_{4+5} extend far proximal of the level of the node, veins R_3 and IR_3 are connected by an oblique crossvein, the discoidal cells have very acute distal angles, the body is metallic green in most species and the legs have long spines. The suture between mesepimeron and metepisternum is visible only in its upper part, as in platycnemidids and coenagrionids. Pruinosity develops in older individuals, particularly males, on the back of the head, the thorax, and on the base and apex of the abdomen.

At rest the wings are held partly open (except in *Sympecma*), not closed vertically over the back as in other damselflies. Lestids are agile and relatively rapid in flight (*leste* is a French word meaning nimble). Females oviposit in branches or stems of bushes or large herbaceous plants growing over or near water, or in floating vegetation, and they do so usually in tandem, the male grasping the support with his legs whilst the female inserts her eggs into the plant tissue. European *Lestes* overwinter as diapausing eggs and larvae which develop to adults the following summer, but *Sympecma* are remarkable in hibernating in the adult stage. Biology of the German species is described by Loibl (1958).

There are two European genera in the family.

Key to genera of Lestidae

1 Dark parts of body brownish bronze, only weakly greenish; abdomen with a distinct dorsal pattern; thorax with broad, pale humeral lines. Posterior margin of pronotum with a large median lobe. Wings (fig. 35) rather pointed with the pterostigma much nearer to the wing apex in the forewing than in the hindwing; discoidal cell of forewing narrower and shorter than that of hindwing .... *Sympecma* (p.65)

– Body mostly metallic green or green-bronze dorsally and yellow laterally, the yellow colour scarcely visible on the abdomen in dorsal view except at the intersegmental incisures; humeral sutures marked by fine, yellow lines. Mature males of most species develop areas of powdery bluish pruinescence at the base and apex of the abdomen, and on the dorsum of the thorax between the wings. Posterior margin of pronotum without a median lobe. Wings (fig. 26) more rounded apically; pterostigmata of fore- and hindwings almost equidistant from apices; discoidal cells of similar shape in both pairs of wings ....
.... *Lestes* (p.58)

LESTES Leach, 1815

Puella Brullé, 1832
Anapetes Charpentier, 1840

A genus of about eighty species, distributed almost worldwide. The six European representatives are of similar general appearance but can be distinguished fairly easily on close examination. They vary little and few subspecies have been described.

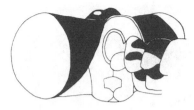

36. *Lestes viridis*

37. *L. barbarus*

Figures 36, 37 Heads in postero-lateral view to show the extent of metallic green coloration (shaded black)

Key to species of *Lestes*

1   Lower posterior part of head, between mouth-opening and neck, yellow (fig. 37) 2
–   Lower posterior part of head metallic green-bronze, sometimes with bluish pruinescence (fig. 36) 3

2(1) Pterostigma unicoloured, brownish. Inferior abdominal appendages of male (fig. 40) very short and broad with long apical setae; superior appendages mainly black. Valvifer of female (below base of penultimate abdominal segment) with a short point on its posterior edge (fig. 46) .... *virens* (p.62)
–   Pterostigma bicoloured, the apical third whitish and the basal part brown. Male inferior abdominal appendages (fig. 39) longer with diverging, finger-like apices and without long setae; superior appendages yellow with the tips dark. Valvifer of female rounded posteriorly (fig. 45) ....
.... *barbarus* (p.61)

3(1) Pterostigma very large and dilated, subtending three or four cells .... *macrostigma* (p.62)
–   Pterostigma smaller, extending above only two cells (fig. 26, p.52) 4

4(3) Pterostigma brown. Inferior abdominal appendages of male short, less than half as long as superior appendages and without long, fringing hairs (fig. 38). Female valvifer rounded posteriorly (fig. 44) .... *viridis* (p.61)

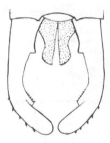

38. *Lestes viridis*

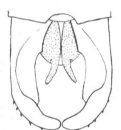

39. *L. barbarus*

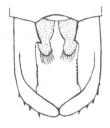

40. *L. virens*

41. *L. sponsa*

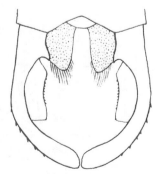

42. *L. macrostigma*

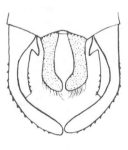

43. *L. dryas*

Figures 38–43 Male terminal abdominal appendages, in dorsal view, of *Lestes* species. The inferior appendages are stippled

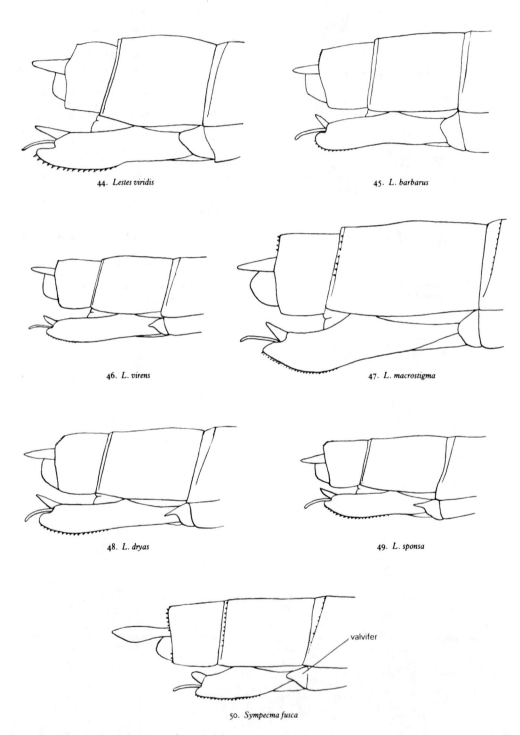

44. *Lestes viridis*

45. *L. barbarus*

46. *L. virens*

47. *L. macrostigma*

48. *L. dryas*

49. *L. sponsa*

valvifer

50. *Sympecma fusca*

Figures 44–50 Apical abdominal segments, from the right, of female Lestidae

– Pterostigma black. Inferior abdominal appendages of male long, exceeding half the length of the superior appendages and with long, fringing hairs (figs 41,43). Female valvifer with a posterior point (figs 48,49) 5

5(4) Pterostigma about 2.0–2.5 times as long as wide. Male inferior appendages with tips bent inwards (fig. 43). Female valve extending distinctly beyond the apex of S10 (fig. 48) .... *dryas* (p.64)

– Pterostigma about three times as long as wide (fig. 26, p.52). Male inferior abdominal appendages almost straight (figs 22, p.47; 41). Female valve not or hardly extending beyond apex of S10 (figs 20, p.47; 49) .... *sponsa* (p.63)

LESTES VIRIDIS (Vander Linden)   Plate 4 (11,12)

Agrion viridis Vander Linden, 1825, *Monograph. Libell. Europ. Spec.* : 36.
Type locality: Bologna, Italy.

   Agrion leucopsalis Charpentier, 1825
F – Le Leste vert, **D** – Weidenjungfer

DESCRIPTION OF ADULT

This species is placed by some authors in the genus *Chalcolestes* Kennedy, 1920, mainly because its larva differs from other *Lestes* larvae, but the adult is generally similar to other European species and it is most conveniently treated with them. It is a large insect, males usually 42–47mm in length and females about 3mm shorter. The body colour is dark green with quite strong coppery reflections dorsally and yellow ventrally. The lower part of the face is yellow, but this colour does not extend to the posterior surface of the head (fig. 36). The clear brown pterostigma (yellowish in immature specimens) is a useful indicator in the field. Perceptible pruinescence is not developed by mature insects so that the reddish brown colour of the thorax between the wing bases is never obscured. The superior male appendages (fig. 38) are yellow with conspicuously darkened apices; other features of the abdominal append-ages and ovipositor (fig. 44) are mentioned in the key.

BIOLOGY

Frequents ponds, lakes, canals and slow-flowing rivers, but overhanging bushes, especially sallows, are required for oviposition. It is the only species of *Lestes* (*sensu lato*) sometimes to develop in running water and it will take advantage of intermittent spring streams. Behaviour and ecology were studied by Jurzitza (1969) and Dreyer (1978); males defend a vertical territory in trees growing at the waters' edge and guard their ovipositing partners against the attentions of other males. Twigs of *Salix* are often selected as an oviposition site, although *Myrica gale* and a variety of tree species have also been recorded. Galls may form about the oviposition site.

FLIGHT PERIOD

L. viridis has a late flight period, chiefly August and September, although in the extreme south of Europe it may appear in June, and occasional individuals persist into November.

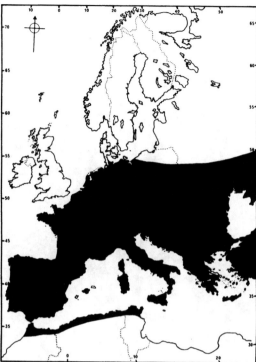

Map 5 – *Lestes viridis*

DISTRIBUTION (map 5)

Southern and central Europe north to the Channel Islands and the Netherlands. It occurs outside Europe in North Africa, Turkey, Georgia and the Middle East.

LESTES BARBARUS (Fabricius)   Plate 5 (14)

Agrion barbara Fabricius, 1798, *Suppl. entomol. syst.* : 286.
Type locality: North Africa (Barbary).

F – Le Leste sauvage, **D** – Südliche Binsenjungfer

DESCRIPTION OF ADULT

The bicoloured pterostigma of this species is diagnostic, although in teneral specimens the two colours may not be distinct. It shares with *L. virens* the feature of yellow coloration on the back of the head below the neck (fig. 37), but *L. barbarus* is a considerably larger insect (body length generally 40–45mm) than *L. virens* and its male abdominal appendages (fig. 39) and female ovipositor (fig. 45) are distinctive. The yellow humeral lines on the thorax are relatively broad. At the apex of the abdomen, yellow coloration extends on to the dorsal surface of S9–10, narrowly bordering the central dark markings, but this may be obscured by bluish pruinescence in mature males.

BIOLOGY

Breeds in stagnant water, sometimes slightly brackish. The female oviposits in *Juncus, Carex, Alisma* and similar plants, as well as in the branches of shrubs. The ecology of

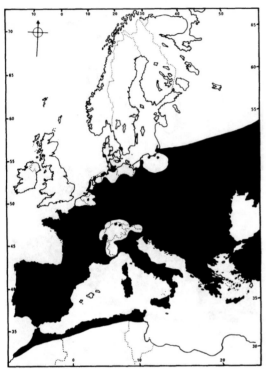

Map 6 – *Lestes barbarus*

L. barbarus in the German Democratic Republic is discussed by Donath (1981). Insects tend to remain faithful to the ponds in which they developed, so that colonization of new breeding sites is slow (Utzeri *et al.*, 1984).

FLIGHT PERIOD

May to the end of October. Peak emergence in Italy about mid-May (Carchini & Nicolai, 1984) but markedly later further north.

DISTRIBUTION (map 6)

In Europe, *L. barbarus* is mainly Mediterranean in distribution, being widespread in Spain, Italy and southern France, and becoming progressively more local northwards to Jersey (Channel Islands). It is uncommon in Germany and Austria, has been recorded only erratically from southern Switzerland, and only a few isolated colonies exist in Belgium, Holland, Denmark (Nielsen, 1979) and Poland (Mielewczyk, 1972). The most northerly record is from Kullaberg in Sweden (Ander, 1963). *L. barbarus* is also found in North Africa, Iran, and the south of the USSR east to India and Mongolia.

LESTES VIRENS (Charpentier)   Plate 5 (13)
Agrion virens Charpentier, 1825, *Horae entomol.* : 8.
Type locality: Lusitania (Portugal, S.W. Spain).
  Agrion paedisca Eversmann, 1836
  Lestes vestalis Rambur, 1842
  Lestes sellatus Sélys, 1862
F – Le Leste verdoyant, **D** – Kleine Binsenjungfer

DESCRIPTION OF ADULT

A delicately-built species and usually rather small (body length seldom exceeding 39mm). The lower, posterior part of the head is yellow, as in *L. barbarus*, but the pterostigma is uniformly brownish and the yellow humeral lines are narrow. Male abdominal appendages and female abdominal apex are illustrated in figures 40 and 46 respectively.

Insects from northern Europe, distinguished as subspecies *L. virens vestalis*, have the sides of the thorax more extensively darkened than those from southern regions. In the nominotypical subspecies, *L. v. virens*, the yellow line bordering the humeral suture is unbroken and runs almost to the base of the forewing, whereas in *L. v. vestalis* this line is usually reduced and does not nearly reach the forewing base. Additionally, in *vestalis* there is a black area of variable extent on the metepimeron between hind leg and hindwing; the metepimeron in *L. v. virens* is entirely yellow. The two forms are not sharply delimited, however, and their separation as subspecies is of questionable merit.

BIOLOGY

Usually found around bushes and rank, herbaceous vegetation on the margins of ponds and lakes. Oviposits above the water level, often in *Juncus* and *Oenanthe* stems.

FLIGHT PERIOD

In southern Europe, *L. virens* may appear as early as April, but its flight period is mainly from July to September or occasionally later. Peak emergence in Italy is in early June (Carchini & Nicolai, 1984).

DISTRIBUTION (map 7)

L. v. virens is a southern insect, found in southern France, Spain, Sardinia and North Africa. *L. v. vestalis* occurs in Italy, northern France, and across central Europe to the eastern Mediterranean, but it does not reach Britain or Scandinavia. Ranges of the two forms, like their morphological differences, are not sharply defined.

LESTES MACROSTIGMA (Eversmann) Plate 5 (15)
Agrion macrostigma Eversmann, 1836, *Bull. Soc. Imp. Moscou* **9** : 246.
Type locality: Orenburg (Chkalov), Russia.
  Lestes picteti Sélys, 1840
F – Le Leste à grands stigmas, **D** – Dunkle Binsenjungfer

DESCRIPTION OF ADULT

Immediately recognizable by its enlarged pterostigma, *L. macrostigma* differs from the other *Lestes* species also in having broader, more rounded wings, and metallic violet and bluish reflections on thorax and abdominal base and apex. Only the central abdominal segments are dark green with extensive coppery reflections. Powdery blue pruin-

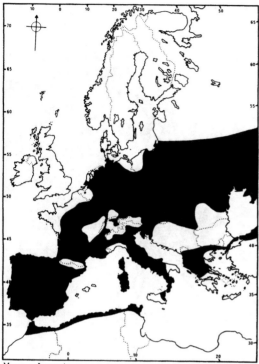

Map 7 – *Lestes virens*

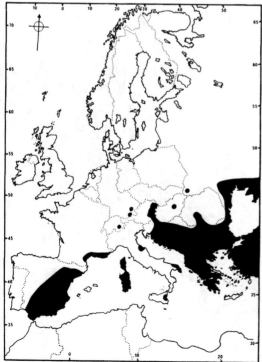

Map 8 – *Lestes macrostigma*

escence develops on the thorax and the base and apex of the abdomen in sexually mature individuals. The form of the male abdominal appendages (fig. 42) and of the female ovipositor (fig. 47) provide additional specific characters. *L. macrostigma* is large, similar in body length to *L. viridis*.

BIOLOGY

Larvae in stagnant, acid and often somewhat brackish water.

FLIGHT PERIOD

Mid-May to September.

DISTRIBUTION (map 8)

In Europe, *L. macrostigma* has a mainly Mediterranean distribution and is found in Spain, south-east France, Corsica, Sardinia, Sicily, peninsular Italy (rare, three localities on the Adriatic coast), Greece and Cyprus. There is a single, ancient record of its occurrence in Switzerland. It is very local in Hungary (Balaton), Austria and south Germany, but becomes more plentiful east through Turkey and southern USSR to Mongolia. Its reputed occurrence in North Africa is based on an old record from Morocco (Martin, 1910).

LESTES SPONSA (Hansemann)　　　Plate 6 (16,17)

Agrion sponsa Hansemann, 1823, *Wiedemann's zool. Mag.* 2: 159.

Type locality: Germany.

> *Lestes autumnalis* Leach, 1815 (in part)
> *Agrion forcipula* Charpentier, 1825
> *Lestes nympha* Stephens, 1835
> *Lestes picteti* Fonscolombe, 1838
> *Lestes spectrum* Kolenati, 1856

E – Emerald Damselfly, **F** – Le Leste fiancé,
D – Gemeine Binsenjungfer

DESCRIPTION OF ADULT

L. sponsa resembles *L. dryas*, but the characters given in the key, and the illustrations of male abdominal appendages (fig. 41) and female abdominal apex (fig. 49), should ensure correct identification of specimens that can be examined in the hand. *L. sponsa* is a smaller, more slender species than *L. dryas*, and the metallic spots on the first abdominal segment are roughly triangular in *L. sponsa* but rectangular in *L. dryas*, although in mature males this feature is obscured by bluish white pruinescence which develops also on the thorax between the wings and at the apex of the abdomen. In male *L. sponsa*, the pruinescence at the base of the abdomen usually completely covers the dorsal surfaces of S1 and S2, but it is less extensive in *L. dryas* occupying only the basal two-thirds of S2.

BIOLOGY

Frequents ponds and lakes with an abundance of tall grasses, sedges, rushes, horsetails and the like about their margins. It is in such plants that the female oviposits, generally above the water surface. The maturation period, spent away from water, is prolonged, lasting from sixteen to thirty days (Corbet, 1956).

FLIGHT PERIOD

Usually mid-June to October, but mature specimens

reported in Austria at the end of May. The life history of *L. sponsa* in Britain is described by Pickup *et al.* (1984).

DISTRIBUTION (map 9)

Widespread in central and northern Europe where it is the most abundant species of lestid. Less common in the south where it is found mostly in mountainous country, and almost absent from southern Spain (only Doñana Biological Reserve: Ferreras Romero & Puchol Caballero, 1984). There are old records from Algeria (Martin, 1910) and Tunisia. Eastwards, *L. sponsa* extends across Asia to Japan.

LESTES DRYAS Kirby   Plate 6 (18)

Lestes dryas Kirby, 1890, *Syn. Cat. Neur. Odon.* : 160.
Type locality: Germany.

    Lestes nympha sensu Sélys, 1840
    Lestes forcipula sensu Rambur, 1842
    Lestes uncatus Kirby, 1890

E – Scarce Emerald Damselfly, **F** – Le Leste dryade,
D – Glänzende Binsenjungfer

Early confusion of *L. dryas* and *L. sponsa* created nomenclatural difficulties, the names *nympha* Sélys and *forcipula* Rambur being unavailable for *dryas* having previously been applied to *sponsa*. The Nearctic *Lestes uncatus* was described on the same page as, and before, *L. dryas*, but reasons for the selection of *dryas* as the valid name of the taxon are put forward by Cowley (1935).

DESCRIPTION OF ADULT

Differences between *L. dryas* and *L. sponsa* are given in the key and mentioned under *L. sponsa*.

BIOLOGY

Habitats of *L. dryas* are shallow ponds, temporary pools and ditches, neutral to slightly acidic, frequently in woodland and often, in the south of the range, at an altitude of about 1000m. There is usually a luxuriant growth of marginal and emergent vegetation in which the insects fly low and inconspicuously. *Sympetrum sanguineum* and *Coenagrion pulchellum* are often found with *L. dryas*. *L. dryas* is declining in parts of northern Europe where heavy applications of nitrogenous fertilizers are causing eutrophication of standing water on arable land.

FLIGHT PERIOD

L. dryas makes its appearance as an adult ten to fifteen days after *L. sponsa* (Aguesse, 1968). Both species are most frequent from July to September.

DISTRIBUTION (map 10)

L. dryas is an Holarctic species, found in North America as well as in Europe and Asia (where it is found as far east as Japan). Like *L. sponsa*, *L. dryas* has a mainly central and northern European distribution, but it is the less common of the two species. Scarce in the British Isles, but rediscovered in south-east England in 1983 after being thought extinct, and surviving in a number of Irish localities. In the Mediterranean region it is found mainly in mountains, and in North Africa it is recorded only from Morocco (Lieftinck, 1966).

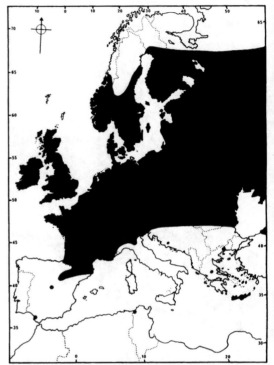

Map 9 – *Lestes sponsa*

Map 10 – *Lestes dryas*

SYMPECMA Burmeister, 1839

Sympycna Charpentier, 1840

The body of *Sympecma* lacks the strong, metallic green coloration characteristic of *Lestes*. The dark areas are black or dark brown with weak bronze to slightly greenish reflections, and the pale areas are light fawn, sometimes slightly pinkish on the sides of the thorax, and more whitish towards the apex of the abdomen and terminal appendages. Pale areas are more extensive than in *Lestes* on the dorsal surface of the abdomen, and the pale stripes on the humeral sutures are broad. The narrow, pointed wings (fig. 35, p.58) with the brown pterostigma distinctly further from the tip in the hindwing than in the forewing, and the very narrow and short forewing discoidal cell, further distinguish *Sympecma* from *Lestes* (fig. 26, p.52). Additionally, the posterior border of the pronotum in both sexes is produced into a broad, low lobe. The abdominal appendages of the female are almost or fully as long as the dorsal surface of S10.

Sympecma species hold their wings together over the abdomen when at rest, in the manner of damselflies in other families. They are the only European Odonata to overwinter as adults. The genus includes only two species and is exclusively Palaearctic.

Key to species of *Sympecma*

1 Dark stripe on mesepimeron entire and with lower edge only weakly sinuate; pale band on humeral suture with upper margin straight (fig. 51). Inferior abdominal appendages of male reaching to the basal teeth of superior appendages (fig. 52). Abdominal appendages of female as long as S10 dorsally (fig. 50, p.60) .... *fusca* (p.65)

– Dark stripe on mesepimeron narrowed and sometimes broken anteriorly, of irregular width, and with lower edge strongly sinuate; pale band on humeral suture with a small but distinct emargination on its upper edge behind the middle (fig. 53). Inferior appendages of male very short (fig. 54). Abdominal appendages of female not quite as long as S10 dorsally ....
.... *annulata* (p.66)

SYMPECMA FUSCA (Vander Linden)   Plate 6 (19)

Agrion fuscum Vander Linden, 1820, *Agrion. bonon. descr.*: 4.
Type locality: Bologna, Italy.
Agrion phallatum Charpentier, 1825
F – Le Leste brun, **D** – Gemeine Winterlibelle

DESCRIPTION OF ADULT

Predominantly pale fawn in colour with the top of the head and thorax, and a median dorsal abdominal stripe, dark brown or black. The dark areas have rather weak metallic reflections. The species blends in colour very well with the

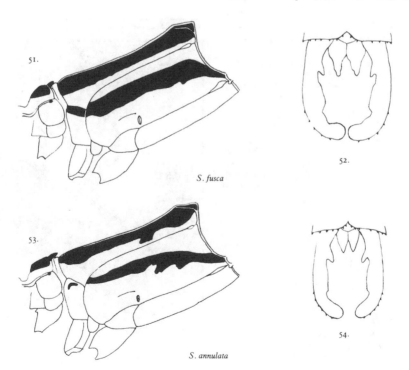

51.

52.

S. fusca

53.

54.

S. annulata

Figures 51–54 *Sympecma* species
(left) thorax from the left
(right) male terminal abdominal appendages in dorsal view

dry grasses and twigs amongst which it overwinters. In spring, there is a darkening of the body colour and the eyes acquire a bluish tint.

BIOLOGY

S. fusca breeds in oligotrophic pools and is tolerant of slight salinity. The female oviposits in floating plant debris such as broken reeds. Adults leave the breeding sites in autumn and move to woods and copses, usually rather open in character with a ground layer of tall grass. Here they overwinter, often a considerable distance from the water to which they return in spring and where they become reproductively active for the first time.

FLIGHT PERIOD

Adults emerge from late July to early September and then hibernate. Oviposition occurs mainly in April and May, and larval development is rapid, being completed in about two months. Adults can therefore be found throughout the year except for a brief period about midsummer. Over-wintering adults may be stimulated to activity by warm spells in the middle of winter.

DISTRIBUTION (map 11)

Widespread throughout southern and central Europe, and reaching the Channel Islands and southern Sweden in the north. Also found in North Africa, the Middle East, and eastwards to the USSR.

SYMPECMA ANNULATA (Sélys)   Plate 6 (20)

Sympycna paedisca Brauer var. *annulata* Sélys, 1887, *Annls Soc. ent. Belg.* **31** : 43.

Type locality: Malatya and Antakya (Antioch), Anatolia, Turkey.

  Sympycna paedisca Brauer, 1882 (*Agrion paedisca* Eversmann, 1836, is *Lestes virens* Charpentier)

  Sympecma annulata braueri Bianchi in Jacobson & Bianchi, 1905

  Sympecma striata St. Quentin, 1963

F – Le Leste enfant, **D** – Sibirische Winterlibelle

S. annulata encompasses a number of named forms and European specimens are referable to the western subspecies *S. a braueri* Bianchi which was described from Switzerland. *S. a. braueri* is quite often accorded full specific status, but Kiauta & Kiauta-Brink (1975) have shown that Dutch and Japanese specimens have identical karyotypes.

DESCRIPTION OF ADULT

S. annulata is very similar to *S. fusca* in general appearance but is slightly smaller, has a different thoracic pattern (fig. 53) and differs in the structure of the male terminal abdominal appendages (fig. 54).

BIOLOGY AND FLIGHT PERIOD

As *S. fusca*.

DISTRIBUTION (map 12)

France (Savoy), Italy (Lombardy, Piedmont and Venezia Tridentina), a number of localities in Holland, Switzerland and Austria, rather more frequent in Germany and Hungary, and becoming increasingly widespread in Russia and Asia (Iraq, Iran, Uzbekistan, Turkestan, Afghanistan, Kashmir, Siberia, Mongolia, Japan) where it exists in a number of forms.

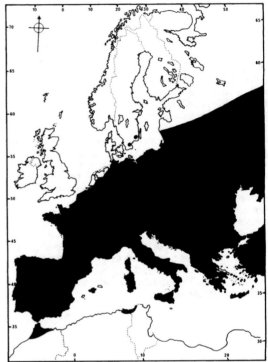

Map 11 – *Sympecma fusca*

Map 12 – *Sympecma annulata*

PLATYCNEMIDIDAE

The European representatives of this Old World family all belong to the genus *Platycnemis* which is easily recognized by the almost rectangular discoidal cell (quadrilateral cell) (figs 14, p.44; 25, p.52) and by the more or less flattened and expanded whitish tibiae (figs 55–57). There are usually only two cells (postquadrilaterals) between the discoidal cell and subnodal vein (the forewing depicted in figures 14 (p.44) and 25 (p.52) is atypical in this respect), the head is broad with a continuous, pale postocular line, and there are two pale stripes on each side of the dorsal surface of the synthorax, a broad antehumeral stripe and a narrower humeral stripe. The suture between mesepimeron and metepisternum is reduced and visible only in its upper part. Expanded, lanceolate tibiae may be present in both sexes, but this development affects only the middle and hind legs and is greater in males than in females.

PLATYCNEMIS Burmeister, 1839

Platyscelus Wallengren, 1894

A large genus of about thirty species, well represented in Asia and Africa but including only three European species. Species and forms occurring in Europe and Asia Minor have been studied by Heymer (1968).

Key to species of *Platycnemis*

1 Male abdomen mostly clear orange in colour, central segments with only incisures dark. Female abdomen dull orange. Tibiae (fig. 57) very weakly expanded in female, and in male only about as strongly as in females of the other species. Superior abdominal appendages of male (fig. 60, p.68) strongly bifid. Posterior margin of female prothorax (fig. 60) with large mid-lateral points .... *acutipennis* (p.70)

– Ground colour of abdomen white, greenish, bluish or, in some females, slightly rusty. Central segments of male abdomen with dark markings in addition to the darkened incisures. Tibiae (figs 55, 56) strongly expanded, especially in males. Superior appendages of males (figs 58,59, p.68) at most moderately bifid apically. Posterior margin of female prothorax (figs 58,59) with at most small mid-lateral points 2

2 Male abdomen white with dark marks on central segments reduced to small, brown, posterior points. Hind tibia (fig. 56) with median black line usually almost absent in males and short in females. Superior appendages of male (fig. 59) with only a weakly-developed dorsal tubercle, in lateral view appearing scarcely bifid. Posterior margin of female prothorax (fig. 59) with small mid-lateral points but strongly-developed lateral lobes .... *latipes* (p.69)

– Male abdomen light blue with dark median markings occupying almost the full length of the central segments (except sometimes S6). Median black line on tibiae (fig. 55) complete or almost so. Immature specimens have the coloration of *latipes*, but the male superior appendages are more distinctly bifid in lateral view (fig. 58) and the posterior margin of the female prothorax has at most a slight indication of mid-lateral points and is without prominent lateral lobes (fig. 58) .... *pennipes* (p.68)

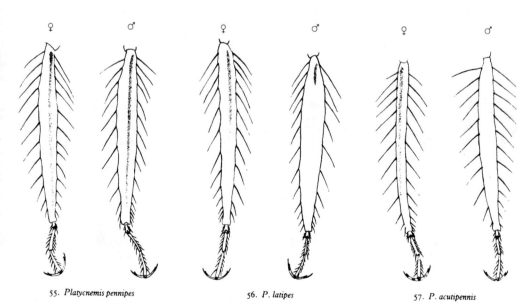

♀ ♂ ♀ ♂ ♀ ♂

55. *Platycnemis pennipes* 56. *P. latipes* 57. *P. acutipennis*

Figures 55–57 Hind tibiae and tarsi of *Platycnemis* species, females on the left and males on the right. The median, black tibial stripe is indicated by stippling

PLATYCNEMIS PENNIPES (Pallas) Plate 7 (21–23)

Libella pennipes Pallas, 1771, *Reisen durch versch. Prov. Rüss. Reiches Jahren 1768–1774* **1** : 469.

Type locality: Crimea, Ukraine, USSR.

Agrion platypoda Vander Linden, 1820
Agrion lacteum Charpentier, 1825

E – White-legged Damselfly, **F** – L'Agrion à larges pattes, **D** – Federlibelle

DESCRIPTION OF ADULT

The mature male has a mainly blue abdomen, although this colour never assumes the intensity of that of a male *Coenagrion*. S7–9 are mostly black with a median, blue line broadening posteriorly, and S10 is entirely blue or with just a pair of small, black, lateral spots; the more anterior segments are blue with black, median markings which are divided by a very fine, blue, central line. The black markings on S6 may be much reduced. The ground colour of the mature female abdomen is greenish yellow with the pale parts of head and thorax slightly rust-coloured. Newly emerged insects are creamy white with the dark marks on S2–6 restricted to a pair of small spots near the posterior margin of each segment. This pale, teneral form (*lactea*) was once thought to be a distinct species, but Schmidt, Er. (1950) concludes that variation in the extent of dark markings is humidity dependent, very pale forms being found in dry regions. Persistence of the white ground colour throughout life is reported in some populations (e.g. in southern Italy). The tibiae (fig. 55) in both sexes are whitish with a black line (rarely absent) running the length of the outer face of the central ridge. Typically the ratio maximum width of hind tibia : length of median tibial spine is 0.7 in males and 0.6 in females, but in *P. p. nitidula* (Brullé, 1832) (= *insularis* Sélys, 1863), a large and pale subspecies from the Balkans, the male ratio is about 1.3 and the black line on the broad hind tibia is vestigial. In these respects, *P. p. nitidula* resembles *P. latipes*. The forms of the male superior abdominal appendages and posterior margin of the female prothorax (fig. 58) are the most reliable features for distinguishing between *pennipes* and *latipes*.

BIOLOGY

Breeds in well-vegetated, slow-flowing streams and rivers, but sometimes also in ponds and lakes. It is the only European *Platycnemis* ever found in completely standing water. Floating vegetation, such as plant debris or water-lily leaves, is required for oviposition. Behaviour of adult *P. pennipes* has been studied by Buchholtz (1956). As in other species of the genus, males execute a fluttering courtship flight. Oviposition occurs in tandem. Males may use their modified tibiae in intraspecific 'threat' display.

FLIGHT PERIOD

Mostly in June and July, declining in numbers through August.

DISTRIBUTION (map 13)

From the south of Scandinavia and southern England over all Europe except Ireland and the Iberian Peninsula. *P. p. nitidula* occurs in Greece, Albania, Yugoslavia and perhaps Bulgaria (Dumont, 1977d). Eastwards, *P. pennipes* extends into the USSR as far as the valley of the Yenisey, and in parts of the Ukraine it is the most numerous species of Odonata. Southwards it reaches Armenia and the Mediterranean coast of southern Turkey.

PLATYCNEMIS SUBDILATATA Sélys, 1849

This is the sole representative of *Platycnemis* in North Africa (Tunisia, Algeria, Morocco). It is closely allied to *P. pennipes* from which it differs in its less dilated tibiae, whitish ground colour with a slight greenish blue tint in the mature male and reddish in the teneral female, male abdomen with median, dorsal black band (finely divided by pale mid-dorsal crest) extending from S1 on to S10, and in small details of male superior appendages and female prothorax.

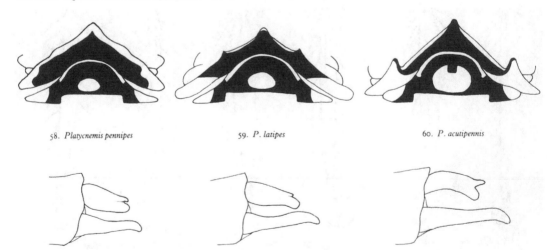

58. *Platycnemis pennipes*   59. *P. latipes*   60. *P. acutipennis*

Figures 58–60 *Platycnemis* species
(above) hind margins of female pronota in posterior view
(below) terminal abdominal appendages of males

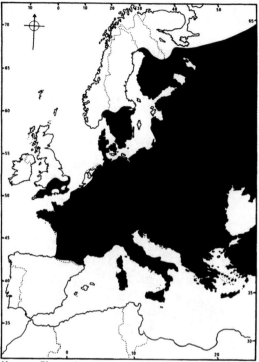

Map 13 – *Platycnemis pennipes*

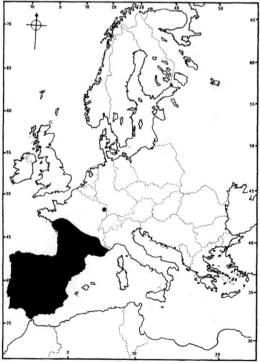

Map 14 – *Platycnemis latipes*

PLATYCNEMIS KERVILLEI (Martin, 1909)

This species occurs in Turkey, Syria, Lebanon and northern Iraq. There is a single record (Dumont, 1977c) from Gemlik on the southern coast of the Sea of Marmara. Teneral specimens have white abdomens with two small, black spots at the apex of each segment, but blue coloration quickly develops so that 'adults are completely dark blue' (Dumont, 1977c). The tibiae are unflattened in both sexes of this predominantly spring species.

PLATYCNEMIS LATIPES Rambur Plate 7 (24,25)

Platycnemis latipes Rambur, 1842, *Hist. nat. ins. Neur.* : 242.

Type locality: Montpellier, France.

D – Weisse Federlibelle, Helle Federlibelle

DESCRIPTION OF ADULT

The ground colour of the mature male is ivory white with a very faint greenish blue tinge developing on head and thorax. The black pattern on the last four abdominal segments is similar to that of *P. pennipes* although rather more extensive on S10. Dark markings are much reduced anteriorly, however, with only a median dorsal spot on the first abdominal segment and paired spots posteriorly on S2–6. These spots are largest on S6 and may be almost obliterated anterior to this segment. The following characters should prevent confusion with immature male *P. pennipes*: the black median line on the posterior tibia (fig. 56) is reduced to a very short basal dash; the tibiae are wider, about 0.9 times as wide as the length of middle tibial spines; the superior appendages are scarcely bifid apically (fig. 59); the pterostigma is shorter, its costal edge being about 2.5 times as long as its width (about 3.0 times in *pennipes*); body length (about 33mm) usually a millimetre or so less than that of *pennipes*. Separating females of *latipes* and *pennipes* is difficult but *latipes* has a slightly more rusty ground colour, the black line on the hind tibia usually extends for no more than the proximal third, and the pterostigma, as in the male, is smaller in *latipes* than in *pennipes*. The most reliable distinguishing feature of female *latipes* is the shape of the posterior pronotal margin (fig. 59) which has more prominent mid-lateral points and more extensive side lobes.

BIOLOGY

Slowly flowing streams and rivers. Behaviour studied by Heymer (1966). Males perform a zigzag flight in courtship.

FLIGHT PERIOD

Rather later than that of *P. pennipes*, remaining plentiful until almost the end of August.

DISTRIBUTION (map 14)

Endemic to Spain, Portugal and the south and west of France extending east to Alsace (Heymer, 1968). Only in France do *P. latipes* and *P. pennipes* occur together.

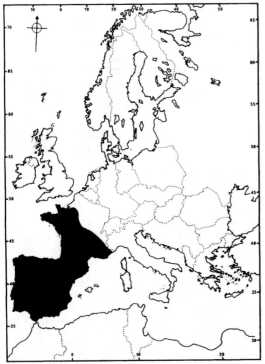

Map 15 – *Platycnemis acutipennis*

distinctly triple-pointed posterior margin of the female prothorax (fig. 60) ensure positive identification.

BIOLOGY

Usually inhabits slow-flowing rivers. Behaviour studied by Heymer (1966). Males have a fluttering courtship flight.

FLIGHT PERIOD

Earlier than *P. pennipes* or *P. latipes*, being abundant by the end of May and almost over by the end of July.

DISTRIBUTION (map 15)

Similar to that of *P. latipes*, occurring in southern and western France and the Iberian Peninsula, but reaching further north, as far as Normandy (Gadeau de Kerville, 1905). The three European species of *Platycnemis* fly together in south-west France.

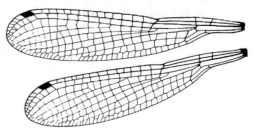

61. *Coenagrion puella* ♀

PLATYCNEMIS DEALBATA Sélys, 1863
(= *syriaca* Hagen in Sélys & Hagen, 1850)

A species closely allied to *P. latipes* and found from north-west India through Kashmir and Afghanistan west to the Caucasus, southern (but not western) Turkey, Iran, Lebanon, Israel, Syria and Jordan. Schneider (1985) figures the prothorax and male genitalia.

PLATYCNEMIS ACUTIPENNIS Sélys
Plate 7 (26,27)

Platycnemis acutipennis Sélys, 1841, *Rev. zool. Guérin Méneville* : 246.
Type locality: Montpellier, France.
   Platycnemis diversa Rambur, 1842

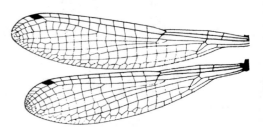

62. *Enallagma cyathigerum* ♀

DESCRIPTION OF ADULT

The abdominal ground colour, clear orange in the mature male and dull orange in the female, makes this a distinctive species. The eyes are bluish. The dark pattern on the male abdomen is limited to S1 and S7–10 where it resembles that of the two preceding species but is much reduced on S10. The tibiae (fig. 57) are only slightly expanded, the hind tibiae being only 0.6 times as wide as the length of a median tibial spine in the male and about 0.3 times as wide in the female. The relatively narrow, pointed wings, after which the species is named, provide a further point of difference between *P. acutipennis* and its European congeners, and the deeply bifid male superior abdominal appendages and the

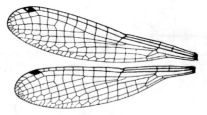

63. *Ischnura elegans* ♂

Figures 61–63 Left wings of Coenagrionidae

COENAGRIONIDAE

Synonyms: Coenagriidae, Agrionidae, Agriidae

(The question of *Coenagrion* versus *Agrion* is discussed under *Coenagrion*)

This is the dominant family of European damselflies and it is world-wide in distribution. It includes ninety-five genera, eight of which are represented in Europe. Males are brightly marked with blue or red, and the black areas, except in *Nehalennia*, are only weakly metallic. The wings (figs 24, p.52; 61–63) are unpigmented with most of their cells rectangular; the discoidal cell is trapezoid and followed by at least three cells before the subnodal vein. The suture between mesepimeron and metepisternum is visible only in its upper part, as in Lestidae and Platycnemididae. European genera are included in three subfamilies:

1 Coenagrioninae — wings shortly petiolate; female without a vulvar spine; frons not angulate. *Pyrrhosoma, Erythromma, Coenagrion, Cercion.*

2 Ischnurinae — wings with rather longer petioles; female with a vulvar spine; frons not angulate. *Enallagma, Ischnura.*

3 Nehalenniinae — wings shortly petiolate; frons angulate. *Nehalennia, Ceriagrion.*

Adults of most species can be found in large numbers amongst vegetation bordering the waters in which they developed, but females especially may sometimes be found far from water.

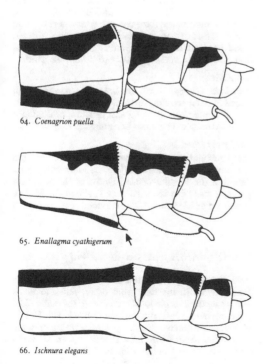

64. *Coenagrion puella*

65. *Enallagma cyathigerum*

66. *Ischnura elegans*

Figures 64–66 Abdominal segments S8–10, in left lateral view, of female Coenagrionidae. Vulvar spines are indicated by arrows

Key to genera of Coenagrionidae

1 Occiput dorsally entirely bronze-black, lacking pale post-ocular marks 2
– Two conspicuous pale spots, or (in *Nehalennia*) a pale line, on the occiput behind the eyes 4

2(1) Pale parts of abdomen blue (male) to greenish (female); eyes of male red in life. Antehumeral stripes reduced or absent. Apical venation of hindwing denser than that of forewing and with row of cells between costa and radius distal to the pterostigma partially divided .... *Erythromma* (p.72)
– Pale parts of abdomen red; eyes of male not bright red 3

3(2) Legs red. Pterostigma not longer than the cell beneath it. Antehumeral stripes absent, or incomplete and very fine. Male superior abdominal appendages (fig. 152, p.97) very short, much shorter than the dorsal length of S10. Frons with anterior and dorsal faces sharply differentiated ....
.... *Ceriagrion* (p.97)
– Legs black. Pterostigma longer than the cell beneath it (fig. 24, p.52). Red or yellow antehumeral stripes present. Male superior abdominal appendages (figs 67,68, p.74) almost as long as the dorsal surface of S10. Frons with anterior and dorsal faces rounded into each other .... *Pyrrhosoma* (p.72)

4(1) Side of synthorax with a short, black stripe only on the dorsal part of the metapleural suture. Male S2 with a mushroom-shaped black mark connected by a short stalk to the posterior black ring (fig. 132, p.89). Female S8 with a well-developed vulvar spine ventrally (fig. 65). Hindwing with a usually partly double row of cells between radius and costa distal to the pterostigma (fig. 62)*. Yellowish to blue antehumeral stripes broad .... *Enallagma* (p.89)
– Side of synthorax with a black line on the upper end of the mesometapleural suture in addition to one on the metapleural suture (fig. 13, p.42). Male S2 differently marked. Female S8 with vulvar spine absent or small (figs 64,66). Hindwing with a single row of cells between costa and radius distal to the pterostigma (figs 61,63) (sometimes partly double in *Cercion*). Antehumeral stripes usually narrower, sometimes absent or incomplete 5

5(4) Female with short vulvar spine on S8 (fig. 66). Male with a dorsal tubercle on S10 (figs 143–145, 147, p.91). Pterostigma of forewing of male (fig. 63) bicoloured, black basally and white apically. Abdomen bronze-black with pale (mainly blue) coloration dorsally on some of the three apical segments only .... *Ischnura* (p.90)
– Female without a vulvar spine (fig. 64). Male with (*Nehalennia*) or without a dorsal tubercle on last abdominal segment. Male pterostigma not bicoloured. Abdomen in most species more extensively pale 6

6(5) Abdomen dorsally strongly metallic greenish with blue coloration only on the three apical segments. Postocular mark linear. Small (22–24mm long) with short, rounded wings having only two or three cells distal to the pterostigma. .... *Nehalennia* (p.95)
– Abdomen dorsally weakly metallic, usually more extensively pale. Postocular marks ovoid to circular. Larger insects with longer wings (fig. 61) having four or more cells distal to the pterostigma .... *Coenagrion* (p.75) and *Cercion* (p.88)

---

* This character is not always reliable. Only one or two cells in the row may be divided, and sometimes the row is partially double in only one hindwing or, occasionally, in neither.

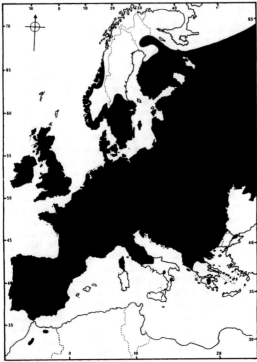

Map 16 – *Pyrrhosoma nymphula*

PYRRHOSOMA Charpentier, 1840

An exclusively Palaearctic genus of two species, represented in Europe by the type species, *P. nymphula*.

PYRRHOSOMA NYMPHULA (Sulzer)

Plate 8 (28–30)

Libellula nymphula Sulzer, 1776, *Ab. Gesch. Ins. Linn. Syst.*: 169.

Type locality: Zurich, Switzerland.

   Libellula minius Harris, 1782
   Agrion sanguinea Vander Linden, 1825

E – Large Red Damselfly, D – Frühe Adonislibelle
F – La Petite Nymphe au corps de feu,

DESCRIPTION OF ADULT

This species with *Ceriagrion tenellum* are the only predominantly red and black damselflies. *P. nymphula* is readily distinguished from *C. tenellum* by its black legs, the presence of red or yellow antehumeral stripes, the pterostigma being longer than the cell beneath it, and by its larger size (body length 33–36mm) and robust form. Postocular spots are absent. On emergence, the antehumeral stripes are yellow, but they soon become crimson (except in one female form); in males, and sometimes also in females, they are narrowly broken posteriorly and have the form of an !-mark. The abdomen of the male is dorsally red with S1 and S7–10 extensively marked with black. Laterally, the thorax and front of the abdomen are yellow-marked.

All black areas have a green to bronze metallic tinge. There are three distinct female colour forms. The most abundant is the homeochrome typical form, coloured rather like the male but with a median, black line on S2–6 which expands posteriorly into a spot on each of these segments. The intersegmental incisures are yellow. Form *fulvipes* Stephens, 1835, (Plate 8(29)) is very male-like with the mid-dorsal, black line on the abdomen fine, and the apical black spots on S2–6 much reduced. The heterochrome form *melanotum* Sélys (Plate 8(28)) is very different from these homeochrome forms; the antehumeral stripes remain primrose yellow throughout life and the abdomen is mainly bronze-black with the apex rust-coloured.

A subspecies, *P. n. elisabethae* Schmidt, 1948a, described from Kalàvrita in Peloponnese, Greece, differs from *P. n. nymphula* in having the male inferior abdominal appendages extending far beyond the apices of the superior appendages (inferiors slightly shorter than superiors in the nominotypical subspecies), and in the female pronotum being more strongly lobed posterolaterally (figs 67, 68, p.74). Some consider *P. n. elisabethae* to be a distinct species, but Buchholz (1954) and Stark (1979) found specimens from Greece and Austria, respectively, to be partly intermediate in character.

BIOLOGY

In the more northerly parts of its range, and in the mountains of southern Europe, *P. nymphula* inhabits peat pools, lochans, ponds, ditches, seepages and sluggish streams, but at low altitude in southern Europe it is an inhabitant of more swiftly-flowing water. In south-eastern Europe, *P. n. elisabethae* is chiefly an insect of the plains, whilst the nominotypical form tends to occur more often in mountains. The maturation period lasts from nine to fifteen days and is spent away from water. Corbet (1952) and Lawton (1970) describe population studies in England.

FLIGHT PERIOD

P. nymphula emerges early in the year, appearing in mid-April in southern England and persisting until almost the end of September. In the south of Europe, the flight period is virtually over by the beginning of August.

DISTRIBUTION (map 16)

A widespread and common species over much of northern Europe but becoming more local in the south and absent from most of the Mediterranean islands. In North Africa, *P. nymphula* has been reported near Ifrane in the Middle Atlas Mountains of Morocco, and it extends eastwards into Asia.

ERYTHROMMA Charpentier, 1840

The two European species in this genus of three species can be distinguished in the field from other blue and black damselflies by the conspicuously red eyes of the males. The male abdomen dorsally is mainly bronze-black with S9+10 bright blue. Pale antehumeral stripes are wanting or much reduced so that from the side the thorax appears contrastingly blackish in front and blue laterally. There are no pale postocular spots; exceptionally two minute points may be detectable. The male has long superior and short inferior anal appendages, and the female is without a spine on S8.

When mature, *Erythromma* are mostly to be seen over water, males spending long periods perched on floating vegetation, often water-lily leaves.

Key to species of *Erythromma*

1 Larger, body length usually more than 30mm. Tibiae and tarsi of male externally black. Antehumeral stripes absent in males and indicated only anteriorly in females. Wings with four or five postquadrilateral cells between the discoidal cell and the subnodal vein. Male superior abdominal appendages (fig. 69, p.74) with broad, internal flanges extending almost to their apices .... *najas* (p.73)

– Smaller, body length less than 30mm. Tibiae and tarsi externally mainly pale. Antehumeral stripes present in both sexes, complete in females but usually interrupted in males. Wings with three postquadrilateral antenodal cells. Male superior abdominal appendages (fig. 70, p.74) with internal flanges broad only basally .... *viridulum* (p.74)

ERYTHROMMA NAJAS (Hansemann)
Plate 8 (31–33)

Agrion najas Hansemann, 1823, *Wiedemann's zool. Mag.* 2 : 158.
Type locality: Germany.

 Agrion chloridion Charpentier, 1825
 Agrion analis Vander Linden, 1825

E – Red-eyed Damselfly, **F** – La Naiade aux yeux rouges, **D** – Grosses Granatauge

DESCRIPTION OF ADULT

A relatively large coenagrionid that can sustain lengthy periods of flight. *Erythromma* and *Enallagma* are genera characterized by the division of some cells in the hindwing series distal to the pterostigma, between costa and radius, and by the general venation of the hindwing apex being finer than that of the forewing (fig. 62). These features are particularly evident in *E. najas*. The first abdominal segment of the male is blue with a rectangular, black dorsal mark in its anterior half, and S9 and S10 (fig. 69) are dorsally almost entirely blue, but almost all of the remainder of the dorsum of the abdomen is bronzed black. The female abdomen is dorsally bronze-black with the lateral yellow colour scarcely extending to the dorsal surface except on the first segment; there are also blue transverse marks dorsally between the posterior abdominal segments, most evident between S8 and S9. The sides of the thorax are blue in the male, yellow in the female, and the female antehumeral stripe, reduced at least to !-shape, is also yellow. Between the wing bases in both sexes are blue thoracic marks. Mature individuals develop a rather thin, whitish pruinosity on thorax and abdomen. The posterior border of the pronotum of the female (fig. 69) is strongly trilobed, deeply indented on either side of the sharply-defined median lobe. Male superior abdominal appendages (fig. 69) are about as long as the dorsal surface of S10, in dorsal view appearing almost obliquely truncated by virtue of the broad, internal flanges reaching almost to the apices.

E. najas has a large size range, specimens from Great Britain and northern France usually exceeding 35mm in length, but specimens from southern France normally not more than about 30mm.

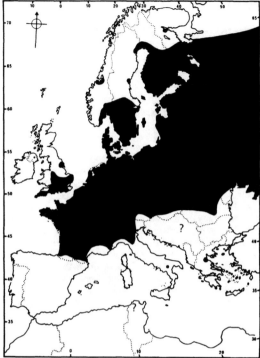

Map 17 – *Erythromma najas*

BIOLOGY

Frequents lakes and large ponds, sometimes slowly-flowing canals and dykes, but the current must not be too great to prevent the growth of water-lilies or other plants with broad, floating leaves which seem to be required as perches by the males. Females oviposit in submerged stems of a variety of plant species, but usually water-lilies, and they are accompanied by their mates. The couple descend to a depth of up to half a metre or more, enveloped in a bubble of air. After egg-laying, which may last for about thirty minutes, the couple regain the surface by floating upwards, in contrast to the usually pedestrian re-emergence of other Coenagrionidae.

FLIGHT PERIOD

Generally May to August, but the end of April to mid-July in northern Italy.

DISTRIBUTION (map 17)

Widespread in northern and central Europe but almost absent from the Mediterranean region. Old records (Martin, 1910) from Tunisia require verification. In Asia it has been recorded from Turkestan, Manchuria, eastern Siberia, and Hokkaido, Japan. Frequent in southern and midland England, but absent from Scotland, and there is only a nineteenth century record for Northern Ireland. Not found in Spain and Portugal, and in Italy known mainly from the north but also from Basilicata. Common in France and Germany, and extending from southern Scandinavia, Belgium, Holland and Poland south to Switzerland, Austria, Czechoslovakia, Hungary and Romania.

ERYTHROMMA VIRIDULUM (Charpentier)
Plate 8 (34,35)

Agrion (Erythromma) viridulum Charpentier, 1840, *Libell. europ. descr. depict.* : 149.

Type locality: Silesia, Poland.

Agrion bremii Rambur, 1842

F – La Naiade au corps vert, **D** – Kleines Granatauge

This is the only included species in *Pseuderythromma* Kennedy, 1920, but establishment of a separate genus for it seems unnecessary.

DESCRIPTION OF ADULT

The pattern of *E. viridulum* in both sexes is similar to that of *E. najas*, but the blue coloration is more intense and the black areas have stronger metallic reflections. S10 of the male abdomen is black dorsally with two large, blue spots (fig. 70). *E. viridulum* has paler legs than *E. najas*, externally only the femora having black, median lines with the tibiae and tarsi whitish and darkened only apically. Also, yellow antehumeral stripes are better developed in *E. viridulum* than in *E. najas*, and it is a much smaller species (body length about 26mm). Eyes of the male are reddish brown rather than crimson. The posterior border of the pronotum of the female (fig. 70) is simple, without a median lobe, and the male superior abdominal appendages (fig. 70) are narrowed and slightly inturned apically.

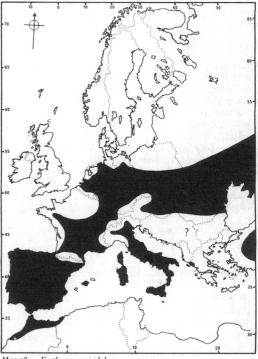

Map 18 – *Erythromma viridulum*

BIOLOGY

E. viridulum flies about dykes, broad ditches, still backwaters of rivers and small lakes. Larvae are tolerant of slightly brackish conditions. Both sexes fly close to the surface of open water in a direct manner, and males will perch upon floating vegetation, like those of *E. najas*, or on horizontally orientated reeds well above the water surface.

FLIGHT PERIOD

Mid-June to mid-September. Two generations a year are reported in Greece (Galletti & Pavesi, 1983).

DISTRIBUTION (map 18)

Mainly the Mediterranean region of Europe and North Africa (Morocco, north-west Algeria) east to Turkestan. It extends northwards, in scattered colonies, to the Netherlands and north Germany, possibly expanding its range, but it does not reach Britain or Scandinavia.

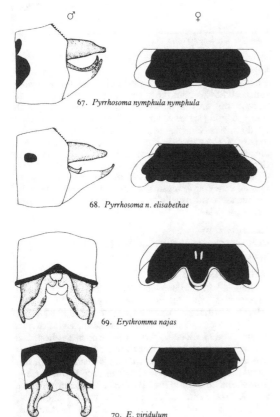

♂   ♀

67. *Pyrrhosoma nymphula nymphula*

68. *Pyrrhosoma n. elisabethae*

69. *Erythromma najas*

70. *E. viridulum*

Figures 67, 68 *Pyrrhosoma nymphula*
(left) male terminal abdominal appendages, in left lateral view
(right) female pronota

Figures 69, 70 *Erythromma* species
(left) male terminal abdominal appendages, in dorsal view
(right) female pronota

COENAGRION Kirby, 1890

Agrion Leach, 1815, *nec* Fabricius, 1775

The employment of *Agrion* versus *Calopteryx* Leach and *Coenagrion* Kirby has generated considerable discussion (e.g. Schmidt, Er., 1948b; Calvert *et al.*, 1949; Montgomery, 1954) because of confusion over Fabricius' intention in defining *Agrion*. The more junior names of *Calopteryx* (and see under this genus) and *Coenagrion* are used in this work and in most other recent publications.

Coenagrion is represented by about thirty-five species in all zoogeographical regions, and it is easily the largest genus of European Zygoptera, comprising eleven species of rather similar general appearance. There are two elliptical, ovoid or triangular postocular spots, and the front of the thorax has two, usually well-defined, antehumeral stripes. The abdomen of the male is blue or greenish blue, dorsally patterned with black. Black areas on the body have very weak bronze reflections. The female has the abdomen more extensively marked with black than the male, and the pale coloration is commonly yellowish to olive green. Females of several species are dimorphic, being of two colour forms. The pale coloration is yellowish or greenish in the heterochrome form, and blue and more extensive, approaching the male condition, in the homeochrome or andromorph form. Males can generally be specifically identified by the black design on the dorsal surface of the abdomen, although this is subject to some degree of intraspecific variation, together with the form of the abdominal appendages. Females are more difficult to distinguish, but the shape of the posterior border of the pronotum (grasped by the male during copulation) provides useful characters. The mesostigmal laminae of females are two small, transverse plates in front of the mesepisterna and just behind the anterior spiracles. These too are engaged by the male and provide diagnostic criteria, but they are difficult to examine without strong magnification and are not used in the key.

Most species of *Coenagrion* occur in profuse colonies so that both sexes can be obtained, and this is a considerable aid to identification. Oviposition occurs whilst the mated pair maintain the tandem position, and the female at least is frequently totally submerged during the process. The species fly amongst rank vegetation at the margins of ponds, lakes, dykes, ditches and, less commonly, streams and rivers. Males, because of their brighter coloration and greater volatility, are more easily found than females. Separate keys are provided for the sexes, and *Cercion lindeni* is included, for convenience, in both.

Key to males of species of *Coenagrion* and *Cercion*

1 Superior abdominal appendages (figs 129,131, p.89) elongated, slightly longer than S10 and fully twice as long as the inferior appendages, curved inwards and downwards. Black design on S2 extending from apex to base (fig. 126, p.89); S8 dorsally black .... *Cercion lindeni* (p.88)

– Superior abdominal appendages (figs 85–95, p.78) shorter than S10, in most species no more than half as long. Black design on S2 not reaching the anterior margin of the segment (figs 71–84, p.77); S8 blue, at most with two small, black points (*Coenagrion*) 2

2(1) Superior abdominal appendages at least slightly longer than the inferior appendages (figs 85–87, 89, p.78) 3

– Superior abdominal appendages shorter than the inferior appendages (figs 88,90–95, p.78) 6

3(2) Black design on second abdominal segment consisting of three isolated marks, the median mark a transverse crescent (fig. 76); black mark on S1 occupying only the anterior half of the segment. Occipital surface of head black with a blue border to the eyes. Pronotum dorsally (fig. 108) without a blue border .... *lunulatum* (p.82)

– Black mark on dorsal surface of S2 a single design connected to the posterior black ring by a median stalk (figs 71,73,74, p.77); black mark on S1 reaching to the posterior margin of the segment. Occiput white to blue with the central black area not extending laterally beyond the mouthparts. Pronotum dorsally (figs 96,99,102, p.79) with a bluish border postero-laterally 4

4(3) Pterostigma (fig. 98) lozenge-shaped, its costal edge not longer than its basal edge, very dark brown centrally but conspicuously pale peripherally. Superior abdominal appendage (fig. 85) with an internal tooth near the base, visible in dorsal view. S2 (fig. 71) with a trifid black design, there being a well-developed median point between the lateral arms .... *mercuriale* (p.80)

– Pterostigma (figs 101,104) elongated, its costal edge much longer than its basal edge, light brown. Superior abdominal appendage (figs 86,87) without an internal tooth. S2 (figs 73,74) with the black design lacking a well-developed median point 5

5(4) Posterior margin of pronotum (fig. 99) with a large, median, obtusely triangular lobe. Pterostigma (fig. 101) almost quadrangular, the posterior edge only slightly shorter than the costal (anterior) edge. Point of superior abdominal appendage (fig. 86) inwardly curved ....
.... *scitulum* (p.80)

– Posterior margin of pronotum (fig. 102) with a small, central tubercle. Pterostigma (fig. 104) triangular, the posterior edge much shorter than the costal edge and curving into the distal edge. Point of superior abdominal appendage (fig. 87) downwardly curved ....
.... *caerulescens* (p.81)

6(2) Inferior abdominal appendages (fig. 90) enormously developed, twice as long as the last abdominal segment. S4 (fig. 77) to S6 black in dorsal view. Thorax without or with the merest vestige of antehumeral stripes ....
.... *armatum* (p.83)

– Inferior abdominal appendages (figs 88,91–95) much shorter than S10. S4 (figs 75,78–84) to S6 broadly blue anteriorly. Antehumeral stripes developed and usually complete 7

7(6) S3+4 with the black design restricted to the posterior thirds of the segments mid-dorsally, but with black, lateral lines running forwards almost to the anterior edges (figs 81,82,84) 8

– S3+4 with the dorsal black design often more extensive (figs 75,78–80) but without black, lateral lines running far forwards except in the dark form of *pulchellum* (fig. 83).
.... 10

8(7) S2 (fig. 84) with black design usually connected with the posterior black ring by a median stalk. Posterior border of pronotum (fig. 123) strongly trilobed. Superior abdominal appendage (fig. 95) without an internal tooth. Antehumeral thoracic stripes narrow, usually broken posteriorly to form an !-mark .... *pulchellum* (part) (p.87)

– S2 (figs 81,82) with black design only exceptionally connected to the posterior black ring. Posterior border of pronotum (figs 119,121) not or weakly trilobed. Superior abdominal appendage (figs 93,94) with an internal tooth. Antehumeral stripes broader and complete 9

9(8) Black, lateral lines on S3+4 very fine; U-shaped mark on S2 complete (fig. 82). Superior abdominal appendage (fig. 94) with a small, internal tooth; inferior appendage only slightly inwardly curved (a widespread and abundant species) .... *puella* (p.86)

– Black, lateral lines on S3+4 broad, extending downwards almost to the ventral mid-line; U-shaped mark on S2 with transverse part semicircular or triangular, often disconnected from the lateral arms (fig. 81). Superior abdominal appendage (fig. 93) lacking an internal tooth; inferior appendage terminating in an inwardly directed hook (known only from the northern Alps in Europe) .... *hylas* (p.85)

10(7) Dorsal black design on S2 (fig. 75) with lateral arms disconnected; S3+4 with dorsal black marks terminating anteriorly in short, median points .... *hastulatum* (p.82)

– Dorsal black design on S2 usually entire (figs 78,80,83); if not (fig. 79), then the black marks on S3+4 terminate anteriorly in long, slender, median points 11

11(10) Dorsal black markings on S3+4 terminate anteriorly in long, slender, median points (figs 78,79). Pale postocular spots behind the eyes with their posterior edges conspicuously denticulate (fig. 125). Abdomen relatively stout. .... *ornatum* (p.84)

– Dorsal black markings on S3+4 terminate broadly in front in three short points (figs 80,83). Pale postocular spots with their posterior edges more regular. Abdomen slender 12

12(11) Dorsal black mark on S2 (fig. 80) not connected to the posterior black ring. Posterior border of prothorax with a median lobe only (fig. 117) .... *johanssoni* (p.85)

– Dorsal black mark on S2 (fig. 83) broadly connected to the posterior black ring. Posterior border of prothorax trilobed (fig. 123) .... *pulchellum* (part) (p.87)

Key to females of species of *Coenagrion* and *Cercion*

1 Pale postocular spots transversely elongated, scarcely broader than the pale median line between them. Pterostigma (fig. 130) elongated, its costal edge more than twice as long as its basal edge, and posterior border of prothorax (fig. 128) only weakly undulating. Abdominal segments with dorsal black markings relatively narrow, encroaching little on to the sides of the segments which in lateral view are mainly bluish green .... *Cercion lindeni* (p.88)

– Pale postocular spots rounded, ovoid or subtriangular. If the pterostigma is twice as long as broad (ratio costal:basal edges), then the posterior border of the prothorax is strongly lobed. Abdominal segments usually with relatively broad, dorsal, black markings (*Coenagrion*) 2

2(1) Posterior border of prothorax (fig. 103) with two, very prominent, pale, tongue-like projections separated by a deep incision .... *caerulescens* (p.81)

– Posterior border of prothorax differently-shaped 3

3(2) S8 bluish with a bilobed black mark in the posterior half only .... *armatum* (p.83)

– S8 mainly black, the design extending forwards to the anterior margin, or (*lunulatum*) nearly so and not bilobed 4

4(3) Back of head between foramen and mouth mainly pale, the black coloration not extending laterally beyond the outer limits of the labium and often restricted to the area immediately adjacent to the foramen 5

– Back of head black with pale border to the eyes not extending inwards as far as the outer limits of the labium 6

5(4) Pterostigma (fig. 101) twice as long as broad. Posterior border of pronotum (fig. 100) trilobed with a pale edge only on the outer lobes .... *scitulum* (p.80)

– Pterostigma (fig. 98) about as long as broad. Posterior border of pronotum (fig. 97) almost continuously pale and with only a small, median lobe .... *mercuriale* (p.80)

6(4) S8 pale anteriorly. Posterior border of pronotum (fig. 109) black with a strongly-developed median lobe .... *lunulatum* (p.82)

– S8 black. Posterior border of pronotum otherwise 7

7(6) Posterior border of pronotum more or less trilobed (figs 115, 122, 124) 8

– Posterior border of pronotum with a median lobe only (figs 106, 118, 120) 10

8(7) Posterior border of pronotum (fig. 115) with median lobe apically incised and pale at least at the tip on either side of the incision. Pale postocular spots with posterior edges denticulate (fig. 125). S3 with dorsal, black, median mark sometimes rather narrow and terminating in a long median point .... *ornatum* (p.84)

– Posterior border of pronotum with median lobe entire. Pale postocular spots with posterior edges more regular. S3 with dorsal, black mark broad, apically trifid or reaching to the anterior of the segment 9

9(8) Posterior border of pronotum (fig. 122) weakly trilobed ... . *puella* (p.86)

– Posterior border of pronotum (fig. 124) strongly trilobed. .... *pulchellum* (p.87)

10(7) S10 pale dorsally .... *johanssoni* (p.85)

– S10 black dorsally 11

11(10) S2 dorsally with a black mark in the posterior third extending anteriorly as a narrow point .... *hylas* (p.85)

– S2 dorsally with a broad, black bar extending forwards from a posterior expansion to reach the anterior margin of the segment .... *hastulatum* (p.82)

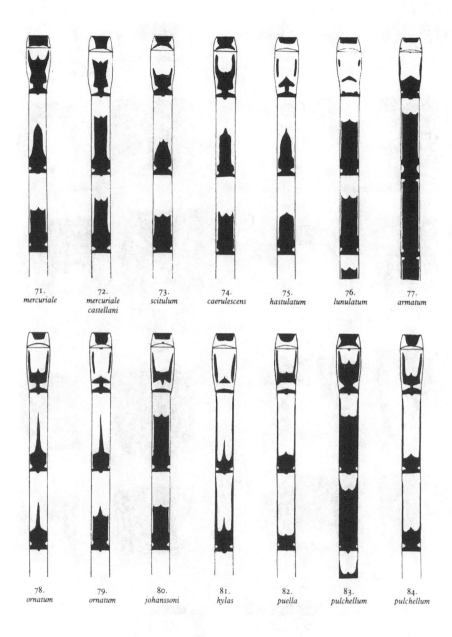

| 71. | 72. | 73. | 74. | 75. | 76. | 77. |
| mercuriale | mercuriale castellani | scitulum | caerulescens | hastulatum | lunulatum | armatum |

| 78. | 79. | 80. | 81. | 82. | 83. | 84. |
| ornatum | ornatum | johanssoni | hylas | puella | pulchellum | pulchellum |

Figures 71–84 Male *Coenagrion* species, abdominal segments S1–4, in dorsal view, to illustrate the black designs. These are variable and the illustrations are of only the more usual patterns

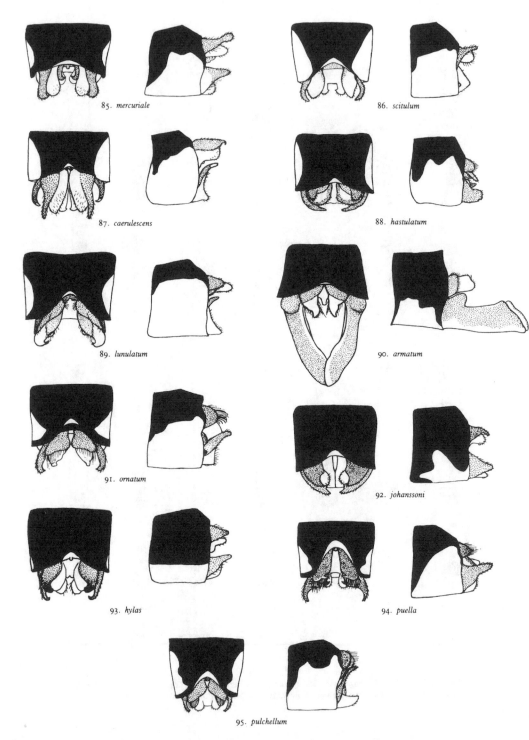

Figures 85–95 Male *Coenagrion* species, abdominal appendages and segment S10 from above (left) and in left lateral view (right). Dark areas of appendages are stippled

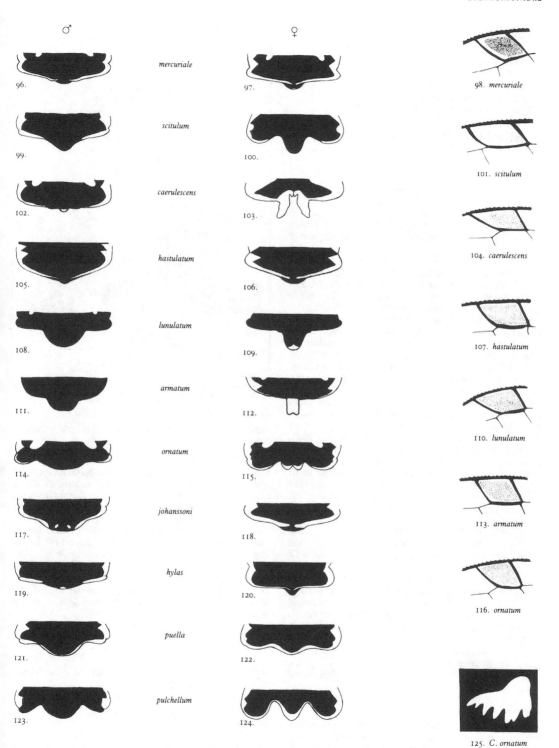

♂ ♀

96. mercuriale 97. 98. mercuriale

99. scitulum 100. 101. scitulum

102. caerulescens 103. 104. caerulescens

105. hastulatum 106. 107. hastulatum

108. lunulatum 109. 110. lunulatum

111. armatum 112. 113. armatum

114. ornatum 115. 116. ornatum

117. johanssoni 118.

119. hylas 120.

121. puella 122.

123. pulchellum 124.

125. C. ornatum

Figures 96–125 (96–124) *Coenagrion* species, posterior pronotal margins of males (left column) and females (middle column), and pterostigmata of left wings (right column) (125) right postocular spot

COENAGRION MERCURIALE (Charpentier)
Plate 9 (36,37)

Agrion mercuriale Charpentier, 1840, *Libell. europ. descr. depict.* : 159.
Type locality: Lüneburg, Germany.

Agrion fonscolombii Rambur, 1842
Agrion hermeticum Sélys, 1872
Coenagrion castellani Roberts, 1948

E – Southern Damselfly, **F** – L'Agrion de Mercure,
D – Helm-Azurjungfer

DESCRIPTION OF ADULT
Coenagrion mercuriale, *C. scitulum* and *C. caerulescens* resemble each other in morphology and distribution; they constitute the Mediterranean species-group (Conci & Nielsen, 1956). In all three species the occiput is mainly pale, and the superior abdominal appendages are longer than the inferiors.

C. *mercuriale* may be distinguished from *C. scitulum* and *C. caerulescens* by its short, lozenge-shaped pterostigma (fig. 98) which is shorter than the cell it surmounts and has a dark centre surrounded by a much paler border. Postocular spots are large and rounded, and the prothorax is similarly-shaped in both sexes with an almost straight posterior margin bearing a small, median tubercle (figs 96,97). The black mark on the first abdominal segment reaches the posterior border and, in the male, S2 (fig. 71) bears a black 'sign of Mercury' design with a prominent median point between lateral arms. Variations in this design are numerous and include the broadening or absence of the stalk connecting it to the posterior black ring, and the separation of the lateral arms. S3–5 of the male abdomen each have a dorsal black mark ending anteriorly as a median point. In the Italian subspecies *C. m. castellani* Roberts, 1948, originally described as a distinct species, the black abdominal marks are broader and those on S3–5 terminate in front in three points, the outer two extending at least as far forwards as the median point (fig. 72). Differences between the two European subspecies in the structure of the male abdominal appendages are as follows: in *C. m. mercuriale* the superiors are only slightly longer than the inferiors and mainly pale with the hooked tip almost closed (fig. 85), but in *C. m. castellani* the superiors are distinctly longer than the inferiors, almost entirely dark, and with the apical hook more open.

The typical heterochrome form of the female has the abdomen olive-green laterally with small, pale marks anteriorly on S3–7. Between S7 and S10 there are bright blue intersegmental rings. In the homeochrome or andromorph form, the pale colour is more extensive and blue as in the male.

C. *mercuriale*, like the other species in the Mediterranean group, is relatively small (body length 27–31mm) with a comparatively short, stout abdomen and rather broad wings.

BIOLOGY
Inhabits streams, runnels and water meadows, often with a moderate rate of flow. *C. mercuriale* tends to be most numerous in areas where the substrate is calcareous, although colonies in Devon, England, occur on boggy heathland with *Drosera* and *Schoenus*.

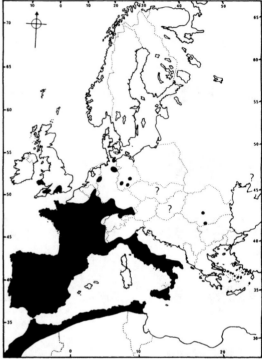

Map 19 – *Coenagrion mercuriale*

FLIGHT PERIOD
Beginning of May to the beginning of September (June to August in the British Isles).

DISTRIBUTION (map 19)
C. *m. mercuriale* is a central and west European insect, more numerous and with a wider distribution than either *C. scitulum* or *C. caerulescens*. It is found in Britain (extreme south of England and near the west coast of Wales), France, Portugal, Spain, Belgium, Holland (two old records), Luxembourg, Germany (very rare in the German Democratic Republic), Switzerland (extinct since 1942 in the west) and Austria. It is recorded also from Romania (Cirdei & Bulimar, 1965) and the Caucasus (Schmidt, Eb., 1978a). *C. m. castellani* occurs in Italy (Conci, 1949). A dark form from Morocco, Algeria and Tunisia, often referred to *C. m. hermeticum* (Sélys, 1872), is probably identical to the nominotypical subspecies (Lieftinck, 1966).

COENAGRION SCITULUM (Rambur)
Plate 9 (38,39)

Agrion scitulum Rambur, 1842, *Hist. nat. ins. Neur.* : 266.
Type locality: near Paris, France.

Agrion distinctum Rambur, 1842

E – Dainty Damselfly, **F** – L'Agrion mignon,
D – Gabel-Azurjungfer

DESCRIPTION OF ADULT
Distinguished from the other species in the Mediterranean group by its very elongated pterostigma (fig. 101) which is

very nearly twice as long as broad. The posterior edge of the prothorax (figs 99,100) is of quite a different shape from that seen in *C. caerulescens* which also has an elongated pterostigma. In the female, it is strongly trilobed with the central lobe entire and black, and in the male it has a large, triangular, median lobe. The black design on the male abdomen (fig. 73) is as follows: S1 with mark reaching posterior margin, S2 typically with a U-shaped mark connected with the posterior black ring, but occasionally this is reduced to an isolated transverse bar lacking lateral arms, S3–5 with posterior marks more or less trifid anteriorly, S6,7,10 dorsally mainly black, S8 and S9 mainly blue. The superior abdominal appendages of the male (fig. 86) are longer than the inferiors, more conspicuously so than in *C. mercuriale*, and they more resemble those of *C. caerulescens* in not having a basal spine visible in dorsal view. The female is very like the female of *C. mercuriale*, but has rather more extensive blue markings anteriorly on the abdominal segments.

BIOLOGY

To be found near ditches, dykes, weedy eutrophic ponds and small streams, often frequenting water with a slight flow. Oviposition is in floating vegetation and takes place in tandem with the male adopting an unusual forward-sloping, not vertical, attitude whilst grasping the female's prothorax and mesostigmatic plates (Boulard, 1981).

FLIGHT PERIOD

End of May to usually the end of July but sometimes lasting until September.

DISTRIBUTION (map 20)

Commonest in the Mediterranean region of Europe and North Africa, extending from Spain (north-east) and Morocco to the Middle East. It is local in central Europe (no records for the Netherlands, Denmark, Poland or Czecho-slovakia; questionable record for Switzerland) and very rare in the north. The only known British colony at Benfleet in Essex was eliminated by North Sea floods in 1953.

COENAGRION CAERULESCENS (Fonscolombe)
Plate 9 (40–42)

Agrion caerulescens Fonscolombe, 1838, *Annls Soc. ent. Fr.* 7 : 568.

Type locality: Aix en Provence, France.

Agrion aquisextanum Rambur, 1842

F – L'Agrion bleuâtre

The specific name is sometimes incorrectly spelt 'coerulescens' as in the *Orthetrum* species.

DESCRIPTION OF ADULT

C. caerulescens closely resembles *C. scitulum* in several respects including the possession of an elongated pterostigma. In *C. caerulescens*, however, the pterostigma (fig. 104) is approximately triangular but in *C. scitulum* it is quad-rangular. Females may be recognized most readily by the form of the posterior pronotal border (fig. 103) which is deeply incised medially, the incision flanked by two pale-coloured lobes clearly visible under a small hand lens. The male has merely a very small, blue-tipped tubercle at the centre of the pronotal border (fig. 102). The occiput is

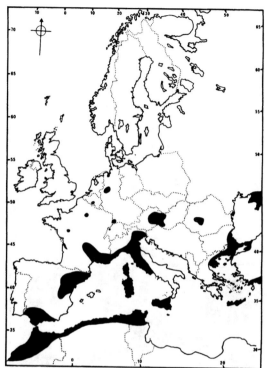

Map 20 – *Coenagrion scitulum*

black only around the foramen, and there is frequently a minute pale spot between the median ocellus and each lateral ocellus. The blue antehumeral stripes are broad, sometimes much broader than the black humeral stripes. The first abdominal segment has a rectangular black mark in its anterior half only. The black design on S2 of the male (fig. 74) is U-shaped with a broad base connected to the posterior black ring, S3–7 have increasing areas of black posteriorly, covering the posterior half of S3 and terminating in front in a median point, and virtually all of S7. Male abdominal appendages (fig. 87) resemble those of *C. scitulum* but the apical point of the superiors is directed ventrally instead of mesially.

Two European subspecies are distinguished (Schmidt, Er., 1959). In *C. c. caerulescens* S9 of the male has a black mark restricted to its postero-lateral part, whilst in *C. c. caesarum* Schmidt this mark extends over most of the segment. Females also differ in abdominal colour pattern, *C. c. caesarum* (Plate 9(41)) being a darker form with S8–10 dorsally almost entirely black, whilst in the typical form they are mainly bluish.

BIOLOGY

Breeds in streams and small rivers where the current is slow.

FLIGHT PERIOD

June to August.

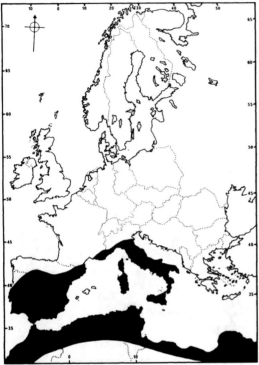

Map 21 – *Coenagrion caerulescens*

DISTRIBUTION (map 21)

West Mediterranean, the nominotypical subspecies occurring in Spain and southern France, and subspecies *C. c. caesarum* in Italy. A large form of the latter is found in Sicily and Sardinia, and a smaller form inhabits mainland Italy (Schmidt, Er., 1959). The subspecies *C. s. theryi* Schmidt, 1959, which occurs in Morocco, Algeria and Tunisia, is considered by Lieftinck (1966) to be inseparable from populations of south-western Europe.

COENAGRION HASTULATUM (Charpentier)
Plate 9 (43,44)

Agrion hastulatum Charpentier, 1825, *Horae entomol.*: 20.
Type locality: Saxony, Germany and Silesia, Poland.
E – Northern Damselfly, **F** – L'Agrion hasté,
D – Speer-Azurjungfer

DESCRIPTION OF ADULT

Separation of the two lateral arms from the posterior black mark on S2 of males is a feature of *C. hastulatum*, *C. lunulatum* and *C. armatum*; in other species it occurs only as a variation. These three species (the boreal species-group) all have a northern or eastern range in Europe and they tend to be greenish blue, in contrast to the sky-blue of males in the preceding Mediterranean group. The eyes of males in life are greenish yellow and the thorax appears green. The posterior surface of the head is extensively darkened, and the black design on the third and following abdominal segments of males is not extended anteriorly as fine lateral lines.

In the male of *C. hastulatum*, the black, spear-shaped mark on S2 (fig. 75) is narrowly connected to the posterior of the segment although there is some variation and the connection may occasionally be absent; exceptionally, the lateral arms are absent or connected with the central mark. S1 has the black mark extending only to the middle in males, but in females it usually reaches the posterior margin. Black marks on S3–5 of the male occupy only the posterior halves of the segments, and they terminate anteriorly in median points; S6 also has a broad, blue anterior area. The black mark on S2 of the female is dilated posteriorly but of even breadth from the dilation to the anterior of the segment (as in *C. lunulatum*); it is not thistle-head shaped, narrower anterior to the dilation than at the anterior margin of the segment. In both sexes, the posterior pronotal edge (figs 105,106) is only weakly lobed medially and antehumeral blue stripes are complete and quite broad. Male abdominal appendages (fig. 88) are short, the superiors projecting less far than the inferiors.

BIOLOGY

A species found at the sheltered edges of acid peat bogs, moorland pools and lochans with broad, sedgy margins, commonly above 1000m altitude in central Europe and frequently associated with *Enallagma cyathigerum* (Schmidt, Eb., 1964a).

FLIGHT PERIOD

End of May to the middle of August, but mostly June and July. Life cycle duration is one or two years in southern Sweden but three or four years in northern Sweden (Norling, 1984a).

DISTRIBUTION (map 22)

Essentially a boreo-alpine species found in northern Europe (Scotland, all Scandinavia, Belgium, Holland, Luxembourg, Germany, Poland and Russia), the Alps (France, Switzerland, Austria, Italy), the Massif Central, Jura and Vosges of France, the Pyrenees (both French and Spanish sides), and in Czechoslovakia, Hungary and Romania. Outside Europe it is widely distributed in northern Asia.

COENAGRION LUNULATUM (Charpentier)
Plate 9 (45,46)

Agrion lunulatum Charpentier, 1840, *Libell. europ. descr. depict.* : 162.
Type locality: Silesia, Poland.
> *Agrion vernale* Hagen, 1839 (*nomen nudum*, see Mielewczyk, 1974)

E – Irish Damselfly, **F** – L'Agrion à lunules,
D – Mond-Azurjungfer

DESCRIPTION OF ADULT

C. lunulatum is in many respects similar to *C. hastulatum* with which it sometimes flies. It is, however, a darker, more robustly-built species with the postocular spots small and, in the male, without a pale, transverse line between them. Antehumeral pale stripes are narrow. The black design on S2 of the male (fig. 76) is divided into three separate parts, the median, transverse, crescentic part suggesting the insect's scientific name. This design is seen only very exceptionally in *C. hastulatum*. Blue coloration on the male

Map 22 – *Coenagrion hastulatum*

Map 23 – *Coenagrion lunulatum*

abdomen is generally more restricted than in *C. hastulatum* with the black design on S3–5 occupying half to three-quarters of the segments and, in each, ending anteriorly in three points. S6 has only a small, basal, blue area. The female's second abdominal segment has a black dorsal design that is not constricted in front of the posterior dilation, resembling that of *C. hastulatum* but differing from the 'thistle head' mark of most other species. In both sexes the last abdominal segment is more deeply excised posteriorly than in *C. hastulatum*. Male superior abdominal appendages (fig. 89) are slightly longer than the inferiors, a feature which *C. lunulatum* shares with the Mediterranean species-group.

BIOLOGY

Inhabits mesotrophic ponds, fens and marshes, acidic in character, usually on peatland. Often associated with *C. hastulatum* in waters sparsely invaded by *Carex* and *Equisetum*.

FLIGHT PERIOD

End of May to end of June.

DISTRIBUTION (map 23)

Northern and eastern Europe through Siberia (including north of the Arctic Circle) and Mongolia to Japan, but in rare, scattered colonies. Recorded from Scandinavia (north to Soroya Island, Finnmark), Holland, Belgium, France (Savoie, Puy-de-Dome, Lozère (Dommanget, 1981)), Germany, Poland, Russia, Switzerland (extinct ?), Austria, Hungary (Lohinai, 1982) and Romania. Discovered in Ireland (Sligo) in 1981 (Cotton, 1982) with further records in 1983 from Westmeath, Offaly, Longford and Galway, and more recently, further north, from Fermanagh and Londonderry.

COENAGRION ARMATUM (Charpentier)
Plate 10 (47,48)

Agrion armatum Charpentier, 1840, *Libell. europ. descr. depict.* : 164.
Type locality: Lüneburg, Germany.
E – Norfolk Damselfly, **F** – L'Agrion armé,
D – Hauben-Azurjungfer

DESCRIPTION OF ADULT

Charpentier named this species after the form of the male's very large inferior abdominal appendages (fig. 90), a feature that makes it impossible to confuse *C. armatum* with any other species. Males are characterized also by the ante-humeral stripes being reduced to four small blue points or absent altogether. The blue coloration of the male has a distinctly greyish or greenish hue, and it is a dark species with pale colour restricted to S1–3,8,9; the middle segments of the abdomen are almost entirely bronzed black. The black design on the male's second abdominal segment (fig. 77) is very broadly connected with the posterior black ring, and its two lateral arms are isolated or almost so. The black mark on S1 does not reach the posterior margin in either sex. The female's body colour pattern is not dissimilar to that of the male, with relatively large greenish blue areas on S1,2 and 8, the latter segment having a bilobed black mark posteriorly. This is the only European species of *Coenagrion* in which the dorsal black design does not extend the full length of S2 in the female. Antehumeral stripes are complete in the female. In both sexes, the posterior border of the prothorax (figs 111,112) has a well-developed, median lobe.

BIOLOGY

Larvae found in stagnant, mesotrophic waters, usually with extensive beds of reeds or sedges amongst which the adults fly.

FLIGHT PERIOD

Mid-May to the beginning of July in Germany, July to August in Finland (Robert, 1958).

DISTRIBUTION (map 24)

C. armatum is a boreal species found in Scandinavia (Hämäläinen, 1984c), Holland (only three records, the last in 1924), Belgium, northern Germany (restricted to northern Schleswig-Holstein in recent years but disappeared after the hot summers of 1975 and 1976 when its habitats dried up (Schmidt, Eb., 1978b)), Poland, Romania and Russia. Widespread in Siberia and Mongolia. It used to occur near Stalham in Norfolk, England, but has not been seen there since 1957 and is presumed to be extinct, its former habitat now being badly polluted (Hammond, 1977) and having a much-lowered water table.

COENAGRION ORNATUM (Sélys)

Plate 10 (49,50)

Agrion ornatum Sélys in Sélys & Hagen, 1850, *Revue des Odonates* : 203.
Type locality: Hildesheim near Hanover, Germany.
F – L'Agrion orné, **D** – Vogel-Azurjungfer

DESCRIPTION OF ADULT

A small, robust species with relatively broad wings. It appears to be most closely allied to the preceding boreal group of species. In the male, the abdomen is very bright blue with the black pattern distinctive; S2 has a U-shaped black mark narrowly connected to the apical black ring (the lateral arms are usually isolated) (figs 78,79), S3–7 have increasingly large, posterior, black marks that terminate in long, forwardly-directed median points with no indication of lateral lines. Antehumeral stripes are complete and broad, and the occiput has a broad blue border to the eyes, approaching the condition of species in the Mediterranean group. A characteristic feature of *C. ornatum* seems to be the shape of the postocular spots (fig. 125); in both sexes they are quite large, ovoid and their posterior edges are invaded by streaks of black so that they appear dentate. The female is dimorphic, pale areas of the homeochrome form being greenish blue and of the heterochrome form yellowish green (Schmidt, Er., 1929). The pale anterior annulations on the female's abdominal segments are sometimes (not always) relatively large and the black, posterior marks then have long, tapering, median points, analogous to those of the male. The posterior border of the female prothorax (fig. 115) is trilobed with the central lobe short, pale at the tip and apically weakly incised; that of the male (fig. 114) is weakly trilobed, like that of *C. puella* but with the central

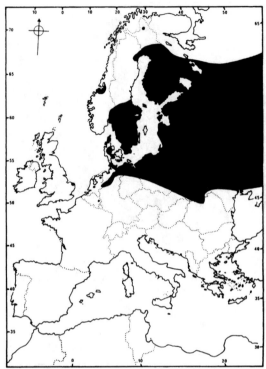

Map 24 – *Coenagrion armatum*

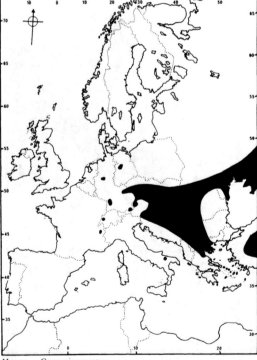

Map 25 – *Coenagrion ornatum*

lobe entirely black. Male abdominal appendages are illustrated in figure 91.

BIOLOGY

Frequents small, shallow, calcareous streams with a slow current. Often associated with *Carex* amongst which its coloration is cryptic. *C. ornatum* does not move far from its breeding sites.

FLIGHT PERIOD

An early summer species, flying from the end of May to mid-July.

DISTRIBUTION (map 25)

An eastern Palaearctic species recorded in Europe from Greece, Romania, Hungary, Czechoslovakia, Austria, Switzerland (extinct ?), Germany, Italy (one locality in the south-east near Foggia (Nielsen & Conci, 1951)) and France (mountains in the south-east, very rare).

COENAGRION JOHANSSONI (Wallengren)
Plate 10 (51,52)

Agrion johanssoni Wallengren, 1894, *Ent. Tidskr.* **15** :267.
Type locality: Norbergs, Västmanland, Sweden.

Agrion concinnum Johansson, 1859 (*nec* Rambur, 1842).

DESCRIPTION OF ADULT

This Scandinavian species is allied to the following three species which constitute the *C. puella* species-group, characterized by slender build, dark occipital head surfaces, and fine, lateral, black lines on the median segments of male abdomens.

The male of *C. johanssoni* has larger posterior black marks on the abdominal segments than have males of the following species (except some *C. pulchellum*), and hence it appears a generally darker insect with the fine lateral lines on the middle abdominal segments largely obscured. S2 (fig. 80) has a U-shaped design not connected to the posterior black ring. The female's abdominal colour pattern is distinctive with all of S10 and the posterior of S9 being blue so that it is almost as blue as the male. In both sexes, the posterior border of the pronotum (figs 117,118) has a broad, median lobe which in the female has an almost complete, pale, thickened margin. Male abdominal appendages are depicted in figure 92; the inferiors are slightly longer than the superiors and have a pale, internal apophysis which makes them appear bifid in profile.

BIOLOGY

Marshes, ponds, lake bays and moorland bogs are the habitat of *C. johanssoni*. Adults fly quite rapidly, keeping to vegetation, and are seldom numerous in Europe. Females especially are difficult to find.

FLIGHT PERIOD

Late June to mid-August.

DISTRIBUTION (map 26)

Scandinavia (except Denmark), Russia and Siberia to north of the Arctic Circle.

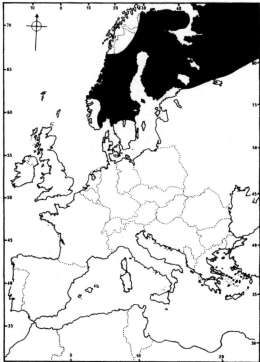

Map 26 – *Coenagrion johanssoni*

COENAGRION HYLAS (Trybőm)   Plate 10 (53)

Agrion hylas Trybőm, 1889, *Bih, K. svenska VetenskAkad. Handl.* **15** : 12.
Type locality: Yenisey, Siberia, USSR.

Agrion freyi Bilek, 1954

F – L'Agrion de Frey, **D** – Sibirische Azurjungfer

C. hylas is the latest species of damselfly to be discovered breeding in Europe. At first, the European population was thought to represent an undescribed species (Bilek, 1954), but the name given to it was later synonymized with *C. hylas* (Lieftinck, 1964). Harz (1978) questions the validity of this synonymy and it is probably best to consider *freyi* Bilek a subspecies of *C. hylas*.

DESCRIPTION OF ADULT

C. hylas is a relatively large species, males measuring 35mm or so in body length. In general appearance the male resembles *C. puella*, but the black design on S2 (fig. 81), which is unconnected with the posterior black ring, is usually disjointed with the median transverse part being an isolated triangle. The fully-developed, lateral, black lines on S3–6 are very broad so that ventrally the abdomen appears black. The median, dorsal points running forwards from the posterior black marks on the abdominal segments are much longer than in *C. puella*, the posterior marks themselves occupy a greater portion of S3–5 than in *C. puella*, and S7 has a pair of anterior blue spots (this segment is normally entirely black in *C. puella*). In the female, S2 has a narrow, parallel-sided, median, black bar extending anteriorly from the posterior black mark (i.e. it is not thistle

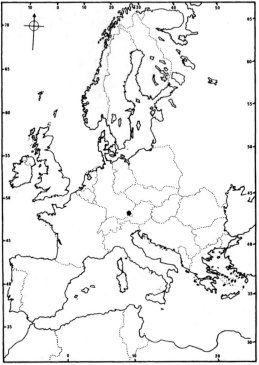

Map 27 – *Coenagrion hylas*

head-shaped as in *C. puella*), S3–6 have large, dorsal, black marks extended forwards to the anterior of the segments as increasingly stout median points, S7 has a pair of very small, anterior blue spots, S8 has a larger pair of anterior blue spots (about half the segment length), and S9,10 are black. Blue, antehumeral thoracic stripes are complete, and the blue postocular spots are large and rounded.

I have not examined European specimens of *C. hylas*, but there are very clear colour photographs of both sexes in the book by Jurzitza (1978: 41). Illustrations of the male abdominal appendages (fig. 93) and the pronota (figs 119,120) are from Siberian specimens.

BIOLOGY

Found in Europe in the vicinity of the mouths and deltas of streams feeding calcareous, montane lakes that have a prolific growth of *Equisetum* and *Potamogeton*.

FLIGHT PERIOD

Mid-June to the end of July.

DISTRIBUTION (map 27)

C. hylas is a north-eastern Asiatic species, occurring in the tundra of Siberia as far west as the Ural Mountains (Belyshev & Haritonov, 1980) and often north of the Arctic Circle. It is recorded also from Mongolia, Manchuria and northern Japan, with a western outpost in south-east Bavaria. The German colony was discovered by Bilek in 1952 near Inzell (Mündung des Zwingsees), but it subsequently disappeared, having suffered from over-collecting and the development of the site as a fish farm and swimming pool in the mid-1960s. Heidemann (1974),

however, reports finding another colony in 1973, the precise location of which in the northern Alps has not been publicized. It should be mentioned here that the collection of any Odonata in the Federal Republic of Germany is prohibited by the Federal Species Conservation Act of 1980.

COENAGRION PUELLA (Linnaeus)
Plate 10 (54–56)

Libellula puella Linnaeus, 1758, *Syst. Nat.* (Edn. 10) **1** : 546.
Type locality: southern Sweden.
  Agrion furcatum Charpentier, 1825
  Agrion irina Brullé, 1832
  Agrion annulare Stephens, 1835
E – Azure Damselfly, **F** – L'Agrion jouvencelle,
D – Hufeisen-Azurjungfer

DESCRIPTION OF ADULT

This is the commonest and most widely-distributed European *Coenagrion*; it is also the type species of the genus. Males may be recognized by the characteristic U-shaped black mark on S2 (fig. 82), isolated (with rare exceptions) from the posterior black ring, in conjunction with the rather small, posterior black marks on S3–5 which occupy barely one- quarter of the lengths of the segments but which have fine, forwardly-projecting lateral lines. The posterior border of the pronotum of the male (fig. 121) has a central lobe with a complete, narrow, pale border; in the female (fig. 122) it is weakly trilobed with a broader pale border. Superior male abdominal appendages (fig. 94) have an internal tooth visible in dorsal view; this tooth is absent in *C. hylas* and *C. pulchellum*.

The commonest form of the female is heterochromic (Plate 10 (54)) with a thistle head design on S2 and S3–10 dorsally black, the very narrow intersegmental incisures between S3–7 being yellowish green, and between S7–10 blue. In the homeochromic andromorph (Plate 10 (55)), the pale areas are bluish and extend over approximately the anterior fifths of S3–6; this form closely resembles the female of *C. pulchellum*.

BIOLOGY

A ubiquitous insect characteristic of small bodies of standing water ranging in character from oligotrophic to eutrophic. Often abundant amongst vegetation bordering lakes, ponds and almost stagnant ditches and dykes. Usually rarely at streams, but confined to lotic waters in Finland (Hämäläinen, 1984b).

FLIGHT PERIOD

End of April to the middle of September, but most numerous in June and July.

DISTRIBUTION (map 28)

Occurs in every European country except Iceland, but found only in the extreme south of Norway, Sweden and Finland. A separate subspecies (*C. p. kocheri* Schmidt, 1960) is found in North Africa. This has black spots on the dorsum of male S8 and a broad and stalked U-shaped mark on S2.

COENAGRION SYRIACUM (Morton, 1924)
(= *ponticum* Bartenef, 1929)

This species differs from *C. puella* in the structure of the male abdominal appendages (superiors longer) and female prothorax (more distinctly trilobed), and in a tendency to be more darkened (U-shaped mark on male S2 usually stalked). *C. syriacum* apparently replaces *C. puella* in the southern Caucasus, Middle East and Turkey north to the Bosporus, although both species are reported from southern Georgia.

COENAGRION PULCHELLUM (Vander Linden)
Plate 11 (57,58)

Agrion pulchella Vander Linden, 1825, *Monograph. Libell. Europ. Spec.* : 38.

Type locality: northern Italy.

Agrion interruptum Charpentier, 1825

E – Variable Damselfly, F – L'Agrion joli,
D – Fledermaus-Azurjungfer

DESCRIPTION OF ADULT

The colour pattern of most species of *Coenagrion* is variable and *C. pulchellum* is perhaps more variable than most, resulting in the naming of a number of forms. There are undoubtedly regional differences, and the normal form of north-west Europe (form *pulchellum* or form *interruptum* (Charpentier), fig. 84), can be distinguished on the male pattern of black markings from the normal form of the Mediterranean region (form *mediterraneum* Schmidt, 1964, fig. 83). It is often possible, however, to see a complete range of variation within a single population, and the value of taxonomically subdividing *C. pulchellum* is debatable. In addition to variation in both sexes in the pattern of black markings, females are dimorphic, being either of a blue homeochrome form or a greenish yellow heterochrome form.

C. pulchellum resembles the common *C. puella* in being a slender damselfly with, in males, fine lateral black lines running forwards from the posterior black marks on the third and following abdominal segments. In eastern and Mediterranean Europe, the dorsal abdominal marks are often very extensive, extending over about the posterior three-quarters of S3–5 and obliterating the lateral black lines. The insects then appear dark and resemble male *C. johanssoni* from which they may be separated by their strongly trilobed pronotal border and the broad attachment of the black mark on S2 to the posterior black ring. This latter character serves to differentiate most male specimens from *C. puella*, the black design on S2 only rarely being unstalked. Female abdominal segments are extensively blue-marked anteriorly in the typical homeochrome form, contrasting with the much darker and more common heterochrome form of *C. puella*. The posterior border of the pronotum (figs 123,124) is trilobed in both sexes, very strongly so in females. Blue antehumeral stripes are narrow in the male and often interrupted in north European insects. Male abdominal appendages are illustrated in figure 95.

BIOLOGY

Lakes, ponds, marshes and slow-flowing ditches are the habitats of *C. pulchellum*. It appears to be more tolerant

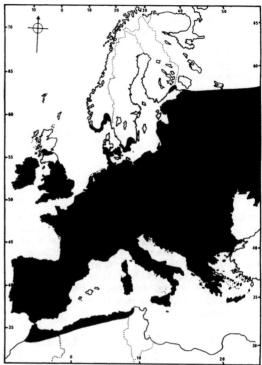

Map 28 – *Coenagrion puella*

Map 29 – *Coenagrion pulchellum*

than *C. puella* of acid conditions. Oviposition is usually into the undersurfaces of floating leaves, the eggs being arranged in a spiral formation.

FLIGHT PERIOD

The first half of April to mid-August in the south of the range but seldom appearing before late May in the north.

DISTRIBUTION (map 29)

Europe and western Asia. A widespread species in Europe with a similar range to *C. puella*, but less common and extending further north in Fennoscandia and absent from North Africa.

CERCION Navas, 1907

Cercion was erected for the single European species, *C. lindeni*, on the basis of its distinctive male abdominal appendages and some other characters which separate it from species of the closely allied genus *Coenagrion*. Some authors retain *C. lindeni* in *Coenagrion*, and Conci & Nielsen (1956) associate it with the Mediterranean group of *Coenagrion*. It is included with *Coenagrion* in the preceding keys. An additional eight species of *Cercion* are found in the eastern Palaearctic.

CERCION LINDENI (Sélys)   Plate 11 (59,60)

Agrion lindenii Sélys, 1840, *Monograph. Libell. Europ.* : 167.
Type locality: Belgium.
F – L'Agrion de Vander Linden, L'Agrion à longs cercoïdes,
D – Pokal-Azurjungfer

DESCRIPTION OF ADULT

A blue and black damselfly, although the blue often has a distinctly green tinge. It is a relatively robust species with body length 30–34mm. The pale postocular spots are small and transversely elongated (absent in form *nigriceps* Navas from southern Spain), and a pair of pale spots usually lies between median and lateral ocelli. The blue antehumeral stripes are broad, each more than one-third as broad as the black mid-dorsal thoracic band. The light brown pterostigma (fig. 130) is twice as long as broad, shaped much as in *Coenagrion caerulescens*. Posteriorly the prothorax is entire and slightly curved in the male (fig. 127), weakly trilobed in the female (fig. 128). On the male's dorsal abdominal pattern (fig. 126) is distinctive; a black median mark extends the full length of S2, black posterior marks on S3–6 are produced forwards as fine, median points, and S8 is dorsally almost all black. In European specimens, S10 is mainly blue with a median, longitudinal, black mark, but in Moroccan examples this mark is often broader and covers most of the last segment (Aguesse, 1968). On the abdomen of the mature female, the pale colour flanking the broad, median dorsal, black band is yellowish to olive-green on the anterior and posterior segments but blue on the central three or four segments. Characteristic male abdominal appendages (figs 129,131), with the superiors twice the length of the inferiors, bluntly pointed, and longer than S10, separate *C. lindeni* from males of all *Coenagrion* species.

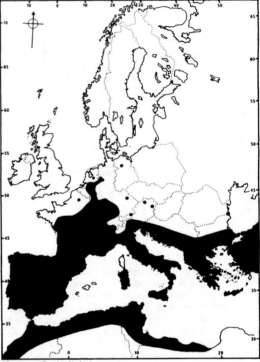

Map 30 – *Cercion lindeni*

BIOLOGY

Inhabits slow-flowing rivers and the larger expanses of standing water. It is often found in gravel pits and other artificial water bodies. Males fly low over the water surface, or wait for females whilst perched on mats of vegetation. Females oviposit in tandem at the water surface, or alone while completely submerged. In the latter case, males will await re-emergence of their mates and reclasp them when they surface to assist them from the water (and to obstruct the attentions of other males) (Heymer, 1973). Males do not accompany their mates in subaquatic oviposition (Utzeri *et al.*, 1983; pers. obs.).

FLIGHT PERIOD

End of May to mid-September.

DISTRIBUTION (map 30)

Widespread and plentiful from North Africa and Portugal across southern Europe to Turkey and Syria. Much less common north of the Alps where it is known from the Netherlands, Belgium, Luxembourg, Germany, Switzerland and Austria (two records only). In Germany it is commonest, and apparently spreading, in the Rhine valley and south-west of the Federal Republic, but it becomes progressively more scarce north-eastwards as far as Mark Brandenburg, German Democratic Republic, where populations may represent a separate subspecies *C. l. lacustre* Beutler, 1985. Reported recently (Bulimar, 1984) from Romania.

ENALLAGMA Charpentier, 1840

As here understood, following Davies (1981), *Enallagma* is a large genus of almost seventy species which occur in all zoogeographical regions except the Australasian and Oriental. *E. cyathigerum*, the type species, is the sole European representative, and it also occurs in North America where *Enallagma* is the best-represented genus of Zygoptera. Adults can be distinguished at rest from other coenagrionid genera by the presence of only one oblique, black stripe on the side of the thorax on the upper part of the metapleural suture. There is no darkening in the position of the obsolete mesometapleural suture.

ENALLAGMA CYATHIGERUM (Charpentier)
Plate 11 (61–64)

Agrion cyathigerum Charpentier, 1840, *Libell. europ. descr. depict.* : 163

Type locality: Silesia, Poland.

> *Agrion pulchrum* Hagen, 1839 (*nomen nudum*)
> *Agrion charpentieri* Sélys, 1840
> *Agrion brunnea* Evans, 1845
> *Agrion annexum* Hagen, 1861

E – Common Blue Damselfly, **F** – L'Agrion porte-coupe, **D** – Becher-Azurjungfer

DESCRIPTION OF ADULT

E. cyathigerum is a common blue damselfly which may at once be recognized in the female by the prominent vulvar spine beneath S8 (fig. 65, p.71), and in the male by the characteristic black hastiform, mushroom- or cup-shaped design on the dorsum of S2 (fig. 132) (*cyathigerum* from *cyathos* (Gk), a wineglass). This latter mark is only occasionally disconnected from the posterior black ring, and lateral arms are very exceptional indeed. The abdomen in both sexes is more robust than in most *Coenagrion*. S8 and S9 of the male are entirely blue dorsally. The median dorsal, black pattern on the female abdomen narrows in front to a point on each segment so that pale coloration is evident anteriorly. In the typical heterochrome female (Plate 11 (64)), the pale colour is yellowish to greenish brown, but a blue homeochrome female form (Plate 11 (62,63)) in which the black pattern is somewhat reduced is quite common. Pale antehumeral thoracic stripes are broad in both sexes. The wings (fig. 62, p.70) are rather pointed, and some of the cells between costa and radius distal to the pterostigma in the hindwing are divided. Apical venation of the hindwing is denser than that of the forewing and it encloses a greater number of small cells. Male inferior abdominal appendages (figs 135,136) are twice as long as the superior appendages but shorter than the last abdominal segment.

BIOLOGY

E. cyathigerum has a broad tolerance of the types of water in which it breeds, occurring in both running and stagnant waters. It may be found in streams and small ponds, but is usually most abundant near larger bodies of water, males often flying over the open surface. It is the most abundant damselfly on Scottish lochs and lochans. Males have the habit during their maturation period of resting on bare ground in full sunlight, sometimes 300 or 400m from

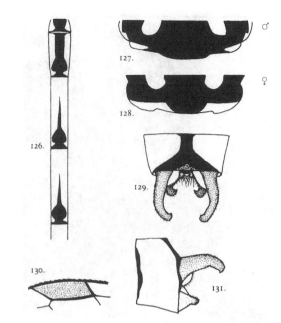

Figure 126–131 *Cercion lindeni*
(126) pattern on male abdominal segments S1–4
(127, 128) pronotal margins of male (above) and female (below)
(129, 131) male abdominal appendages in dorsal view (above) and left lateral view (below)
(130) left forewing pterostigma

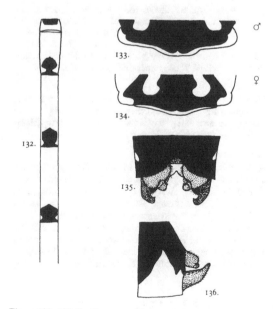

Figure 132–136 *Enallagma cyathigerum*
(132) pattern on male abdominal segments S1–4
(133, 134) pronotal margins of male (above) and female (below)
(135, 136) male abdominal appendages in dorsal view (above) and left lateral view (below)

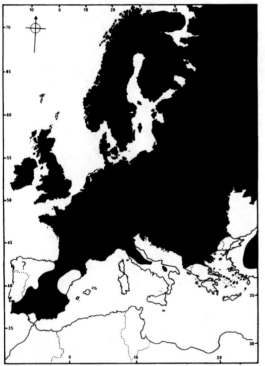

Map 31 – *Enallagma cyathigerum*

breeding sites. Oviposition is in tandem if the female lays eggs whilst standing above the water level, but if she submerges the pair disengage and the male awaits the female's re-emergence. Copulatory activity is studied in detail by Miller & Miller (1981), and Parr (1976) presents a study of population ecology. Larval development takes from two to four years, depending partly on latitude.

FLIGHT PERIOD

Mostly from the beginning of May to early in September, but appearing earlier in southern, lowland localities, and there are records of it persisting well into October in Britain and to the beginning of November in Switzerland.

DISTRIBUTION (map 31)

A circumboreal species found in North America (Florida everglades, and everywhere between 40°N and the Arctic Circle), Europe and Asia east to Mongolia. *E. cyathigerum* and *Lestes dryas* are the only Zygoptera found on both sides of the Atlantic. It is widespread in Europe north of 45°N, recorded from every country except Iceland, but local in the extreme south (to Potenza, Basilicata in Italy (Bucciarelli, 1972) and Cadiz in Spain) where it is found mostly at altitude. It does not occur in North Africa.

ENALLAGMA DESERTI (Sélys, 1871b)

This species is described from the Algerian Sahara and it is reputed to extend east across the Middle East and southern USSR to Japan. *E. deserti* resembles *E. cyathigerum* in pattern but the pale colour is typically more extensive.

Postocular spots are larger and in females not closed posteriorly, the clypeus is pale with two lateral black spots (all black in *cyathigerum*), wing veins are yellowish and the mid-dorsal thoracic crest is pale. Lieftinck (1966) has studied material of *deserti* from Morocco (Middle Atlas) and Algeria. The Moroccan specimens are not especially pale, but males from both countries have terminal abdominal appendages differently-shaped from those of *cyathigerum*. Asian material attributed to *deserti* requires further study.

ISCHNURA Charpentier, 1840
Micronympha Kirby, 1890

The species of *Ischnura* are of delicate build, relatively short-winged, and with blue areas limited in extent and on the abdomen confined to the apical segments. Females are polymorphic. Males have a characteristically bicoloured pterostigma in the forewing (fig. 63, p.70), the proximal part blackish and the apex white to yellowish. Females have a small vulvar spine on S8 (fig. 66, p.71) and males have a more or less bifid, dorsal tubercle on S10 (figs 143–145, 147). The frons is rounded, postocular spots are rather small and round, and the antehumeral stripes are fine. Both of the two latter features may be obliterated.

Ischnura is a genus of world-wide distribution and of similar size to *Enallagma*, represented in Europe by four species and with five other species occurring outside Europe but within the western Palaearctic faunal region.

Key to species of *Ischnura*

1 Males 2
– Females 5

2(1) Posterior part of S8 and almost all of S9 blue. Pterostigma of forewing distinctly larger than that of hindwing. Superior abdominal appendage (fig. 143) without an internal finger-like process; tubercle on dorsum of S10 in posterior view not as high as broad .... *pumilio* (p.92)

– S8 all blue but S9 blue at most only anteriorly. Pterostigma of forewing not obviously larger than that of hindwing. Superior abdominal appendage (figs 144,145,147) with a downwardly-directed finger-like process arising from the mesial surface; tubercle on dorsum of S10 in poster or view slightly higher than broad 3

3(2) Posterior edge of pronotum (fig. 139) medially elevated as a low dome. Superior abdominal appendage (fig. 145) with downwardly-directed process about as long as the rest of the appendage; inferior appendage without a well-developed internal projection. Second abdominal segment (fig. 142) with a distinct transverse ridge on the dorsal surface posteriorly. Postocular spots and antehumeral stripes subobsolete. Spain, Portugal and North Africa .... *graellsi* (p.94)

– Posterior edge of pronotum (figs 138,146) medially elevated as a tongue-like structure. Superior abdominal appendage (figs 144,147) with internal process shorter; inferior appendage sometimes with a well-developed internal process. Second abdominal segment (fig. 141) with dorsal surface in profile almost flat. Postocular spots and antehumeral stripes usually developed 4

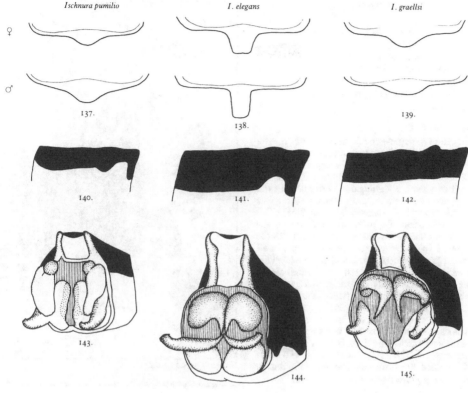

Ischnura pumilio   *I. elegans*   *I. graellsi*

♀

♂

137.   138.   139.

140.   141.   142.

143.   144.   145.

4(3) Posterior edge of pronotum (fig. 138) with median elevation longer than broad, with parallel sides and apex truncate. Superior abdominal appendages (fig. 144) with internal processes not crossing; inferior appendage with a well-developed internal process extending to the point of the process of the superior appendage. Widespread in Europe ... . *elegans* (p.92)

– Posterior edge of pronotum (fig. 146) with median elevation slightly broader than long, with converging sides and apex incised. Superior abdominal appendages (fig. 147) with internal processes crossing; inferior appendage with a weak internal process. Corsica, Sardinia, Sicily and other central Mediterranean islands .... *genei* (p.94)

5(1) Posterior edge of pronotum (fig. 138) with a median tongue-like projection which is parallel-sided and truncate or only very weakly incised apically .... *elegans* (p.92)

– Posterior edge of pronotum (figs 137,139,146) with median elevation less prominent and either dome-shaped, triangular or apically incised 6

6(5) Posterior edge of pronotum (fig. 146) with median elevation incised apically .... *genei* (p.94)

– Posterior edge of pronotum (figs 137,139) with median elevation entire 7

7(6) Dorsal surface of abdomen entirely black in homeochrome form and with pale colour only on S1,2 and the anterior part of S3 in the heterochrome form .... *pumilio* (p.92)

– Pale colour on the dorsal surface of S8, as well as on the anterior segments, although in the heterochrome form the pale areas may be obfuscate .... *graellsi* (p.94)

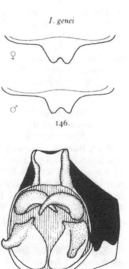

I. genei

♀

♂

146.

147.

Figures 137–147 *Ischnura* species
(137, 138, 139, 146) pronota of female (above) and male (below)
(140, 141, 142) abdominal segment S2 in lateral view
(143, 144, 145, 147) male abdominal appendages and segment S10 in posterolateral view

ISCHNURA PUMILIO (Charpentier)
Plate 12 (65–67)

Agrion pumilio Charpentier, 1825, *Horae entomol.* : 22.
Type locality: northern Italy.

   Agrion xanthopterum Stephens, 1835
   Agrion rubellum Curtis, 1839
   Agrion cognata Sélys, 1841

E – Scarce Blue-tailed Damselfly, **F** – L'Agrion nain,
D – Kleine Pechlibelle

DESCRIPTION OF ADULT

This species, the type species of *Ischnura*, stands a little apart from the other three European representatives, and may be readily recognized by its characteristic colour patterns. The male has the subapical blue colour on the abdomen shifted posteriorly compared with males in the *elegans* group, and S9 is predominantly blue with just a pair of small, black spots (in *I. elegans*, S8 is all blue). The female has no indication of a pale, subapical, dorsal abdominal mark in either of its forms. The homeochrome female (Plate 12 (66)) has yellow postocular spots and antehumeral stripes, the latter broad, and brownish yellow marks laterally on the first two and last two abdominal segments. The heterochrome female (form *aurantiaca* Sélys, 1837) is a beautiful and distinctive insect ((Plate 12 (67)) with pale areas of the body more extensive and a bright orange in colour. Postocular spots are large and antehumeral stripes very broad so that the thorax is only narrowly lined with black. The dorsum of S1 and S2, the anterior quarter of S3, S9 laterally and almost all of S10 are orange. Wing veins are brownish orange. Both female forms have all wing pterostigmata pale brownish.

Males of *I. pumilio* differ from those of the other species in having a relatively large pterostigmata in the forewing, a lower tubercle on the dorsal apex of the last abdominal. segment and mainly pale superior abdominal appendages (fig. 143) which lack an internal finger-like process. In both sexes, the posterior edge of the pronotum is medially raised as a low dome (fig. 137). *I. pumilio* is small, its body length usually 26–30mm in males and 28–31mm in females.

BIOLOGY

Breeds in shallow ditches, swampy meanders, the boggy edges of pools and streams, and marshes, usually where the rate of flow is very slow. The larvae can survive in slightly brackish water and are tolerant of a wide pH range. *I. pumilio* will rapidly colonize newly-formed biotopes, but colonies tend to die out after a few years. In Switzerland, it has been found at altitudes in excess of 2000m. Aspects of the ecology of the species are described by Krieger & Krieger-Loibl (1958), Dapling & Rocker (1969), Jurzitza (1970), Zimmerman (1973) and Rudolph (1979).

FLIGHT PERIOD

From the end of May to mid-September, with a well-defined peak about the end of June.

DISTRIBUTION (map 32)

Found over much of Europe in scattered colonies, but most numerous in the south. It occurs only in the south and west of the British Isles, and was unrecorded from Scandinavia until Hämäläinen (1985b) reported finding it in Helsinki and very recently it has been cited from Gotland. Outside Europe, *I. pumilio* is recorded from the Azores, Madeira

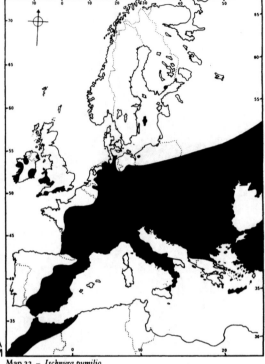

Map 32 – *Ischnura pumilio*

and western North Africa (Morocco, Algeria), the Middle East and from the southern USSR to west and central Siberia.

ISCHNURA ELEGANS (Vander Linden)
Plate 12 (68–74)

Agrion elegans Vander Linden, 1820, *Agrion. bonon. descr.* : 6.
Type locality: Bologna, Italy.

   Agrion pupilla Hansemann, 1823
   Agrion tuberculatum Charpentier, 1825
   Agrion rufescens Stephens, 1835
   Ischnura lamellata Kolbe, 1885

E – Blue-tailed Damselfly, **F** – L'Agrion élégant,
D – Grosse Pechlibelle

DESCRIPTION OF ADULT

I. elegans and the two following species are very similar in appearance with S8 in males all blue and females existing in two major forms, one of which is more or less orange in colour. However, the two following species both have restricted ranges, scarcely overlapping with that of the common and widespread *I. elegans*, so that locality is a good guide to identity. The main morphological characters separating *I. elegans* from its allies are to be found in the form of the posterior pronotal edge and the male abdominal appendages. In both sexes of *I. elegans* but most strongly developed in the male, there is a tongue-like, raised, median projection posteriorly on the pronotum (fig. 138). The male abdominal appendages viewed from behind (fig.

144) have four processes terminating near to each other centrally, the two elongated, internal processes of the superior appendages being subparallel and directed ventrally, and the shorter internal processes of the inferior appendages curving inwards and upwards to almost meet them. Antehumeral thoracic stripes and postocular spots are most often well-defined, although specimens can be found with both of these reduced or absent. *I. elegans* is normally larger than *I. pumilio*, males from Britain measuring 31–33mm and females 31–34 mm in body length, but in southern France it is generally two or three millimetres shorter and in north-east Spain (*I. e. mortoni*) males may be only 26mm long.

Female *I. elegans* exhibit an interesting array of colour variation (Killington, 1924), and the following named forms are usually present in any one population in western Europe (in the Ukraine I failed to find any heterochrome females amongst hundreds examined):

homeochrome forms

typica (Plate 12 (70))	Very like the male in coloration.
violacea (Plate 12 (71))	This is an immature form that develops into either *typica* or *infuscans* and is seldom found in tandem (see Plate A, fig. 3 (p. 33)). Blue on the thorax is replaced by violet.
infuscans (Plate 12 (72))	Differs from the typical form in having the pale parts of head and thorax yellowish brown to olive, and S8 dorsally dull brownish becoming almost black with age.

heterochrome forms

rufescens (Plate 12 (69,74))	The thorax has a mid-dorsal black stripe only, the rest of the thorax being pink. Postocular spots are orange-brown, occasionally blue. S8 is dorsally blue. This is an immature form of the following.
infuscans-obsoleta (Plate 12 (73))	Resembles *rufescens* but the thorax is more brownish and the dorsal surface of S8 is dark as in form *infuscans*.

A number of allopatric subspecies of *I. elegans* from different parts of its broad geographical range have been described (see Schmidt, Er., 1938, 1968). These include *I. e. mortoni* Schmidt, 1938, from the north-east of Spain, *I. e. pontica* Schmidt, 1938, from eastern Austria (Neusiedlersee), Hungary, and the northern Balkans east to Georgia, the Caspian coast of Iran and to Afghanistan, and *I. e. ebneri* Schmidt, 1938, from Crete, the Cyclades, Rhodes, Cyprus, Turkey and the Middle East. This latter subspecies has the tips of the male superior appendages crossed.

BIOLOGY

An ubiquitous species of ditches, canals, pools, ponds, lakes and slow-flowing streams. The species is tolerant of both brackish conditions and moderate pollution. *I. elegans* is biologically unusual in several respects: there is no evidence of movement away from water during the very brief (about five days) maturation period, it is active in dull and relatively cool weather, immature adults and females spend much time feeding near water, and it has a prolonged

Map 33 – *Ischnura elegans*

copulation time of five or six hours (Parr, 1973; Miller, 1987). Females oviposit unaccompanied late in the day and at this time are much less molested by males than are other Coenagrionidae.

FLIGHT PERIOD

Long and usually without a clearly-defined peak of abundance, in contrast to *I. pumilio*. In the Mediterranean region it appears in March and has two or perhaps even three generations of adults in a year, the last persisting to the end of October. In northern Britain, in contrast, adults do not usually emerge before early May, and larvae require two years to complete their development. In southern England, *I. elegans* is univoltine.

DISTRIBUTION (map 33)

I. elegans ranges over all of Europe except Iceland, northern Scandinavia, Corsica, Sardinia, Sicily and Malta. In the Iberian Peninsula it is found only in Gerona, Cadiz (Ferreras Romero & Puchol Caballero, 1984) and on the Balearic Islands. Formerly considered absent from North Africa but establishment of conspecificity between the Algerian *I. lamellata* Kolbe and *I. elegans* extends its range to that continent. *I. elegans* is widespread in the Middle East, much of the Soviet Union and in China.

ISCHNURA GENEI (Rambur)

Agrion Genei Rambur, 1842, *Hist. nat. ins. Neur.* : 276.
Type locality: Sardinia.

DESCRIPTION OF ADULT

I. genei is closely allied to *I. elegans*, and considered by
Aguesse (1968) to be a subspecies of it. It is, however,
separable by the structure of the posterior border of the
pronotum (fig. 146), which in both sexes has the apex of the
median lobe distinctly incised, and by the form of the male
abdominal appendages (fig. 147). The superior appendages
have elongated processes whose tips cross each other, and
each inferior appendage has a short internal tooth which is
much smaller than the corresponding structure in *I.
elegans*. *I. genei* is a small insect, of similar size to *I. pumilio*.

BIOLOGY

Larvae inhabit small ponds, slowly-flowing streams and
drainage channels.

FLIGHT PERIOD

Specimens examined in collections have been taken in May
and June, but the actual period of adult activity probably
extends beyond these months.

DISTRIBUTION (map 34)

Restricted to Mediterranean islands. Recorded from Corsica, Sardinia, Sicily and Malta, and by Capra (1937, 1976)
from Capraia and Giglio off the Tuscany coast of western
Italy. The last-named island is the only locality where *I.
genei* and *I. elegans* are reported to coexist.

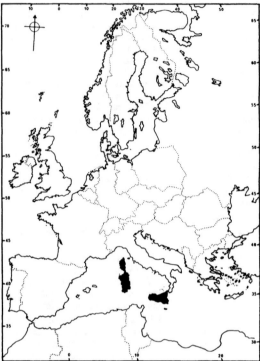

Map 34 – *Ischnura genei*

ISCHNURA GRAELLSI (Rambur)   Plate 13 (75)

Agrion Graellsii Rambur, 1842, *Hist. nat. ins. Neur.* : 275.
Type locality: Barcelona, Spain.

DESCRIPTION OF ADULT

The shape of the posterior edge of the pronotum (fig. 139),
which is entire and only slightly raised medially, distinguishes *I. graellsi* in both sexes from both *I. elegans* and
I. genei. The male abdominal appendages (fig. 145) are also
characteristic; the superior appendages have long, slender,
internal processes whose tips diverge and are uncrossed,
and the inferior appendages are without internal processes.
I. graellsi is no larger than *I. pumilio*. The two postocular
spots are scarcely visible or entirely obliterated, and the
antehumeral thoracic stripes are very narrow or incomplete
in males.

BIOLOGY

The larva has recently been described (Ferreras Romero,
1981b). It inhabits standing, sometimes slightly brackish,
water.

FLIGHT PERIOD

March to September.

DISTRIBUTION (map 35)

Spain, Portugal and North Africa (Mediterranean littoral of
Morocco, Algeria and Tunisia). Reports of *I. graellsi* from
the extreme south-west of France require confirmation; the
Pyrenees are generally considered to mark its north-eastern
limits.

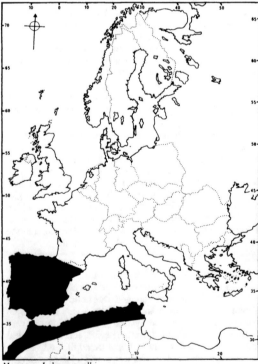

Map 35 – *Ischnura graellsi*

The following species of *Ischnura* are extra-limital but occur within the western Palaearctic region close to the frontiers of Europe.

ISCHNURA SENEGALENSIS (Rambur, 1842)

A large species with a long vulvar spine in the female and a weakly-developed dorsal tubercle on S10 in the male. Found over all Africa (including Madagascar but not the Canary Islands or Madeira) south of about 35°N (i.e. not north of the Atlas Mountains). It also ranges from Egypt, Yemen, Oman and Iraq to the Caucasus, Pakistan and thence east to Japan.

ISCHNURA SAHARENSIS Aguesse, 1958

Allied to *I. genei* and found in North Africa mostly south of the Atlas Mountains, from the Canary Islands and Morocco (south to Agadir) to Libya (Lieftinck, 1966).

ISCHNURA FOUNTAINEI Morton, 1905

The male differs from all other *Ischnura* species herein mentioned in having superior abdominal appendages longer than the inferiors, the black spines of the latter being directed inwards. The posterior lobe of the pronotum is not prominent and very weakly incised in males. The heterochrome female has a yellow-orange thorax with only a median black stripe, a pale spot in front of each lateral ocellus, and a mid-dorsal, metallic black stripe on S2–9. It is a species particularly associated with oases in desert regions and is often found in company with *I. saharensis*. It is found in North Africa from southern Algeria and Tunisia through Libya, Egypt and Sinai to Iraq, Iran, Saudi Arabia and Azerbaijan to Uzbekistan and Tadzhikistan, USSR.

ISCHNURA FORCIPATA Morton, 1907

Not known from west of the Caucasus.

ISCHNURA INTERMEDIA Dumont, 1974

Found in August 1973 on Nemrut Mountain in Malatya, Turkey at an altitude of about 2000m. It is intermediate in character between *I. pumilio* and *I. forcipata*.

NEHALENNIA Sélys in Sélys & Hagen, 1850
Argiallegma Calvert, 1907

A small genus of six described species, five of which occur in the New World which is probably the centre of origin (Belyshev & Haritonov, 1977). It is represented in Eurasia by *N. speciosa*, the type species. They are small damselflies, predominantly greenish black with brassy or metallic reflections, the frons is sharply, transversely ridged, postocular spots are absent and the eyes are widely separated.

NEHALENNIA SPECIOSA (Charpentier)
Plate 13 (76)

Agrion speciosum Charpentier, 1840, *Libell. europ. descr. depict.* : 151.
Type locality: Lüneburg, Germany.
  Agrion sophia Sélys, 1840
F – La Néhalennie précieuse, **D** – Zwerglibelle

DESCRIPTION OF ADULT
This is the smallest European odonate with a body length usually in the range 22–24mm. Thorax and abdomen are very narrow, and the abdomen is relatively long. Wings are short and broad, only 0.6 to 0.7 times as long as the abdomen, and their venation is reduced with only two or three cells between costa and radius distal to the pterostigma. The pterostigma (fig. 148) is rhomboidal, light brown centrally with a broad, clear margin, and only as long as the cell it surmounts. Vein R_2 forks only about four (forewing) or three (hindwing) cells after the node. The pecten on the front tibia consists of a row of seldom more than four pale spines. The occiput dorsally is black with a continuous, blue, postocular line (no indication of paired spots). The frons has a transverse, right-angled ridge or crest, and the clypeus projects forwards horizontally. Eyes in life in both sexes are blue becoming browner with age. The thorax dorsally is metallic green, powdery blue between the wing bases in the mature male, and pale antehumeral stripes are absent. The legs are whitish with narrow, dorsal, dark lines on the femora. In the male, the abdomen dorsally is metallic green with blue colour confined to the terminal three segments; the apical half of S8, most of S9, and all of S10 except for two small, dark points, are blue. The female has at first a similar abdominal coloration to that of the male, but at maturity the abdomen becomes more orange-brown. Thorax and abdomen in both sexes are ventrally whitish.
  The last abdominal segment of the male bears a pair of small, contiguous, spinous, median projections dorsally on

Figure 148 *Nehalennia speciosa* male left forewing pterostigma

its posterior border (fig. 149). The superior abdominal appendages are about half the length of S10, much longer than the very short inferiors (fig. 149), obliquely truncated and appearing apically rounded when viewed laterally. There is no vulvar spine in the female. The posterior pronotal edge (figs 150,151) is simply rounded, although raised and very thin in the female.

BIOLOGY

N. speciosa is locally abundant, but only in very scattered colonies found near shallow, mesotrophic pools and marshy places on upland moors with 'prairies' of sedges, fescue grass, or sometimes *Phragmites*. An account of some aspects of the biology of the species is provided by Schiess (1973).

FLIGHT PERIOD

From the end of May to about the end of July.

DISTRIBUTION (map 36)

Northern and eastern Europe, east through Siberia to Japan, but not the British Isles. In Europe, *N. speciosa* has been recorded from south-east Sweden, Finland (very local in the south and not reported for twenty-seven years prior to 1981 (Hämäläinen, 1981)), a few localities in Belgium and the Netherlands, from Switzerland (Canton Zurich only (De Marmels & Schiess, 1977)), Austria (Puschnig (1935) records observations made in Karinthia), Germany, Poland, Czechoslovakia, northern Italy (Balestrazzi & Bucciarelli, 1971b; Bucciarelli, 1976; Pecile, 1981) and Romania.

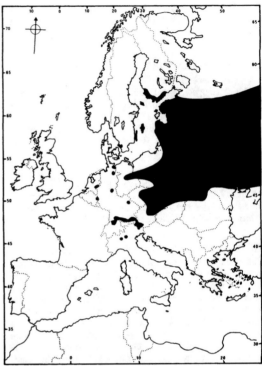

Map 36 – *Nehalennia speciosa*

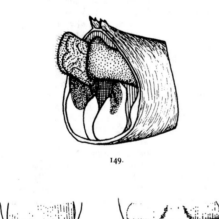

149.

150.   151.

Figures 149–151 *Nehalennia speciosa*
(149) male abdominal appendages and segment S10 in postero-lateral view
(150, 151) prothorax of male (left) and female (right)

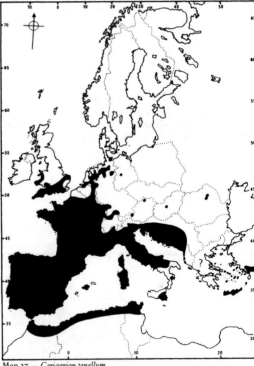

Map 37 – *Ceriagrion tenellum*

CERIAGRION Sélys, 1876
Palaeobasis Kennedy, 1920

Small, delicate, red damselflies represented in Europe by a single species, but with forty-seven species distributed in all parts of the Old World and in Australasia. The frons is angulate, as in *Nehalennia* to which *Ceriagrion* is allied, with widely-separated eyes, absence of postocular spots, and slightly metallic sheen on the black areas of the body.

CERIAGRION TENELLUM (Villers)
Plate 13 (77–79)
Libellula tenella Villers, 1789, *Caroli Linnaei Entomol.* : 15.
Type locality: France.
  Agrion rubella Vander Linden, 1823
  Agrion rufipes Stephens, 1835
E – Small Red Damselfly, **F** – L'Agrion délicat,
D – Späte Adonislibelle

DESCRIPTION OF ADULT

The red coloration of *C. tenellum* gives it a superficial resemblance to *Pyrrhosoma*, in which genus it was at one time placed. *C. tenellum* is, however, a very much smaller and more delicate damselfly than *P. nymphula* and its legs are yellowish brown to reddish brown rather than black, it has very short male abdominal appendages (fig. 152), the

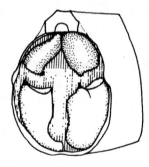

Figure 152 *Ceriagrion tenellum*
male abdominal appendages and segment S10 in posterolateral view

frons is angulated with dorsal and anterior faces sharply separated, and other differentiating characters are included above in the key to genera of Coenagrionidae. The mature male has red eyes, frons, clypeus, antennal bases and abdomen, and reddish brown legs and pterostigmata; the vertex is bronzed black and there are no pale postocular spots. The thorax dorsally is black with at most the merest trace of pale antehumeral stripes. Yellowish antehumeral stripes are better developed in the female but they are very narrow and usually broken. The face of the female is cream-coloured with the postclypeus and base of the labrum blackish. Wings are apically rounded with small, rhomboidal pterostigmata (posterior edges angular rather than curved) and there are three postquadrilateral antenodal cells in all wings.

Females of *C. tenellum* are polymorphic, four colour forms being recognizable:

typica (Plate 13 (79))	S1–3 are dorsally red except for a black spot at the base of S1 and black apex to S3, S4–8 are dorsally mainly black and S9,10 are mainly reddish.
erythrogastrum Sélys	The homeochrome form, in which the abdomen is as red as in the male. The head, also, is coloured like that of the male.
intermedium Sélys	The abdomen is dorsally red but S6–8, sometimes also S3–5, are apically black. The head is coloured as in the typical female with postclypeus and base of labrum black, the remainder of face cream-coloured.
melanogastrum Sélys (Plate 13 (78))	The dorsal surface of the abdomen, except for intersegmental incisures, is almost entirely bronzed black. The characteristic yellow to reddish leg colour remains.

The typical form of the female is the most plentiful throughout Europe, form *melanogastrum* is not uncommon, but the remaining two forms are generally rare.

C. tenellum tenellum is small, male body length 25–32mm, female body length 29–34mm. A not very well-differentiated subspecies, *C. t. nielseni* Schmidt, 1953, is rather larger (males 29–34mm, females 31–35mm) and in the typical form has female S8 dorsally red with a black median line and black sides. No female form corresponding to *intermedium* has been recorded from populations of *C. t. nielseni*, although the three other forms are represented.

BIOLOGY

Found at ponds, marshes, upland peat bogs and sometimes slowly flowing ditches. It is tolerant of very shallow water, and in Britain is found only in acid conditions. Adults flutter weakly amongst grasses and rushes fringing the breeding sites. Females oviposit in tandem, usually in vertical, submerged stems in shallow water. *C. tenellum* often forms very dense colonies. Parr & Parr (1979) give an account of the species in England.

FLIGHT PERIOD

Appears rather later than most other coenagrionids: mid-May to the end of August or early September in southern Europe, June to September in more northerly parts of the range.

DISTRIBUTION (map 37)

Most plentiful in south-west Europe and widespread in Spain (including Majorca), south and west France (including Corsica), Italy and Sardinia. Local in southern England and Wales (north to Anglesey), Belgium, Holland, West Germany, Switzerland (five localities known in cantons Zurich, Thurgau and Schwyz; extinct in the west), Austria, Hungary, Romania and Yugoslavia. It is not found in Scandinavia, is known only from Magdeburg in East Germany (Müller, 1984) and does not extend east to Poland or the USSR. *C. t. nielseni* is reported to replace the typical subspecies in Sicily (type locality: Lentini), Yugoslavia, Crete and in North Africa. *C. t. georgfreyi* (Schmidt, 1953) occurs in Turkey and in Syria north of the Jordan valley.

Suborder: **Anisoptera** (Dragonflies)

AESHNIDAE

Aeshnidae are large to moderately large dragonflies. As in Gomphidae and Cordulegastridae, the first antenodal crossvein and another more distally situated are strongly thickened; these primary antenodals are usually the only ones in exact alignment across the costal and subcostal spaces. Aeshnids further resemble cordulegastrids in having the discoidal cells similarly shaped in both pairs of wings (fig. 16, p.45) and elongated in the long axis of the wing, but the eyes of aeshnids are contiguous usually over a considerable length on top of the head (only just touching in Cordulegastridae) and most aeshnids are patterned with yellow, green or blue on a dark brown ground colour, unlike the contrasting black and yellow coloration of cordulegastrids. Among Anisoptera, females of only Cordulegastridae and Aeshnidae have an ovipositor but its structure is very different in the two families. In Aeshnidae the ovipositor is short but it is provided with serrated valves, as in damselflies, enabling the female to insert eggs into plant tissue.

Aeshnidae are very volatile, much of their time being spent in flight, and their size and attractive coloration make them conspicuous and familiar insects of ponds and lakesides. At rest in vegetation, they adopt a vertical position. The family includes six genera with European representatives.

Key to genera of Aeshnidae

1   Median (or basal) space between veins Cu and R+M before the arculus crossed by one or more veins (figs 158,159) 2

–   Median space before the arculus open (figs 153–157) 3

2(1)   Pterostigma long, extending over three or four cells; vein IR_3 unforked; tips of all wings usually infumate (fig. 159) .... *Boyeria* (p.119)

–   Pterostigma very short, extending over two cells only; vein IR_3 bifurcates symmetrically just over half-way between node and base of pterostigma; wings clear (fig. 158) .... *Caliaeschna* (p.118)

3(1)   Vein R_3 sharply curved towards vein R_2 beneath distal part of pterostigma and arculus with veins arising from its anterior half (figs 155,156). Hindwing of male basally rounded and without an anal triangle; male without auricles on S2 4

–   Vein R_3 gently curved towards vein R_2 and arculus with veins arising near to its centre (figs 153,154,157). Hindwing of male basally angulate and with well-defined anal triangle; auricles present on male S2 5

4(3)   Hindwing (fig. 156) with field of cells below fork of anal vein laterally elongated; anal vein (A_1) far removed from cubital vein (CuP) beneath the discoidal cell and separated from it by an irregular group of cells three deep; radial supplementary vein deeply curved. Male inferior appendage in dorsal view (fig. 186, p.113) subtriangular and with its apex slightly incised .... *Hemianax* (p.116)

–   Hindwing (fig. 155) with field of cells below fork of anal vein rounded; anal vein subparallel to cubital vein and separated from it beneath the discoidal cell by two regular rows of cells; radial supplementary vein shallowly curved. Male inferior appendage in dorsal view (figs 184,185, p.113) quadrangular or trapezoidal, broadly truncated .... *Anax* (p.112)

5(3)   Radial and median supplementary veins each subtend a single row of cells; usually only two cubito-anal veins basal to the discoidal cell; pterostigma very long and narrow (fig. 157) .... *Brachytron* (p.116)

–   Radial and median supplementary veins separated respectively from vein IR_3 and median vein by several rows of cells; usually four to six cubito-anal veins; pterostigma less elongated and broader (fig. 154) .... *Aeshna* (p.100)

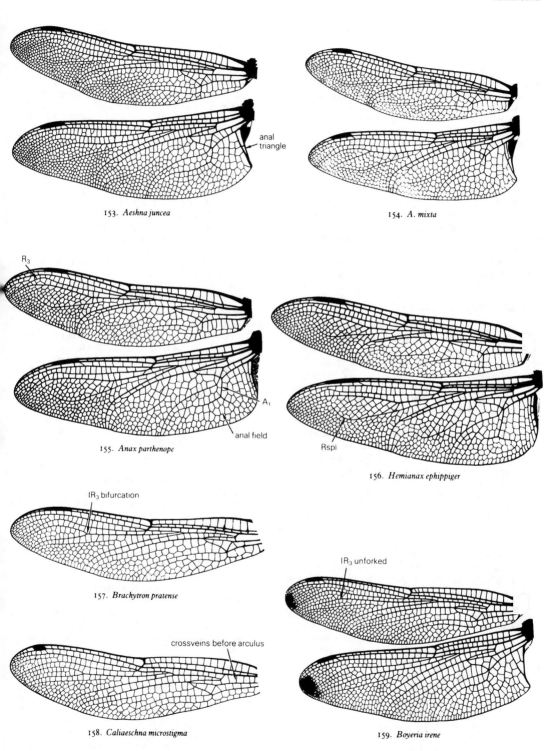

153. *Aeshna juncea*

anal triangle

154. *A. mixta*

R_3

155. *Anax parthenope*

A_1

anal field

Rspl

156. *Hemianax ephippiger*

IR_3 bifurcation

157. *Brachytron pratense*

crossveins before arculus

158. *Caliaeschna microstigma*

IR_3 unforked

159. *Boyeria irene*

Figures 153–159 Wings of male Aeshnidae
(forewings only of *Brachytron pratense* and *Caliaeschna microstigma*)

AESHNA Fabricius, 1775

Aeschna auctt.

The name of this genus is often spelt *Aeschna*, this probably being what Fabricius intended, but since he actually wrote *Aeshna* it is this form which should be adopted (Opinion 34, International Commission on Zoological Nomenclature) and the emendation is invalid. The family name derived from this genus should likewise be spelt without a 'c'.

Aeshna species are mostly brightly marked with green or blue in a fairly constant pattern which makes field identification at a distance often a matter of difficulty. They are very strong fliers, usually hawking at a height in excess of two metres, and they tend to move far from water during maturation. They breed in stagnant waters, often acidic, and adults of the majority of species are mid- to late summer insects, remaining active well into autumn. Females, with the exception of those of *A. affinis*, oviposit unaccompanied by males.

The genus is widespread and the eleven European species represent about one-sixth of the presently-known world fauna.

Key to species of *Aeshna*

1 Membranule long, dark, reaching to the anal angle (fig. 182, p.110); a small area at the base of the wing marked with yellow. Abdomen reddish brown with a yellow triangle on S2 the most conspicuous pale mark on the dorsal surface, there being no green or blue coloration ...... *isosceles* (p.111)

– Membranule shorter, extending two-thirds of the way or less towards the anal angle (figs 180,181); wings clear to uniformly brownish. Abdomen extensively patterned with blue, green or yellow, or if mainly brown (*grandis*), then lacking yellow triangle on S2 and membranule whitish 2

2(1) Abdomen reddish brown with reduced patterning (no pale spots dorsally on S8–10). Wings uniformly tinted brownish amber with reddish brown venation. Frons above with an ill-defined, crescentic brown mark .... *grandis* (p.111)

– Abdomen dark brown to black, patterned on all segments with blue, green or yellow spots. Wings hyaline or (*viridis*) weakly brownish, venation (except costa) black. Frons above with a well-defined crescentic, black mark which is (except in *viridis*) connected by a stalk to a black mark in front of the ocelli forming a T-shaped mark 3

3(2) Eyes contiguous for a distance scarcely longer than the occipital triangle (fig. 171, p.103). Apical venation dense and the bifurcation of vein IR₃ under the pterostigma poorly defined, its anterior branch weak and often rudimentary .... *caerulea* (p.103)

– Eyes contiguous for a distance much greater than length of occipital triangle (fig. 172, p.104). Apical venation less dense and the bifurcation of vein IR₃ always distinct (fig. 154) 4

4(3) Black mark on top of frons unstalked. Sides of thorax almost entirely green with scarcely any black marking. Suture between frons and postclypeus not lined with black. Wings weakly but distinctly brownish; anal triangle of male hindwing two-celled .... *viridis* (p.110)

– Black T-shaped mark on frons complete (e.g. figs 171,172). Sides of thorax usually brown with two oblique, yellow to bluish stripes, but if mainly greenish (*cyanea*, *affinis*) then black sulcal lines are clearly visible at least above the leg

insertions. Suture between frons and postclypeus often lined with black. Wings hyaline or almost so; anal triangle of male hindwing sometimes with more than two cells 5

5(4) S9 and S10 each with a complete blue band dorsally. Antehumeral stripes broad and green. Pterostigma short, not exceeding 3mm in length. Anal triangle of male hindwing with three or more cells .... *cyanea* (p.108)

– S9 and S10 each with a pair of separated pale spots. Antehumeral stripes narrower, sometimes vestigial or absent. Pterostigma relatively longer, often more than 3mm. Anal triangle of male with two or three cells 6

6(5) Smaller insects, usually about 60mm in length. Seven to nine postnodal veins in forewing; pterostigma subtends at most three cells; space between cubital (CuP) and anal (A₁) veins in hindwing (fig. 154, p.99) narrowing abruptly below the distal end of the discoidal cell and thence occupied by just a single row of cells except sometimes at the wing margin; anal triangle of male three-celled 7

– Larger insects, usually exceeding 65mm in length. Usually eleven or more postnodal veins in forewing; pterostigma subtends usually four or more cells; space between cubital and anal veins in hindwing (fig. 153, p.99) not narrowing abruptly and occupied centrally usually by at least a partially double row of cells; anal triangle of male two-celled ...8

7(6) Sides of thorax posterior to humeral sulcus green to bluish with narrow, black sulcal lines. Terminal abdominal appendages shorter, particularly in female where they are less than combined length of S9+10; male superior abdominal appendage with a small, ventral tooth near its base visible in profile (fig. 166) .... *affinis* (p.108)

– Sides of thorax extensively brown with oblique, pale stripes and black sulcal lines. Terminal abdominal appendages longer, in female longer than the combined length of S9+10; male superior abdominal appendage without a subbasal tooth (fig. 165) .... *mixta* (p.107)

8(6) Membranule mainly white. Male inferior appendage not reaching the mid-point of a superior appendage, the latter in profile with dorsal surface markedly concave and with the dorsal crest terminating in a series of small denticulations (fig. 164). Female with complete (always ?) antehumeral stripes .... *serrata* (p.107)

– Membranule infuscate except basally. Male superior appendages less than twice as long as the inferior appendage, less concave dorsally and apically dentate only in *crenata* (figs 161–163). Female with incomplete antehumeral stripes ... 9

9(8) Superior anal appendages of male (fig. 163) with two or three strong teeth dorsally at their apices; female anal appendages (fig. 178, p.106) apically pointed (always ?) .... *crenata* (p.106)

– Superior anal appendages of male (figs 161,162) not apically dentate; female anal appendages (fig. 179, p.106) apically rounded 10

10(9) Occipital surface of head with a small, yellow spot behind the middle of each eye; mid-lobe of labium pale. Side of thorax with two yellow, epimeral stripes but with little trace of metepisternal stripe between them. Male superior anal appendage (fig. 161) narrower .... *juncea* (p.104)

– Occipital surface of head entirely black; mid-lobe of labium with two apical, contiguous, dark spots. Side of thorax with a pale stripe on metepisternum between the two epimeral stripes. Male superior anal appendage (fig. 162) broader .... *subarctica* (p.105)

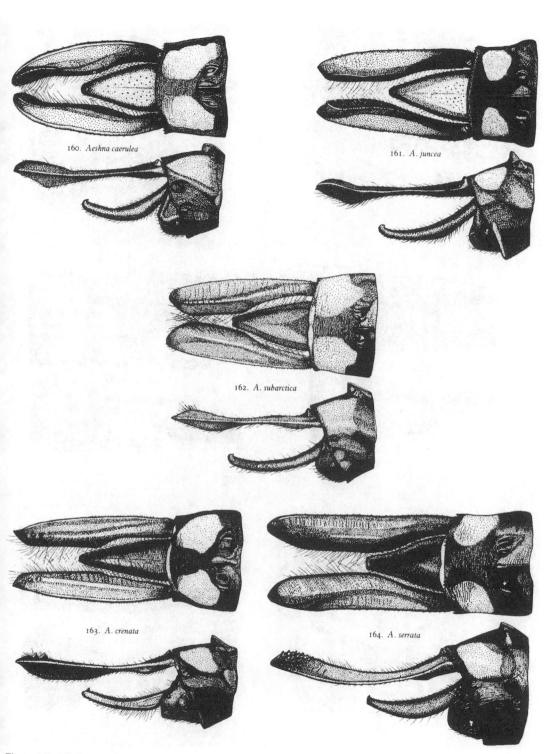

160. *Aeshna caerulea*

161. *A. juncea*

162. *A. subarctica*

163. *A. crenata*

164. *A. serrata*

Figures 160–164 *Aeshna* species
showing male abdominal appendages and segment S10 in dorsal view (above) and right lateral view (below)

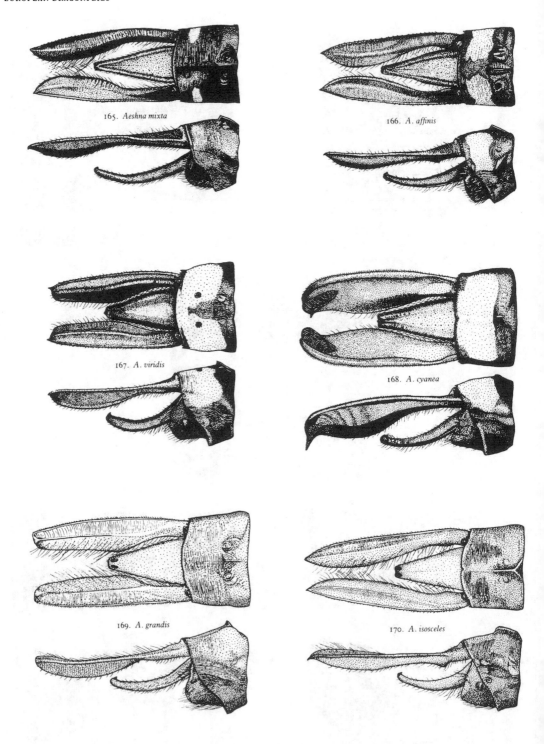

165. *Aeshna mixta*

166. *A. affinis*

167. *A. viridis*

168. *A. cyanea*

169. *A. grandis*

170. *A. isosceles*

Figures 165–170 *Aeshna* species
showing male abdominal appendages and segment S10 in dorsal view (above) and right lateral view (below)

AESHNA CAERULEA (Ström)   Plate 14 (80)

Libellula caerulea Ström, 1783, *Nye Saml. Kong. Danske Vidensk. Selsk. Skr.* **2** : 90.
Type locality: Norway.

> ?*Aeschna septentrionalis* Burmeister, 1839
> *Aeschna borealis* Zetterstedt, 1840

E – Azure Hawker, **F** – L'Aeschne azurée,
D – Alpen-Mosaikjungfer

DESCRIPTION OF ADULT

Mature males are very blue in appearance, lacking any green or yellow coloration, and the extent of the black abdominal ground colour is more reduced than in other species of *Aeshna*. Females are dorsally darker than males, and they, as well as immature males, have the intense blue colour of the mature male replaced by yellowish to lavender blue.

In both sexes, the head is relatively small and the eyes are contiguous over a much shorter distance than in any other European species (see key, p.100 and fig. 171). The black T-shaped mark on top of the frons is strongly defined. Antehumeral stripes are much reduced, and the two pale stripes on each side of the thorax are sinuous and very narrow. Wing venation is very dense apically, the membranule is uniformly infumate, the anal triangle in the male hindwing is two-celled, and vein IR$_3$ does not obviously bifurcate under the pterostigma. McLachlan (1889a) finds this latter character, although somewhat variable, to be a useful supplementary feature for distinguishing *A. caerulea* from other species of *Aeshna* in which the anterior branch of the IR$_3$ furcation is well-defined. Male superior anal appendages (fig. 160) in dorsal view have their outer edges rather strongly convex. *A. caerulea* is one of the smaller species, total body length usually being 54–64mm.

BIOLOGY

Morton (1895), writing on the habitat of *A. caerulea* in Scotland, states that it 'affected the sunniest glades and openings, both in the little birch woods which mark the course of the burns down the hill-sides, and in the larger woodlands on the lower ground. It is a sun-loving thing, only flying freely when the weather is really warm; it is fond of basking on light-coloured stones, but when so resting is shy, and flies off at once if any attempt is made to approach it.' *A. caerulea* also rests in full sunlight on boulders, posts and tree trunks, and it may be located on sunny afternoons by scanning such situations with binoculars. It is an insect of acid *Sphagnum* moorland, flying at low altitudes in Scotland and Scandinavia (*circa* 200m) but normally at 1000–2000m in the Alps. Larvae are found in shallow, acid bog pools and lochans; the ovipositing female requires a thick *Sphagnum* carpet with very little open water.

Male *A. caerulea* exhibit a reversible temperature-related colour change (Sternberg, 1987). With decreasing temperature, the blue eyes and abdominal spots change to a greyish colour. This occurs at nightfall in males roosting on tree trunks and it is said to enhance their concealment.

FLIGHT PERIOD

Mainly in the latter half of June (sometimes as early as the end of May) and virtually over by the end of July in Scotland, much earlier than in most species of *Aeshna*. In more southerly parts of Europe, it flies mostly in July and August, and there are records of it persisting into September in Germany and Switzerland (Ander, 1950).

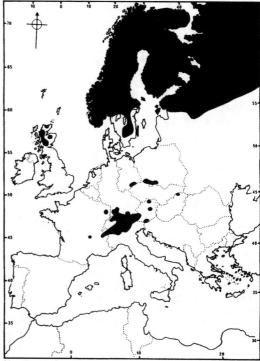

Map 38 – *Aeshna caerulea*

DISTRIBUTION (map 38)

Usually considered a circumboreal species, occurring in Europe, northern Asia and North America. Its distribution is discussed by Ander (1950). New World material (*A. c. septentrionalis*) is at least subspecifically distinct (Valle, 1952), having broader male superior appendages and rather different thoracic pattern and venation (Walker, 1958). In Europe, *A. caerulea* is found throughout Scandinavia except Denmark, in Scotland, the Alps of France, Switzerland (Ris, 1916), Austria and northern Italy (Trentino, Veneto, Lombardy), in Germany (Black Forest, Bavarian Alps, Riesengebirge), Poland (Silesia), Hungary and France (Massif Central, Vosges). There is an isolated population in the Caucasus.

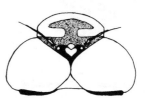

Figure 171 *Aeshna caerulea* head in dorsal view

AESHNA JUNCEA (Linnaeus) Plate 14 (81,82)

Libellula juncea Linnaeus, 1758, *Syst. Nat.* (Edn. 10) **1** : 544.

Type locality: south Sweden.

> *Libellula quadrifasciata* var.e *ocellata* Müller,1764
> *Aeschna rustica* Zetterstedt, 1840
> *Aeschna picta* Charpentier, 1840
> *Aeschna caucasica* Sélys in Sélys & Hagen, 1850

E – Common Hawker, **F** – L'Aeschne des joncs, **D** – Torf-Mosaikjungfer

DESCRIPTION OF ADULT

A. juncea is very similar to the following three species, all four being essentially northern in distribution and referable to the *A. juncea* group.

The black T-shaped mark on top of the frons is well defined and extends in front of the transverse carina (fig.172), and the sulcus between frons and postclypeus is lined with black. There are two yellow spots on the occiput, small but distinct, and in this character *A. juncea* differs from the other species in the group in which the occiput is entirely black or bears only vestiges of spots. Antehumeral stripes are narrow and yellow, reduced in the female to short streaks, and there are two yellow stripes on each side of the thorax. Wings are hyaline, although in old females they may be lightly suffused with brownish yellow. Membranules are whitish basally but heavily infuscated distad, short, and in the male, as in all four species of the group, not reaching to the vein dividing the two-celled anal triangle (fig. 173). The pterostigma is long, especially in females, blackish in males and brownish in females. Venation is black except the costa which is brightly yellow. Legs are mainly black, but reddish brown basally. Ground colour of abdomen and thorax is dark brown and pale markings are edged with black. S3–10 each bear a pair of large, blue (yellow to green in female) spots posteriorly, and additionally, on S3–8, there is a pair of small, subtriangular spots, broadest medially and tapering laterally, in the anterior half of each segment. The mid-dorsal yellowish triangle on S2 is very narrow and reduced almost to a longitudinal line. Male superior anal appendages (fig. 161) are relatively narrow and lack denticulation of the dorsal carina. Female anal appendages are flat and leaf-like, about 2.5 times as long as S10. Male accessory genitalia (fig. 174) in *A. juncea* (and *A. subarctica*) are unusual in having the mesially-directed processes of the anterior hamules very long and separated from the hamular fold. Body length of *A. juncea* is usually within the range 65–74mm.

Figure 172 *Aeshna juncea* head in dorsal view

Map 39 – *Aeshna juncea*

BIOLOGY

Larvae develop in ponds and lakes, preferentially oligotrophic and slightly acid. Females oviposit in submerged and more or less vertical stems and leaves of such plants as *Typha* and *Iris*; usually the entire abdomen of an ovipositing insect is immersed. Adults hawk for insects in open situations, often far from water and seldom close to trees. They may fly until dusk on warm evenings. Larval development extends over four (sometimes three) years (Münchberg, 1930).

FLIGHT PERIOD

Most numerous from late July to the end of September, although at low altitudes some adults survive well into October. Exceptionally early individuals are occasionally encountered; Robert (1958) observed a specimen with fully mature coloration on 26 May 1920 in Switzerland.

DISTRIBUTION (map 39)

The most widespread and abundant of the *A. juncea* group of species, with an Holarctic distribution encompassing North America, Europe and northern Asia to Japan. In Europe, *A. juncea* is commonest in north and central Europe, although in northern Finland it is outnumbered by *A. caerulea*. In the south it is confined to mountainous country and its most southerly recorded locality is Avila, Spain (Benitez Morera, 1950). In the New World it 'is the most abundant *Aeshna* in Alaska and in the Mackenzie Delta beyond the Arctic Circle' (Walker, 1958), and it extends south to a few of the northern United States.

AESHNA SUBARCTICA Walker Plate 14 (83,84)

Aeschna subarctica Walker, 1908, *Can. Ent.* **40** : 385.
Type locality: Canada.

> *Aeshna elisabethae* Djakonov, 1922
> *Aeschna subarctica interlineata* Ander, 1944

F – L'Aeschne subarctique, **D** – Hochmoor-Mosaikjungfer

In 1922 Djakonov described as new, under the name *A. elisabethae*, an *Aeshna* collected in Finno-Karelia. Further material from Lüneburg Heath in northern Germany was recognized (Ris, 1927) as conspecific with Nearctic *A. subarctica*, and Lieftinck (1929), Morton (1927) and Schmidt, Er. (1929) distinguished the species from *A. juncea* with which it had been previously confused. The Palaearctic form of the species is subspecifically distinct and should be known as *A. subarctica elisabethae* Djakonov.

DESCRIPTION OF ADULT

A. subarctica is similar to *A. juncea* and about the same size, but in mature individuals the occiput is all black, lacking the two yellow spots present in *A. juncea*. There may be indications of these spots in immature specimens. The black T-shaped mark on the dorsal surface of the frons does not extend on to the anterior face, and the mid-lobe of the labium has a pair of apical dark spots (entirely pale in *A. juncea*). Male antehumeral stripes diverge more strongly ventrad than in *A. juncea* and they are constricted, sometimes broken, near the top, the apical portion extending almost the entire breadth of the mesepisternum. The ventral surface of the thorax has relatively large yellow spots; in *A. juncea* these are more or less reduced. In both sexes, a narrow, pale line runs along the front of the humeral sulcus (rarely traceable in *A. juncea*), the dorsal half of the anterior lateral thoracic stripe is set back from the humeral sulcus (dorsal third in *A. juncea*), and the top of the stripe is produced anteriorly and far posteriorly below the forewing insertion (only slightly posteriorly in *A. juncea*).

The thoracic stripes are yellow to bluish in *A. subarctica*, yellowish in *A. juncea*. Between the epimeral stripes there is a well-defined pale mark, across or just posterior to the spiracle, which reaches well into the dorsal half of the metepisternum. At most, only vestiges of this stripe are discernible in *A. juncea*, and it is apparently also much reduced in specimens of *A. subarctica* from northern Scandinavia since Ander (1944) uses its presence to distinguish the form from southern Sweden and central Europe, which he names *interlineata*, from more northerly occurring insects. Membranules are more extensively darkened than in *A. juncea* with the small, basal pale area suffused with grey. The male abdominal pattern is less blue than in *A. juncea*, the paired posterior segmental spots on S3 to S6 or S7 often remaining yellow. In females, the lateral spots are whitish. Male superior abdominal appendages (fig. 162) resemble those of *A. juncea*, having a smooth dorsal carina, but they are slightly broader. Male secondary genitalia are also similar in the two species, but in *A. subarctica* (fig. 175) the anterior spinous processes are longer, more slender and almost parallel, whereas in *A. juncea* they are clearly divergent apically. Female anal appendages (fig. 179, p.106) are flat, leaf-like, and symmetrically rounded apically, about three times the dorsal length of S10, relatively slightly longer and broader than those of *A. juncea*.

BIOLOGY

Breeds in acid pools on peat moors with *Sphagnum*. Females oviposit in *Sphagnum*. Eutrophication renders water unsuitable for *A. subarctica*.

FLIGHT PERIOD

Schmidt, Eb. (1964b) records the emergence of several hundred adults in 1962 at a site in Schleswig-Holstein. Emergence commenced about mid-July, reached a peak in mid-August, and declined through September.

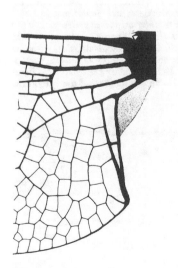

Figure 173 *Aeshna juncea*
base of male left hindwing showing angled anal margin and two-celled anal triangle

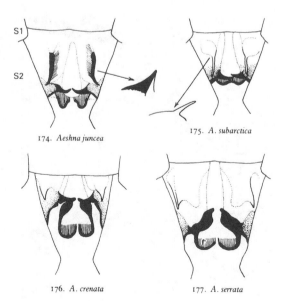

174. *Aeshna juncea*

175. *A. subarctica*

176. *A. crenata*

177. *A. serrata*

Figures 174–177 *Aeshna* species
male accessory genitalia in ventral view

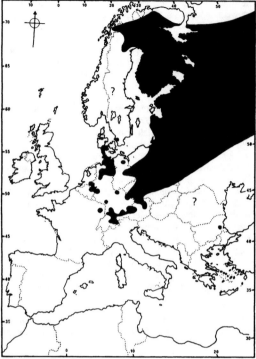

Map 40 – *Aeshna subarctica*

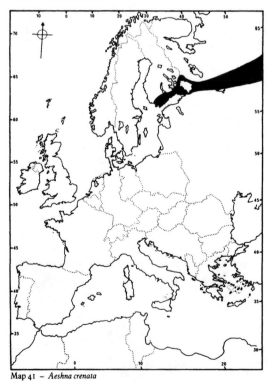

Map 41 – *Aeshna crenata*

DISTRIBUTION (map 40)

A circumboreal species found in North America, Siberia and northern and central Europe. Its discovery in Japan (Hokkaido) is reported by Asahina (1973). In northern Europe, *A. subarctica* is widespread in Scandinavia including Denmark, and in northern Russia. It occurs at a few localities in the south of Holland (Limburg), Belgium, France (Vosges), and in northern Germany (especially Schleswig-Holstein, Harz Mountains and Lüneburg Heath). In central Europe it is recorded from the Black Forest, about ten sites in Switzerland, the Austrian Tyrol and the eastern Balkans. It has not been found in the British Isles, but might perhaps be expected in Scotland. The southernmost known breeding population in North America is on Mount Desert Island, Maine.

AESHNA CRENATA Hagen   Plate 14 (85)

Aeschna crenata Hagen, 1856, *Ent. Ztg. Stettin.* **17** : 369.
Type locality: Siberia.

Martin (1908) places *A. crenata* in synonymy with the North American and Siberian species *A. clepsydra* Say, 1839 (=*arundinacea* Sélys, 1872), but this synonymy has not been accepted by subsequent authors.

DESCRIPTION OF ADULT

A. crenata resembles *A. juncea*, although it is a larger insect varying from about 71–86mm in body length. Males tend to be larger than females with the abdomen rather strongly constricted on the third segment. *A. crenata* differs from *A. juncea* in that the occiput is entirely black without two yellow spots, and the black T-shaped mark on the frons does not extend significantly over the transverse carina separating the dorsal and anterior faces. The most distinctive characters of *A. crenata*, however, are in the anal appendages. The male inferior appendage (fig. 163) is long and relatively narrow, more than half the length of the superior appendages, and the dorsal carina of the superior appendage terminates in three to five teeth. The female anal appendages (fig. 178) in those specimens examined each terminate in an asymmetrical point, differing from the regularly rounded form seen in other species of the group (fig. 179). Male secondary genitalia (fig. 176) are of rather different structure from those of the two previous species.

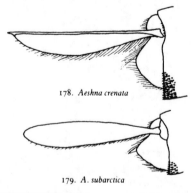

178. *Aeshna crenata*

179. *A. subarctica*

Figures 178, 179 *Aeshna* species
female abdominal appendages in right lateral view

BIOLOGY

A species of moorland bog and taiga.

FLIGHT PERIOD

Probably July to September.

DISTRIBUTION (map 41)

A Siberian species extending into northern Russia and southern Finland. Also reported from Mongolia and Japan (Hokkaido).

AESHNA SERRATA Hagen   Plate 14 (86)
Aeschna serrata Hagen, 1856, *Ent. Ztg. Stettin.* **17** : 370.
Type locality: Kirghizia (USSR).
   Aeschna osiliensis Mierzejewski, 1913
   Aeschna fennica Valle, 1927
F – L'Aeschne dentée

DESCRIPTION OF ADULT

Closely allied to the preceding species and, like it, a very large dragonfly. Males exceed 80mm in total body length and almost match in size the larger species of *Cordulegaster*. *A. serrata* is the fourth and final European species belonging to the *A. juncea* group. It differs from the others in having an all-white (or very nearly so) membranule and complete, well-defined antehumeral stripes in females as well as males. Mature males are very blue; females are greenish blue. Male anal appendages (fig. 164) are distinctive, the inferior being less than half the length of the superiors and the latter have the dorsal carina terminating in a series of small teeth (many more than in *A. crenata*). The dorsal surface of the superior appendage is distinctly concave. Female anal appendages are bluntly rounded, flat and leaf-like, resembling those of *A. juncea* and *A. subarctica* rather than those of *A. crenata*. Male secondary genitalia (fig. 177), however, are most similar to those of *A. crenata*.

BIOLOGY

Larvae develop in acid moorland pools or brackish, estuarine standing water. They are tolerant of periods of exposure to the air.

FLIGHT PERIOD

July to September.

DISTRIBUTION (map 42)

A. serrata, like *A. crenata*, is a Siberian species, but it has an eastern rather than a northern distribution. It is at the edge of its range in Europe, although it penetrates rather further than *A. crenata* having been found sporadically in eastern Sweden and on several of the Baltic islands. European material is referable to *A. s. osiliensis*. In the USSR, *A. serrata* is widespread in Siberia, Mongolia and northern Russia and it is found also in the Caucasus, Armenia and Caspian region. Also recorded from Turkey (Asian).

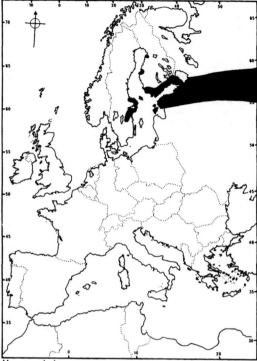

Map 42 – *Aeshna serrata*

AESHNA MIXTA Latreille   Plate 15 (87,88)
Aeschna mixta Latreille, 1805, *Hist. nat. crust. insect.* **13** : 7.
Type locality: Paris, France.
   ?*Libellula coluberculus* Harris, 1782 (*nomen dubium*)
   Aeschna affinis Stephens, 1835 *nec* Vander Linden, 1823
   Aeschna alpina Sélys, 1848
E – Migrant Hawker, **F** – L'Aeschne mixte,
D – Herbst-Mosaikjungfer

DESCRIPTION OF ADULT

Aeshna mixta and the following species, *A. affinis*, have a rather similar colour pattern to species in the *juncea* group, but they are smaller dragonflies and have a more southerly distribution. They can be recognized by the wing characters given in the key to species, in particular by the three-celled anal triangle of the male hindwing and by the relatively long membranule which reaches posterior to the transverse crossvein in the male hindwing anal triangle (fig. 180, p.110).

In *A. mixta* the T-shaped black mark on the frons is complete, the sulcus between postclypeus and frons is not strongly blackened, and the back of the head is all black. Antehumeral thoracic stripes are much reduced or absent, and laterally on the thorax are two yellowish green to blue stripes contrasting with the general dark brown ground colour. Legs are mainly black but reddish brown at their bases. Wings are hyaline, very slightly tinted in females, and the membranules are greyish brown apically, whitish basally. The ground colour of the abdomen is dark brown, blackish around the spots; there is a posterior pair of dorsal blue spots (yellow to green or bluish green in the female, greyish to lilac in immature individuals) on S3–10, a pair of narrowly triangular, dorsal, transverse, yellow streaks just anterior to the middle of S2–8, and lateral bluish spots on S1–8. S2 has a narrow, longitudinal, mid-dorsal yellow streak which reaches to the blue band across the posterior part of the segment. Male superior anal appendages (fig. 165) are long, rather more than three times the length of the dorsal surface of S10. No subbasal ventral tooth is visible in lateral view. Female abdominal appendages are also long, fully as long as the combined length of the two terminal abdominal segments. Total body length 56–60mm.

BIOLOGY

Breeds in lakes and ponds and is tolerant of brackish conditions (Aguesse (1968) gives the limit as 5–6g/l salt, Lieftinck (1926) quotes 2 g/l). Adults are migratory and large congregations may be found far from water, hawking at the edges of woods or in tree-lined lanes.

FLIGHT PERIOD

The usual flight period is from July to the end of October, but precocious emergences in May and even April are on record.

DISTRIBUTION (map 43)

Widespread and common throughout southern and central Europe but absent from Scandinavia (except Denmark) and in the British Isles found only as far north as the English Midlands and South Wales. Also occurs in North Africa, the Middle East, Caucasus region and from the Caspian east to China and Japan.

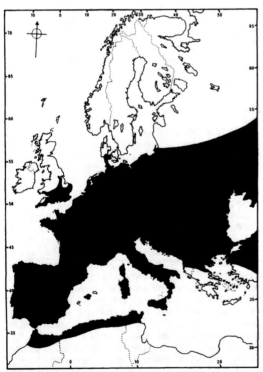

Map 43 – *Aeshna mixta*

AESHNA AFFINIS Vander Linden   Plate 15 (89,90)

Aeshna affinis Vander Linden, 1823, *Opusc. Scient.* **4** : 102.
Type locality: Bologna, Italy.
   Aeschna landoltii Buchecker, 1876
F – L'Aeschne affine, **D** – Südliche Mosaikjungfer

DESCRIPTION OF ADULT

A small *Aeshna* of similar size to *A. mixta* and resembling that species in wing venation. It differs from *A. mixta*, however, in the coloration of the sides of the thorax which are greenish to bluish traversed by black sulcal lines but without a brown metepisternal stripe, and in the male abdominal pattern which appears a generally brighter blue, the anterior pair of dorsal segmental spots being subquadrate instead of almost linear as they are in *A. mixta*. Paired spots on S10 are large in *affinis* but small in *mixta*. Male superior abdominal appendages (fig. 166) have a low but distinct subbasal tooth, absent in *mixta*, visible on the ventral surface in lateral view. Female anal appendages are shorter than the combined length of the two terminal abdominal segments.

BIOLOGY

A migratory species, like *A. mixta*, and therefore liable to turn up in a variety of habitats. Larvae develop in standing waters, sometimes slightly brackish, but *A. affinis* is less tolerant of salinity than *A. mixta*. Schiemenz (1953) reports that oviposition occurs in tandem, contrary to the habit of other species of *Aeshna*. Utzeri & Raffi (1983), in a behavioural study of the species, found females ovipositing both in tandem and alone and they report that the former are much more difficult to approach; wheel formation is adopted exclusively in flight.

FLIGHT PERIOD

Emerges usually earlier than *A. mixta*, flying from June or late May to the end of August.

DISTRIBUTION (map 44)

Southern Europe, North Africa and from the Middle East to China. In Europe, *A. affinis* extends less far to the north and is less common in central Europe than *A. mixta*. It is absent from Scandinavia and Britain, and known from just one locality in the Netherlands. There are scattered records from north Germany, the northernmost from Mark Brandenburg, German Democratic Republic (Beutler, 1980). Even in southern Europe, *A. affinis* appears to be a much scarcer insect than *A. mixta*.

AESHNA CYANEA (Müller)   Plate 15 (91,92)

Libellula cyanea Müller, 1764, *Fauna Insect. fridrichsdalina*: 61.
Type locality: Denmark.
   Libellula aenea Sulzer, 1761
   Libellula anguis Harris, 1782
   Aeshna maculatissima Latreille, 1805
   Libellula varia Shaw, 1806
E – Southern Hawker, **F** – L'Aeschne bleue,
D – Blaugrüne Mosaikjungfer

DESCRIPTION OF ADULT

Like most other species in the genus, *A. cyanea* has a dark

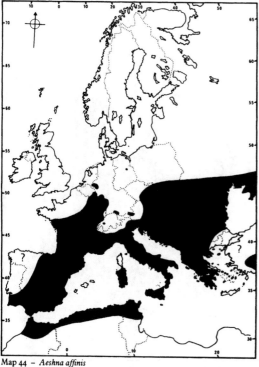

Map 44 – *Aeshna affinis*

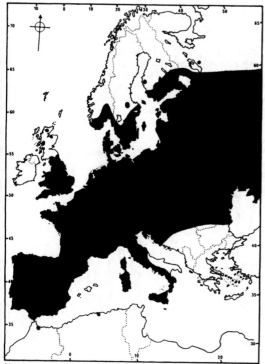

Map 45 – *Aeshna cyanea*

brown body patterned with yellowish green to blue marks, but it has two diagnostic features, seen in both sexes, which simplify its recognition. The yellow or green antehumeral stripes are very broad and oval in shape, and S9,10 each have dorsally an entire band of blue or green across the posterior half, the paired spots seen in other species being medially fused. *A. viridis* is the only other European species to share these features, although fusion of the posterior abdominal spots is less complete.

The T-shaped black mark on the frons is complete in *A. cyanea* (unstalked in *A. viridis*), and the back of the head is mainly black but with a relatively large yellow spot behind the middle of each eye. The sides of the thorax are green with strongly-marked black sulcal lines (ill-defined in *A. viridis*). Wings are hyaline with the membranules whitish, only slightly infuscate apically, and the pterostigmata are short, black in the male and brown in the female. The anal triangle of the male hindwing (fig. 181, p.110) is three- (sometimes four- or five-) celled (two-celled in *A. viridis*). The dorsal crest at the base of S10 in the male is less developed than in any of the other European species of *Aeshna*. Superior male abdominal appendages (fig. 168) have a unique form. They are broad with their apices down-turned so that each, in profile, resembles the shape of a bird's head, and there is a boss bearing setae on the inner face at the apex. The female anal appendages have the flat, foliaceous form usual in *Aeshna* and they are short, less than the length of S9+10 together and about 2.5 times as long as the dorsal surface of S10. Total body length 67–76mm.

BIOLOGY

The common *Aeshna* of farm ponds, small lakes, reservoirs and canals, and often seen hawking along lanes and hedgerows. It is essentially a lowland insect and frequently breeds in the proximity of human habitation, although seldom more than one or two individuals are seen together. Females oviposit at the edges of ponds, often in damp moss. Kaiser (1968, 1969, 1974a, 1974b) gives accounts of behaviour and ecology.

FLIGHT PERIOD

From the beginning of July until October or, exceptionally, the third week of November in England. Further south, emergence may commence early in June.

DISTRIBUTION (map 45)

Widespread across Europe from the Mediterranean to the more southerly parts of Scandinavia. In the British Isles, *A. cyanea* is common in England and Wales but there are only a few scattered records from Scotland, and it has not yet been reported from Ireland. There appears to be an absence of records from the southern Balkans. Also found in North Africa, Asia Minor and east to the Caucasus.

AESHNA VIRIDIS Eversmann   Plate 16 (93,94)

Aeschna viridis Eversmann, 1836, *Bull. Soc. Imp. Moscou* 9 : 242.

Type locality: Russia.

Aeschna virens Charpentier, 1840

F – L'Aeschne verte, **D** – Grüne Mosaikjungfer

DESCRIPTION OF ADULT

Resembles *A. cyanea* in having broad, green antehumeral stripes but it is usually smaller, not exceeding 70mm in length, the wings are weakly suffused yellowish brown (particularly in old individuals), and the black mark on the frons is not connected with the black line in front of the ocelli (that is, it is *not* T-shaped).

The sides of the thorax are almost entirely green and the legs are mainly brown. The membranules are whitish and the wing pterostigmata are brown, especially pale in females, and also long (about as long as the hypotenuse of the discoidal cell). The male anal triangle is two-celled. Pale spots on the last abdominal segment are partially confluent and they each enclose a small, black point in males. Male superior anal appendages (fig. 167) are almost flat dorsally as seen in profile and they have no subbasal tooth. The female appendages are short, just slightly more than twice as long as S10.

BIOLOGY

Breeds in standing, acid to neutral, moorland water bodies of the type often frequented by *A. juncea* but only at low altitude. According to Schiemenz (1953), females oviposit, usually at dusk, in submerged leaves of *Stratiotes aloides*, and although *Typha* and *Sparganium* may also sometimes be used, the preference for *Stratiotes* is very strong. Other species of *Aeshna* show little preference for particular plants. The life history of *A. viridis* in southern Sweden (Norling, 1971) extends over two or three years and is regulated mainly by photoperiod. Adults may form feeding swarms at dusk.

FLIGHT PERIOD

July to September.

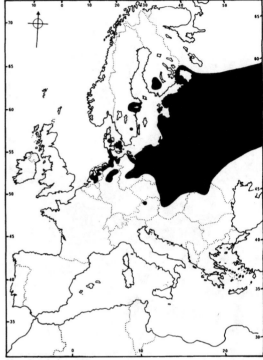

Map 46 – *Aeshna viridis*

DISTRIBUTION (map 46)

A Siberian species extending into north Europe (south Sweden, Finland, Denmark, Holland, north Germany (Schmidt, Eb., 1975), Poland, Austria, Hungary, Romania and Russia) and the Carpathians (but not the Alps), south to the northern Balkans. The northernmost record is from Vyartsilya, Karelia (Hämäläinen, 1983). Generally rare with very scattered records in Europe, and becoming scarcer in parts of Germany. Widespread in southern Asia where it is found as far east as the Pacific seaboard.

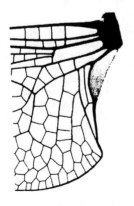

180. *Aeshna mixta*

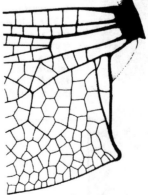

181. *A. cyanea*

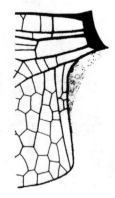

182. *A. isosceles*

Figures 180–182 *Aeshna* species
base of male left hindwing

AESHNA GRANDIS (Linnaeus) Plate 16 (95,96)

Libellula grandis Linnaeus, 1758, *Syst. Nat.* (Edn. 10) **1** : 544.

Type locality: Sweden.

Libellula quadrifasciata Müller, 1764
Libellula nobilis Müller, 1767
Libellula flavipennis Retzius, 1783

E – Brown Hawker, **F** – La grande Aeschne,
D – Braune Mosaikjungfer

DESCRIPTION OF ADULT

The type species of *Aeshna. A. grandis* is an unmistakable insect with a predominantly brown body and wings suffused all over with a brownish tint.

The frons above has a dark brown crescent, rather ill-defined and almost imperceptibly stalked. The thorax is brown, without antehumeral stripes but with two broad, yellow stripes on each side. Sclerites at the bases of the wings are blue, these being the only blue parts on the dorsal surface of the female. Wing veins and pterostigmata are brownish and the membranules are whitish. The anal triangle of the male is two-celled. Legs are mainly reddish brown. The abdomen of the male is brown with two large, blue spots posteriorly on the dorsal surface of the second segment, a pair of blue basal spots on S3, and narrow, transverse, yellow streaks dorsally on S2–8. The rest of the dorsal surface is entirely brown, although there are lateral spots on S2 (yellow) and S3–8 (bluish). The female abdomen dorsally is brown with only a pair of small, yellow spots on S2 and transverse, yellow streaks on S2–8; the sides of the female abdomen bear yellowish marks on S2–8. Male superior abdominal appendages (fig. 169) are slightly shorter than the combined length of S9 and S10, without a subbasal tooth, and in dorsal view apically broadly rounded. The female anal appendages are about 2.5 times as long as the dorsal surface of S10. As its name indicates, *A. grandis* is a large dragonfly, usually attaining 69–76mm in total body length.

BIOLOGY

Females oviposit in submerged parts of plants growing in shallow water at the margins of ponds and lakes. *A. grandis* is a very powerful flier, usually to be seen hawking alone at a height of several metres over water or along lanes. It remains active until dusk and is a species that not infrequently penetrates deeply into urban areas.

FLIGHT PERIOD

End of June to early October, but principally from late July to mid-September. Larval development takes from two to four years.

DISTRIBUTION (map 47)

A western Palaearctic range from Ireland and the French Pyrenees in the west across all of central and northern Europe to about the Baikal region in Siberia. Absent from North Africa, Portugal, Spain, the islands of the Mediterranean, most of Italy (cited only from Piedmont and Trentino), Greece, and also Scotland (although it extends to the far north of Scandinavia).

Map 47 – *Aeshna grandis*

AESHNA ISOSCELES (Müller) Plate 16 (97)

Libellula quadrifasciata, var. 36. *isoceles* Müller, 1767, *Nova Acta Leopold. Carol.* **3** : 125.

Type locality: Denmark.

Aeshna rufescens Vander Linden, 1825
Aeschna chrysophthalmus Charpentier, 1825
Anaciaeschna isosceles auctt.

E – Norfolk Hawker, **F** – L'Aeschne isocèle,
D – Keilflecklibelle

DESCRIPTION OF ADULT

Although differing in a number of respects from other European species of *Aeshna*, there seems little justification for including this species in the genus *Anaciaeschna* which is allied to *Anax* and in which male hindwings are basally rounded and there are no auricles.

Like the preceding species, *A. isosceles* is a predominantly brown dragonfly, but its almost hyaline wings and green eyes distinguish it at a distance from *A. grandis. A. isosceles* and *A. grandis* both lack a black, T-shaped mark on the upper part of the frons, and in both species the thorax is brown without antehumeral stripes but with two yellow stripes on each side. These are, however, much shorter in *A. isosceles* than in *A. grandis*, failing to reach the subalar carina. Basal wing sclerites are yellow, not blue, in *A. isosceles*. Legs are mainly black with only the bases of the femora reddish brown. Wing venation is blackish (brown at the base of the wing) and the pterostigmata are brownish. The wing membrane is clear or very slightly brownish, but there is a distinct saffron patch at the base. This saffron

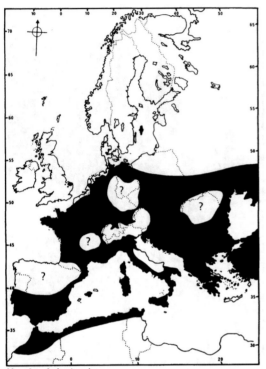

Map 48 – *Aeshna isosceles*

mark, which extends along the anal margin of the hind-wing, distinguishes *A. isosceles* from all of its European congeners; so too does the membranule which is dark and very long, reaching almost to the anal angle (fig. 182, p.110). The anal margin of the male hindwing is much less sharply angled than in the foregoing *Aeshna* species (but auricles on S2 are present), and the anal triangle is divided into from three to seven cells (fig. 182). In both sexes the abdomen is brown with a conspicuous yellow mark on the dorsal surface of the second segment. This mark is shaped like an isosceles triangle, its base on the anterior margin of the second segment, and it gives the insect its trivial name. There are no other pale abdominal markings, but a mid-dorsal black keel extends from S3 to S9, widening on S8,9. There are also transverse, blackish streaks across the anterior halves of S3–7, a broader brown streak interrupted by the yellow mark on S2, and a pair of small, blackish spots posteriorly on S3–8. Male superior anal appendages (fig. 170) are rather narrow and each has a subbasal tooth visible in profile. The female anal appendages are small, only about 1.5 times the dorsal length of the last abdominal segment. Total body length 62–68mm.

BIOLOGY

Larvae are found in small lakes, fens, ditches and dykes at low altitude. Adults do not seem to fly far from their breeding sites. Females have been observed ovipositing in *Stratiotes*. The species is intolerant of pollution. Settling is more frequent than in other species of *Aeshna*.

FLIGHT PERIOD

A relatively early species flying mainly between mid-May (mid-June in England) and the end of July. Aguesse (1968) records a precocious appearance on 12 April 1957 on the Mediterranean coast, and there are some August records.

DISTRIBUTION (map 48)

Widely distributed in the Mediterranean region and central Europe, but becoming very local in northern Europe. In Great Britain it is found only in Norfolk where it is legally protected under the Wildlife and Countryside Act 1981. Not recorded north of the Isle of Gotland in Scandinavia. A number of sites in Mecklenburg, East Germany are reported by Königstedt (1980). *A. isosceles* also occurs in North Africa and, as subspecies *A. i. antehumeralis* Schmidt, 1951, in northern Iran, the Caucasus, Armenia and Turkey.

ANAX Leach, 1815

A genus of some thirty species which are distributed in temperate and tropical latitudes throughout the world but chiefly in southern Asia and Africa. They are very large dragonflies, mostly brightly marked with enamel-like blues and greens. Males have no anal triangle, the anal angle of the hindwing (fig. 183) being rounded as in females, and auricles are absent from S2. The two sectors of the arculus arise from the anterior half and vein R_3 is angled anteriorly beneath the distal end of the pterostigma (figs 155, p.99; 183). Two species are represented in the western Palaearctic region, one of which, *A. imperator*, is the type of the genus.

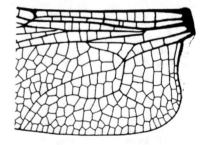

Figure 183 *Anax parthenope* base of male left hindwing

Key to species of *Anax*

1 Thorax green. Male abdomen marked with bright blue on S2–10. Male inferior anal appendage (fig. 184) about one-third as long as a superior, apically truncated in dorsal view; superior appendage much expanded medially, only about three times as long as its maximum breadth, apically rounded without a clearly-defined subbasal tooth. Female without occipital tubercles (fig. 190, p.115) .... *imperator* (p. 114)

– Thorax purplish brown. Male abdomen marked with bright blue only on S2 and base of S3, the pale markings thereafter being duller and darker. Male inferior anal appendage (fig. 185) less than one-fifth as long as a superior, apically bilobed in dorsal view; superior appendage about four times as long as its maximum breadth in dorsal view, apically pointed, and with a strongly projecting subbasal tooth. Female with two occipital tubercles (fig. 191, p.115) .... *parthenope* (p.115)

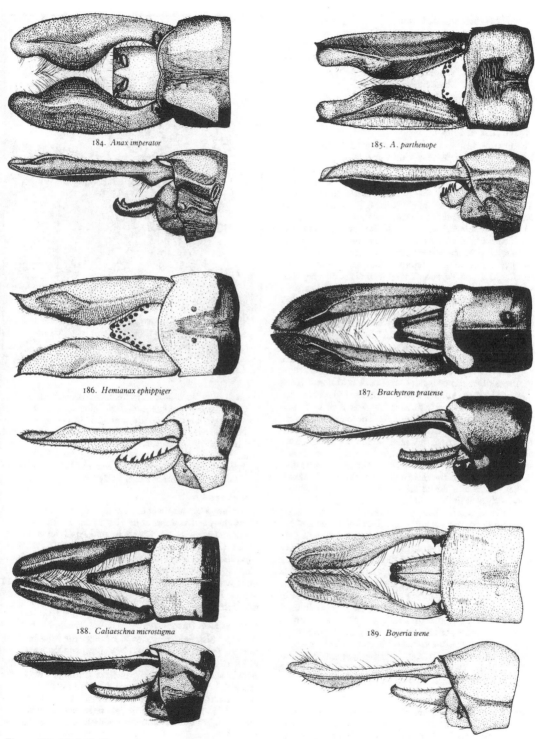

184. *Anax imperator*

185. *A. parthenope*

186. *Hemianax ephippiger*

187. *Brachytron pratense*

188. *Caliaeschna microstigma*

189. *Boyeria irene*

Figures 184–189 Aeshnidae species
showing male abdominal appendages and segment S10 in dorsal view (above) and right lateral view (below)

113

ANAX IMPERATOR Leach Plate 17 (98,99)

Anax imperator Leach, 1815, *Edinb. Encyclop.* **9** : 137.
Type locality: England (probably New Forest, Hampshire).

    Aeshna formosa Vander Linden, 1823
    Aeschna azurea Charpentier, 1825

E – Emperor Dragonfly, **F** – L'Anax empereur,
D – Grosse Königslibelle

DESCRIPTION OF ADULT

The mature male is a brilliantly-coloured dragonfly with bright blue predominating. Frons and clypeus are mainly yellow with a small, brown line on the front edge of the dorsal surface of the frons and a broader, blue one just behind it. These apical marks are separated from a black, triangular spot in front of the vertex. Eyes are blue and touch for a considerable distance. The thorax is apple-green without antehumeral stripes, with two very fine, oblique, black sulcal lines on each side. The mesopleural spiracle and a small point above it are black and conspicuous. Wings are hyaline in males, slightly yellowish brown in old females, the costa is yellow edged with black but the other veins are black, the pterostigma is brown, long and narrow, and the membranule is infuscate apically, whitish basally. The hindwing is rounded at the anal angle, similarly shaped in both sexes. Legs are black with the bases of the femora reddish brown and the ventral surface of the front femur yellow. The abdomen is mainly blue although the first segment and anterior part of the second are greenish. There are two black spots at the front of the first segment and three transverse black lines on the second. A mid-dorsal black line runs from the middle of S2 to the apex of S10, of uneven width with two lateral expansions in each of S3 to S7 and one in the remaining three segments. The lateral abdominal carinae on S4–10 are strongly blackened. The pattern of the female resembles that of the male, but the pale colour on the abdomen is greenish instead of blue, the eyes are green, and the mid-dorsal abdominal line is brownish and broader than in the male. S10 of the male abdomen (fig. 184) is almost without trace of a dorsal crest and is strongly produced posteriorly over the bases of the anal appendages. Its apical margin is weakly bilobed. The superior male appendages (fig. 184) in dorsal view are of massive construction, about as long as the two apical abdominal segments and only about three times as long as broad. On the ventral surface basally is a low prominence, not developed into a tooth. The dorsal surface of the superior appendage bears a median longitudinal ridge, sharp only for a short distance in the apical part of the appendage, and the appendage carries long and dense hairs on its inner face. The inferior appendage is rectangular in dorsal view, about one-third as long as a superior, terminating in strong, upturned teeth, two of which are visible in exact side-view. The female anal appendages are leaf-like, apically rather pointed, as long as S9+10. Males are usually about 76mm and females 73mm long, although males can reach 84mm in length.

BIOLOGY

Males are conspicuous insects patrolling over open water or around reed beds in ponds and lakes, both natural and artificial, from sea level to an altitude of 1500m or more. In flight the abdomen is bent slightly downwards, a characteristic that distinguishes *Anax* from the larger species of

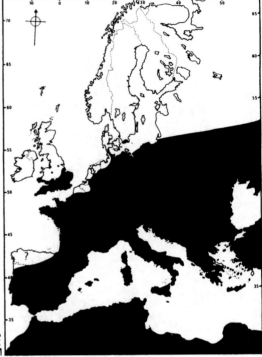

Map 49 – *Anax imperator*

Aeshna. Sometimes large prey items such as butterflies (e.g. *Pieris brassicae*, *Maniola jurtina*) or other dragonflies (e.g. *Libellula quadrimaculata*) are captured. In cloudy weather *A. imperator* retires to reeds, bushes or cornfields, but it will fly until dusk on mild evenings. Females oviposit unaccompanied in mats of aquatic vegetation (*Myriophyllum*, *Ceratophyllum*, *Elodea*, *Potamogeton*) (Robert, 1958) or floating reed debris. For a behavioural study see Consiglio (1976).

FLIGHT PERIOD

A summer dragonfly over most of Europe, flying from June to August, but it appears in May in the south and fully mature specimens were seen in mid-April on Madeira. Corbet (1957b) showed that, in England, some larvae of *A. imperator* develop rapidly and produce adults in the year after egg deposition whilst others attain only their penultimate larval instar in the second year and spend their second winter in diapause to produce adults two years after egg-laying.

DISTRIBUTION (map 49)

A widely-distributed species in Europe, the Middle East, western Asia to the Aral Sea, Uzbekistan and Pakistan, and in Africa where it occurs both in the north from Suez to the Azores and, as a doubtfully distinct (Pinhey, 1961) subspecies *A. i. mauricianus* Rambur, 1842, over much of southern Africa to the Cape of Good Hope and in Madagascar and Mauritius. In Europe, *A. imperator* extends from the Mediterranean to southern Lancashire in England and the extreme south of Sweden, but it has not been recorded from Scotland, Denmark, Norway or Finland, and it may be absent from parts of the Balkans.

ANAX PARTHENOPE (Sélys)   Plate 17 (100)

Aeshna parthenope Sélys, 1839, *Bull. Acad. Belg.* **6** : 389.
Type locality: Lago d'Averno, Naples, Italy.

   Anax parisinus Rambur, 1842

F – L'Anax parthénopéen, L'Anax napolitain,
D – Kleine Königslibelle

DESCRIPTION OF ADULT

A. parthenope may be distinguished from *A. imperator* in the field by its brownish violet thorax, greener eyes and darker abdomen with only S2 and S3 a bright blue colour. The pale colour on the rest of the abdomen is greenish blue but suffused greyish brown. The pattern is similar to that of *A. imperator*. Wings are tinted yellowish brown in both sexes and the membranule is evenly pale grey. The female quite closely resembles the male; it may be readily identified by the two small, black, tubercular callosities behind the occiput (fig. 191), no traces of which are present in female *A. imperator* (fig. 190). S10 of the male is not produced posteriorly as it is in *A. imperator*. Superior anal appendages (fig. 185) are not so broad, they have an apical point, and each bears a long process basally on the ventral surface. The inferior appendage is very short, scarcely one-fifth the length of a superior appendage, and apically it is more rounded, slightly incised medially, and with a conspicuous series of black teeth around its posterior margin dorsally; in lateral view, five or more of these teeth can be seen. *A. parthenope* is on average slightly smaller than *A. imperator*, body length 66–75mm.

BIOLOGY

Frequents ponds and small lakes, like *A. imperator*, but its flight is not so rapid as that of its larger congener, and when males of the two species are at the same pond, territorial *A. imperator* repeatedly harass *A. parthenope* causing them to move continually from one area to another. Ovipositing *A. parthenope* are held in tandem by their mates, in contrast to *A. imperator*, and this difference between the two species provides a helpful field character although tandem oviposition is not invariable in *A. parthenope* (Robert, 1958).

FLIGHT PERIOD

End of May until August, slightly earlier than *A. imperator*. Flies in March in North Africa.

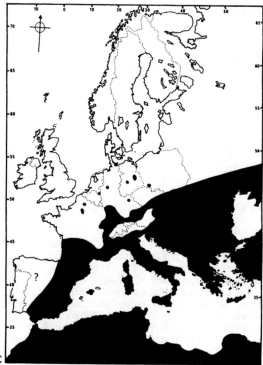

Map 50 – *Anax parthenope*

DISTRIBUTION (map 50)

From the Atlantic (including Canary Islands) to the Pacific. Found in North Africa south to the Somali Republic, southern Europe, the Middle East, Asia Minor and east through the Caucasus to Kashmir, Pakistan, China and Japan (Hokkaido). Isolated populations occur in the Ural Mountains and Altai, USSR. It is local in central Europe and rare in the north where it extends as a vagrant to Holland (one record in 1938: Geijskes & van Tol, 1983) and northern Germany (e.g. Schleswig-Holstein and the island of Sylt: Schmidt, Eb., 1974). The European range of *A. parthenope* is encompassed by that of *A. imperator* and almost everywhere it appears to be the scarcer of the two species. Only in the Camargue have I seen *A. parthenope* abundant. Buchholz (1955) has described a number of geographical forms.

ANAX IMMACULIFRONS Rambur, 1842

This is a large, black and yellow *Anax* found in India, the Middle East, Turkey and on the Isle of Karpathos in the Aegean. Larvae have been located in the latter locality in small, montane water courses.

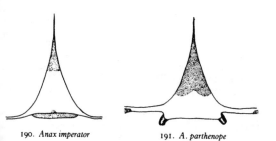

190. *Anax imperator*   191. *A. parthenope*

Figures 190, 191 Female *Anax* species
occipital triangles, showing tubercular callosities of *A. parthenope*

HEMIANAX Sélys, 1883

Bears a general resemblance to *Anax* (cf. figs 183,192) but differs in venational characters given in the key (p.98). Only two species have been described, one from Africa and the southern Palaearctic, the other from Australia and New Guinea.

HEMIANAX EPHIPPIGER (Burmeister)
Plate 17 (101)

Aeschna ephippigera Burmeister, 1839, *Handbuch Entomol.* 2 : 840.
Type locality: Madras, India.

   Aeshna mediterranea Sélys, 1839
   Anax senegalensis Rambur, 1842
F – L'Anax porte-selle, **D** – Schabrackenlibelle

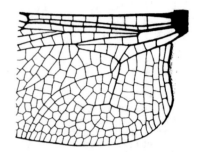

Figure 192 *Hemianax ephippiger* base of male left hindwing

DESCRIPTION OF ADULT

A sandy-coloured dragonfly with bright blue coloration restricted in males to the second abdominal segment and extending slightly on to the base of the third. In females the blue is replaced by a weak violet tinge.

   The sulcus between frons and postclypeus is pale (as in *Anax*), and there is a dark brown, crescentic mark on the anterior edge of the upper surface of the frons which is also darkened posteriorly in front of eyes and ocelli. Eyes are brown above and pale green below. In dorsal view, the eyes in both sexes describe an almost perfect semicircle compared with a semi-ellipse in *Anax* and *Aeshna*. The thorax is yellow-brown with the only blackish lateral markings being the spiracles and a minute spot above them. Wings are relatively broad reaching posteriorly almost to the level of the apex of S3, and their venation antero-basally is yellowish. The pterostigma is light brown, the membranule is whitish with an infuscate outer border, and the hindwings are yellowish centrally in females but clear in mature males. Legs are black with the bases reddish brown and the front femora yellow ventrally. The brownish yellow abdomen has dark brown intersegmental divisions and a median dark line of uneven width edging the dorsal carina, as in *Anax*. The abdomen is slimmer than that of *Anax* and without lateral, black carinae. The posterior edge of S10 of the male is produced back in an even curve. Male superior anal appendages (fig. 186) are brown and with apical, outwardly-turned points, clothed with only short hairs, without either a subbasal ventral tooth or a basal dorsal condyle, but with a sharply-raised dorsal carina in the apical third. The inferior appendage is about one-third as long as the superiors, pale, with a long series of marginal black teeth dorsally; in dorsal view it is subtriangular with its apex weakly incised. Accessory male genitalia are described in detail by Ali Khan (1973). Body length 61–66mm.

BIOLOGY

Breeds in small, standing water bodies, sometimes of a temporary nature and sometimes brackish. A strongly migratory species that may turn up almost anywhere, although it seems to avoid woodland. Several were watched hawking along cliff tops near Novorossiysk on the Black Sea coast in July 1983, and in April 1982 specimens were seen flying over coastal dunes, arable land and heavily-grazed inland valleys on the island of Kos. Flies at dusk and

may be attracted by light. Oviposition occurs in tandem as in *Anax parthenope* and *Aeshna affinis*.

FLIGHT PERIOD

Observed in Europe mainly between the end of March and October, although away from the Mediterranean coast it is usually seen only in autumn. A female was captured, however, at Devonport, England on 24 February, 1903 (McLachlan, 1903), and Icelandic specimens were found in September, October and November.

DISTRIBUTION (map 51)

Essentially a species of arid parts of Africa, the Middle East and south-west Asia to Pakistan, but breeds sporadically in southern Europe. The capture of teneral adults is reported from the Camargue (Jurzitza, 1964) and Huelva, Spain (Belle, 1984). Larvae have recently been found in the south of France, central Italy and Sicily. The species is strongly migratory and has turned up at various places in Europe (Heymer, 1967), more especially in the Mediterranean region but also as far north as England (e.g. Cambridge and Leeds in recent years) and even Iceland (1941, 1964 and three in 1971) in which country it is the only dragonfly recorded (Norling, 1967; Mikkola, 1968; Tuxen, 1976). The appearance of migrants in Piedmont, Italy is reported by Ghiliani (1874) and in Belgrade, Yugoslavia by Mihajlovic (1974). The species is not uncommon in the east Mediterranean (e.g. Kos), Turkey, Iraq and along the east coast of the Black Sea.

BRACHYTRON Sélys in Sélys & Hagen, 1850

The generic name, descriptive of the relatively short body, was used in 1845 in a privately circulated work, *British Libellulinae or Dragon Flies* by M. W. F. Evans, but first published by Sélys (in Sélys & Hagen, 1850:113) five years later. Only one species is known. It is a small aeshnid which appears early in summer and is characterized by its extensively hairy body, long and narrow pterostigma, male hindwing only bluntly angled with a broad anal triangle, and small membranule.

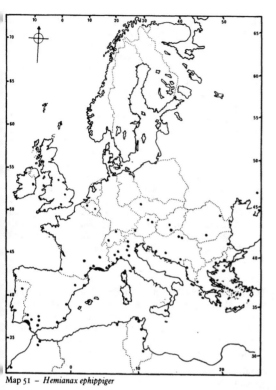

Map 51 – *Hemianax ephippiger*

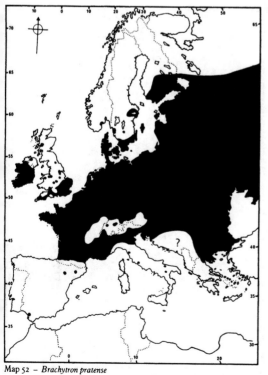

Map 52 – *Brachytron pratense*

BRACHYTRON PRATENSE (Müller)
Plate 18 (102,103)

Libellula pratensis Müller, 1764, *Fauna Insect. fridrichs-dalina* : 64.

Type locality: Denmark.

Libellula hafniensis Müller, 1764
Libellula aspis Harris, 1782
Aeschna teretiuscula Leach, 1815
Aeshna vernalis Vander Linden, 1820
Aeschna pilosa Charpentier, 1825

E – Hairy Dragonfly, **F** – L'Aeschne printanière,
D – Kleine Mosaikjungfer

Müller (1764) failed to associate the two sexes and described the male as *pratensis* and the female as *hafniensis*. Since he described the female earlier in his work on page 61, the name *hafniensis* has page priority, but *pratense* is so well established in the literature that it is desirable to retain it, especially as no change of authorship is involved.

DESCRIPTION OF ADULT

Eyes of *B. pratense* are in contact over a rather short distance, the sulcus between postclypeus and frons is black, and the upper surface of the frons has a black T-shaped mark. Antehumeral stripes are complete and conspicuous in the male but short and narrow in the female. The sides of the thorax are greenish yellow with three oblique black lines, and the mesepisternum is brown (apart from the antehumeral stripes). The thorax is conspicuously pilose. Legs are black and the wings are hyaline or slightly yellowish in mature females, with the pterostigmata brown, very narrow and long (about 4mm). Membranules are whitish. The costa is marked with yellow but other longitudinal veins are black. Venational characters (fig. 157, p.99) are mentioned in the key to aeshnid genera; in addition, the male hindwing has a broadly-based anal triangle of two or, more usually, three cells and the anal margin is not sharply angled. The abdomen of the male is only weakly constricted at the third segment and hairy at the base. It is black dorsally with a yellowish median spot on the first segment, with large, paired, greenish to blue spots in the posterior parts of S2–9, and a medially interrupted, transverse, yellow line in the anterior quarter of each of S2–8 with, just anterior to it on S3–7, a medially interrupted bluish transverse line arising from a pale lateral spot. Laterally there are two pale spots on the central abdominal segments, becoming larger and fusing towards the front of the abdomen but decreasing and eventually disappearing posteriad. The tenth abdominal segment has its posterior margin lined with blue. The female abdomen is broad and has yellow instead of blue spots; S10 is elongated, longer than broad, with a dorsal carina extending almost its full length. Male superior anal appendages (fig. 187) are black, fully as long as the combined length of S9+10, without a subbasal ventral tooth but with a very sharply elevated dorsal carina in the distal third. The inferior appendage is about one-quarter the length of a superior appendage, apically truncated and bilobed. Female anal appendages are about 6mm in length, longer than the last two abdominal segments. Body length 54–60mm.

BIOLOGY

Larvae develop under floating mats of dead vegetation in water of canals, large ditches, dykes and ponds that is stagnant or has but a feeble flow. Half-decomposed floating rushes and sedges are often used as oviposition sites. Males fly low and quite slowly in a zig-zag path amongst reeds. Both sexes, but particularly females, may be encountered in woods quite far from water.

FLIGHT PERIOD

This is brief and very early, commencing in April or even late March in southern Europe, in May in northern Europe, and being virtually finished by the end of June or, in Britain, mid-July.

DISTRIBUTION (map 52)

Europe east to the Caucasus and Caspian Sea. Quite widespread in Europe but rare in the south and known in Spain and Italy mainly from northern localities, although in the latter country it has been recorded from Basilicata. Also found in Greece, including Lesbos. In the north, *B. pratense* has been recently discovered in Scotland (Smith & Smith, 1984), but it does not penetrate beyond the southern parts of Norway, Sweden and Finland.

CALIAESCHNA Sélys, 1883

Small aeshnids, marked with yellow, green and blue. The genus has only one included species and is easily recognized by the small pterostigma and the presence of crossveins in the median space between veins R+M and Cu basal to the arculus (fig. 158, p.99). Vein IR_3 is symmetrically forked, the male hindwing is only moderately angled, and the anal triangle is very broad.

CALIAESCHNA MICROSTIGMA (Schneider)
Plate 18 (104)

Aeschna microstigma Schneider, 1845, *Ent. Ztg. stettin.* **6** : 113.
Type locality: Gelemish (Kellemisch), Turkey.

DESCRIPTION OF ADULT

The smallest European aeshnid with a body length of only 50–58mm. Eyes touch for a distance only about twice the length of the occipital triangle. The upper surface of the frons has a black crescentic mark on the front edge. The thorax is dark brown with complete antehumeral stripes in both sexes, and there are two very broad yellowish to green stripes on each side. Legs are mainly brown. Wings are rather broad and their venation is not dense. The median space between veins R+M and Cu before the arculus is crossed by three to five crossveins; the other European aeshnid to have veins in this position is *Boyeria irene*, but other venational and wing features in *Caliaeschna* and *Boyeria* differ (cf. figs 158 and 159, p.99). The pterostigmata are very short, 1.5–2.0mm long and 2.0–2.5 times as long as broad. Vein IR_3 is symmetrically forked not far distad of midway between node and pterostigma. Radial and median supplementary veins subtend just a single row of cells. In the male, the anal angle of the hindwing is approximately right-angled and the anal triangle is 3-celled and very broad. The membranule is whitish and exceptionally small, adjacent to merely half a cell. The male abdomen is moderately contracted at the third segment with conspicuous auricles on the second. The ground-

colour of the abdomen is dark brown marked with yellow to greenish blue dorsally as follows: a median spot and two posterior lines on S1; a median longitudinal line on S2 and S3 becoming reduced on S4 and represented on S5–8 by a small anterior triangle; a transverse line broken by the dorsal carina so that it appears as two transversely orientated narrow triangles on S3–8 and as two oblique lines on S2, situated near the centres of S2 and S3 but thereafter becoming progressively more anterior in position; paired posterior spots on S2–7 becoming larger posteriad, partially fused on S8 and S9, completely fused on S10 to occupy almost the entire dorsal surface. The last abdominal segment of the male has its dorsal surface moderately convex in profile and a median carina running about two-thirds of its length. Male superior anal appendages (fig. 188) are long, equalling the length of S9+10, and apically rounded. The inferior appendage is triangular in dorsal view, half the length of the superiors. Female anal appendages (fig. 193) are diminutive, no longer than the tenth abdominal segment and just reaching to the apex of the ovipositor.

BIOLOGY

Inhabits relatively fast-flowing and rocky streams and small rivers. Males fly in shaded situations, slowly and very low over the water surface, in a manner resembling that of *Boyeria irene*, from about noon until late evening (Kemp & Kemp, 1985). Females oviposit in moss on boulders in streams.

FLIGHT PERIOD
May, June.

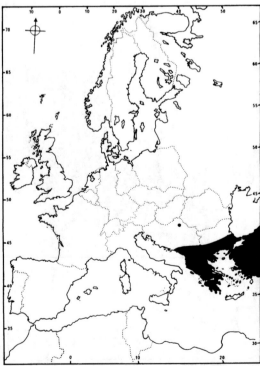

Map 53 – *Caliaeschna microstigma*

DISTRIBUTION (map 53)

Asia Minor to the Caucasus, extending westwards into the Balkans of Europe. Recorded from Albania (Petrela), Yugoslavia (Belgrade, Montenegro), Greece (Halkidiki, Thessaloniki, Peloponnese, Corfu) and also Cyprus, Turkey (Dumont, 1977c), Lebanon, Iraq and Iran.

BOYERIA McLachlan, 1896

Fonscolombia Sélys, 1883

A small genus (five species) with an Holarctic distribution but with only the type species in Europe. There are crossveins in the median space before the arculus, the anal margin of the male hindwing is angled and vein IR_3 is unforked.

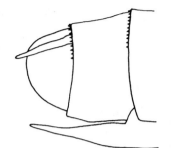

193. *Caliaeschna microstigma*

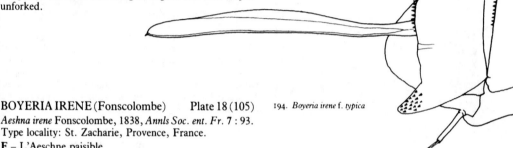

194. *Boyeria irene* f. *typica*

BOYERIA IRENE (Fonscolombe)   Plate 18 (105)

Aeshna irene Fonscolombe, 1838, *Annls Soc. ent. Fr.* 7 : 93.
Type locality: St. Zacharie, Provence, France.
F – L'Aeschne paisible

DESCRIPTION OF ADULT

The presence of two to four crossveins in the median space (fig. 159, p.99) distinguishes *B. irene* from all other species of European Aeshnidae with the exception of *Caliaeschna microstigma*, but the latter has a range which does not overlap with that of *B. irene*. As in *Caliaeschna*, the venation is not dense, the anal margin of the male hindwing is moderately angled and the anal triangle is broad (but less so than in *Caliaeschna*) and includes three to five cells (fig. 196). *Boyeria* differs from *Caliaeschna* in having vein IR_3 unforked, the radial and median supplementary veins each subtend a double row of cells, and the pterostigmata are long.

The body of *Boyeria* is dull green to olive with a rather intricate pattern of brown marks on the abdomen. This brown colour tends to merge with the greenish colour so that the pattern lacks sharpness. The mature male is more distinctly green than the female, its third abdominal segment is quite strongly constricted, and each wing has a small, brownish patch at its apex. The female and immature male lack this infumation of the wing tips but otherwise resemble the male in pattern. Antehumeral thoracic stripes are present in both sexes, legs are brown, and the membranules are very short and whitish. The last abdominal segment of the male is strongly convex dorsally and has no dorsal carina, and the superior anal appendages (fig. 189) each have a small apical point, a low dorsal crest in the apical quarter, and a subbasal tooth. The inferior appendage is apically truncated and about one-third as long as a superior. The anal appendages of the female are usually short, about 2mm long and about as long as the last abdominal segment (form *brachycerca* Navas, fig. 195), but in the less common typical form (fig. 194) they are about 6mm long, fully three times the length of S10. There do not

195. *Boyeria irene* f. *brachycerca*

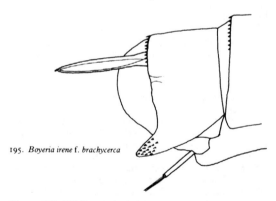

Figures 193–195 Female Aeshnidae
terminal abdominal appendages in right lateral view

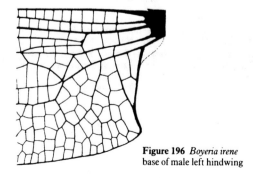

Figure 196 *Boyeria irene*
base of male left hindwing

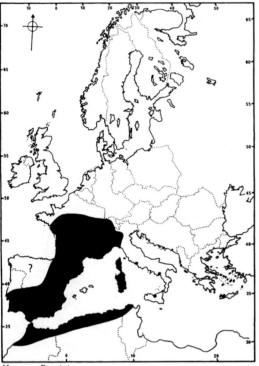

Map 54 – *Boyeria irene*

appear to be intermediate states and both forms may occur at the same localities (Wenger, 1959). Body length 63–67mm.

BIOLOGY

A river-dwelling insect. Males fly low and relatively slowly, back and forth along shaded stretches of rivers, keeping close to the banks when the rivers are broad. They often fly in deep shade and always where there are overhanging trees and bushes. Nooks and cavities in the bank just above the water level, particularly under tree roots, are constantly investigated. The species continues to fly late in the evening and occasionally enters lighted buildings. Miller & Miller (1985a) noted a period of intense crepuscular feeding which coincided with sexual activity.

FLIGHT PERIOD

From July to the beginning of September.

DISTRIBUTION (map 54)

Essentially a west Mediterranean species, known from North Africa, Portugal, Spain, France as far north as Le Mans but most numerous in the south-west, Corsica, Sardinia, western Italy, Sicily (once seen: Bucciarelli, 1977) and Switzerland (rare and extinct in the west). Also recorded from Crete (Cowley, 1940; Schmidt, Er., 1965).

GOMPHIDAE

Gomphidae are immediately distinguished from other families of Anisoptera by the wide separation of the eyes on the dorsal surface of the head. They are yellow or, in just a few European species, partly-green dragonflies, marked with black or brown. Most European species are of medium size, although *Lindenia* is large and *Paragomphus* small. The wings have very reduced membranules (except in *Lindenia*), the discoidal cells are of similar shape in fore- and hindwings, the two veins running from the arculus are well separated (figs 197,198), and the supplementary veins are not developed. Hindwings in males have sharply-angled anal margins. The legs are short. The abdomen is cylindrical, more or less expanded posteriorly and sometimes with foliaceous outgrowths ventrally. In males there are two auricles on the second segment. The female ovipositor is not developed.

Most gomphids have a relatively brief flight period and are seldom found far from their breeding places which are mostly in running water. Males in particular may be seen flying low and, if not disturbed, quite slowly near to the shore. They perch for long periods on sand or shingle in full sunlight, seldom resting on vegetation. Males usually seize females in flight, but copulation is normally completed on the ground. The female oviposits in flight, not in tandem, dipping the tip of her abdomen into the water. Larval development is typically slow, adults of several species not emerging until from three to five years after egg-laying. Eclosion of the adult insect is often on a horizontal surface such as on top of a flat rock.

Key to genera of Gomphidae

1   Discoidal cells divided by crossveins. S7,8 with foliaceous ventral expansions in both sexes .... *Lindenia* (p.135)
–   Discoidal cells undivided. Abdomen either without ventral expansions or (male *Paragomphus*) these are on S8,9 2

2(1) Hindwing without an anal field (fig. 197) 3
–   Hindwing with an anal field of two or three cells (fig. 198) 4

3(2) S8,9 with foliaceous ventral expansions, large in the male (fig. 218, p.129) but small in the female. Five or six crossveins between the node and pterostigma. Male superior anal appendages more than twice as long as the last abdominal segment, down-curved and pointed (fig. 218) .... *Paragomphus* (p.128)
–   S8,9 lacking ventral expansions. At least seven and usually eight to ten crossveins between the node and pterostigma. Male superior anal appendages only about as long as the last abdominal segment, not curved downwards and often truncated (figs 205–210, p.123) .... *Gomphus* (p.121)

4(2) Male superior anal appendages stout, weakly curved and hardly longer than the tenth abdominal segment (fig. 222, p.130). Female with two yellow prominences on the occiput, each with a crown of black teeth (fig. 221, p.130) .... *Ophiogomphus* (p.130)
–   Male superior anal appendages slender and strongly curved, much longer than the tenth abdominal segment (figs 230, 233, 236, p.133). Female without dentate occipital prominences (simple prominences in *O. forcipatus*) .... *Onychogomphus* (p.131)

GOMPHUS Leach, 1815
Diastatomma Burmeister, 1839

Species of *Gomphus* are distinguished from other genera by the absence of an anal field in the hindwing together with the absence of ventral outgrowths on the terminal abdominal segments. The pterostigma of the forewing is slightly shorter than that of the hindwing. The six species known to occur in Europe quite closely resemble each other, being medium-sized, yellow dragonflies patterned with black. The extent of black markings on the thorax is specifically fairly constant and therefore of taxonomic value. Basically, five oblique, black, thoracic lines or fasciae can be identified when the thorax is viewed from the side. These fasciae are numbered and named, from anterior to posterior, as:

1 median dorsal
2 antehumeral
3 humeral or mesepisternoepimeral
4 mesometapleural
5 metepisternoepimeral

Their disposition is illustrated in figure 199, p.122. Structure of the male anal appendages is also specifically characteristic. The inferior appendage is uniformly broadly emarginate with the two arms strongly divergent and terminating in upturned points outside the apices of the superior appendages, but the superior appendages differ in shape between species. The female vulvar scale is likewise diagnostic.

Gomphus ranges across the northern hemisphere from North America to south-east Asia with altogether about sixty species.

Key to species of *Gomphus*

1 Thoracic fascia 2 midway between fasciae 1 and 3, the intervening yellow stripes almost equally broad (fig. 199). Male superior anal appendages not truncated, simply pointed apically (fig. 205, p.123). Female vulvar scale only one-quarter the length of the ninth sternum (fig. 211, p.124) .... *flavipes* (p.123)
– Fascia 2 nearer to fascia 3 than to fascia 1. Male superior anal appendages obliquely truncated with an apical point that is often upwardly-directed and (except in *vulgatissimus*) a ventral or lateral subapical tooth. Female vulvar scale usually longer 2

2(1) Black thoracic fasciae very narrow, fascia 2 sometimes partly obliterated and much narrower than the yellow band between fasciae 2 and 3; fascia 4 extends as a thin line to the dorsal surface of the thorax (fig. 202, p.122). Tarsi mainly yellowish .... *pulchellus* (p.127)
– Fasciae broader, fascia 2 as broad as or broader than the yellow band between fasciae 2 and 3; fascia 4 extends dorsally only as far as the level of the metathoracic spiracle. Tarsi mainly dark brown to black 3

3(2) S8–10 without median dorsal yellow marks 4
– S8–10 with median dorsal yellow marks 5

4(3) Abdomen in both sexes stout and clubbed. Fasciae 2 and 3 very close to each other, separated by only a thin yellow line (fig. 200, p.122) which anteriorly is less than half as broad as fascia 3. Legs with small yellow marks only on coxae and internal face of front femur. Back of head mainly black. Ventral surface of thorax posterior to hind coxae mainly black. Widespread in Europe .... *vulgatissimus* (p.125)
– Abdomen rather more slender and less clubbed. Black markings on body rather less extensive than in *vulgatissimus*; fasciae 2 and 3 separated by a yellow band that is anteriorly as broad as fascia 3. Ventral surface of thorax posterior to hind coxae mainly yellow. South-eastern Europe .... *schneideri* (part) (p.126)

5(3) Fascia 2 much broader than yellow band between fasciae 2 and 3, dorsally extending medially to almost or quite contact fascia 1 (fig. 203, p.122). Male superior anal appendages with a large lateral tooth (fig. 210, p.123). Central and south-western France and Iberian Peninsula only ... *graslini* (p.127)
– Fascia 2 not usually distinctly broader than yellow band between fasciae 2 and 3, dorsally separated from fascia 1 by yellow band (fig. 201, p.122). Male superior anal appendage with a small ventral tooth (fig. 207, p.123) 6

6(5) Species with extensive black markings (approaching *vulgatissimus*). Tibiae black. South-eastern Europe ....
.... *schneideri* (part) (p.126)
– Species with black markings more reduced. Tibiae striped with yellow. South-western Europe .... *simillimus* (p.125)

Figures 197, 198 Male Gomphidae bases of left hindwings.
Note anal loop present in *Onychogomphus forcipatus* (below) but absent in *Gomphus pulchellus* (left)

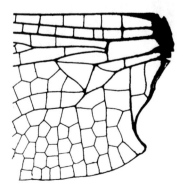

197. *Gomphus pulchellus*

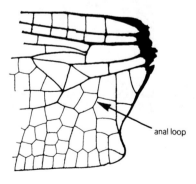

198. *Onychogomphus forcipatus*

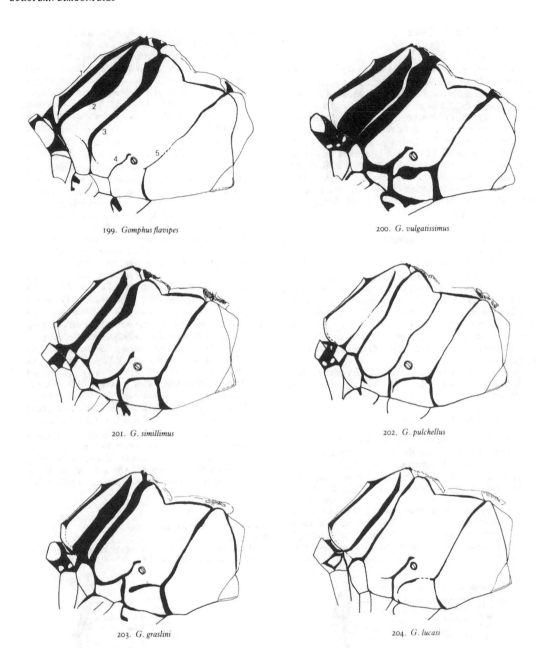

199. *Gomphus flavipes*

200. *G. vulgatissimus*

201. *G. simillimus*

202. *G. pulchellus*

203. *G. graslini*

204. *G. lucasi*

Figures 199–204 *Gomphus* species
thoraces in left lateral view to show designs of black markings.
Numbers on *G. flavipes* (199) refer to the fasciae described in the text

GOMPHUS FLAVIPES (Charpentier)
Plate 19 (106,107)

Aeschna flavipes Charpentier, 1825, *Horae entomol.* : 24.
Type locality: Silesia, Poland.

Petalura selysii Guérin-Méneville, 1838
Aeshna cognata Eversmann, 1836

F – Le Gomphus à pattes jaunes,
D – Asiatische Keiljungfer

Distinct from other species of *Gomphus* in thoracic pattern
and other features, it is suggested by Eb. Schmidt (1987)
that *flavipes* be assigned to the genus *Stylurus* Needham.

DESCRIPTION OF ADULT

G. flavipes has a total body length of 50-53mm and is
slightly larger than the other European species of *Gomphus*
(body length 48–50mm). It is most easily recognized by the
anterior position of the black antehumeral fascia (fascia 2)
(fig. 199), this being far removed from the humeral fascia
and extended dorsally to reach the top of the mesepisternum. S7–9 of the male abdomen are more expanded than in
all other species except *G. vulgatissimus*, but the abdomen
anteriorly is much more slender and more extensively
yellow than in that species. The legs are yellow, striped with
black, but not more yellow than those of *G. simillimus* or *G.
pulchellus* so that the specific name of *G. flavipes* could be
misleading. A more useful distinguishing feature is the
wholly yellow, median, dorsal crest of the thorax between
the black median dorsal fasciae. In other species, this crest
is darkened posteriorly and often anteriorly as well. The
simple, untoothed, dark brown, male superior appendages
(fig. 205) are semicircular in section, flattened below, and

Map 55 – *Gomphus flavipes*

with a sinuate lateral beading on the outer edges, and each
ends in a quite attenuated point. In lateral view, the inferior
appendage is rather sharply angled ventrally, and its apical
emargination is almost semicircular, deeper than in the
other species. The vulvar scale of the female (fig. 211,
p.124) is very short, barely one-quarter the length of the
ninth segment, and divided to its base into two, triangular
lobes.

BIOLOGY

Immature stages are passed in running water and the adults
frequent sandy banks along the lower courses of large
rivers.

FLIGHT PERIOD

Middle of June to the first half of September.

DISTRIBUTION (map 55)

An eastern Palaearctic species extending from Siberia
(Amur), Manchuria, Russia, Turkey, Iran, Iraq and Syria
west into Europe where it is found mainly in the east but
with scattered records from Holland (last observed in
1902), Belgium and France west to Saumur in Maine-et-
Loire (Dommanget, 1981). In Italy it has been found in
Lazio, Veneto, Lombardy and Piedmont (Consiglio, 1950;
Bucciarelli, 1976), and is apparently not uncommon along
the lower reaches of the River Po. There are records of
single captures in England (Hastings in 1818: Stephens,
1835) and Switzerland (mid-19th century: Dufour, 1982).
In Europe the species seems to be declining in numbers and
it is now believed extinct in West Germany (Schmidt, Eb.,
1977).

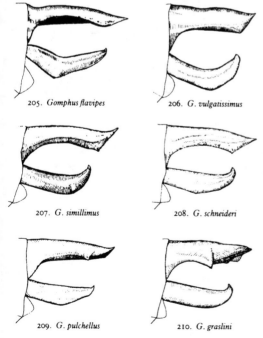

205. *Gomphus flavipes*

206. *G. vulgatissimus*

207. *G. simillimus*

208. *G. schneideri*

209. *G. pulchellus*

210. *G. graslini*

Figures 205–210 *Gomphus* species
male abdominal appendages in left lateral view

123

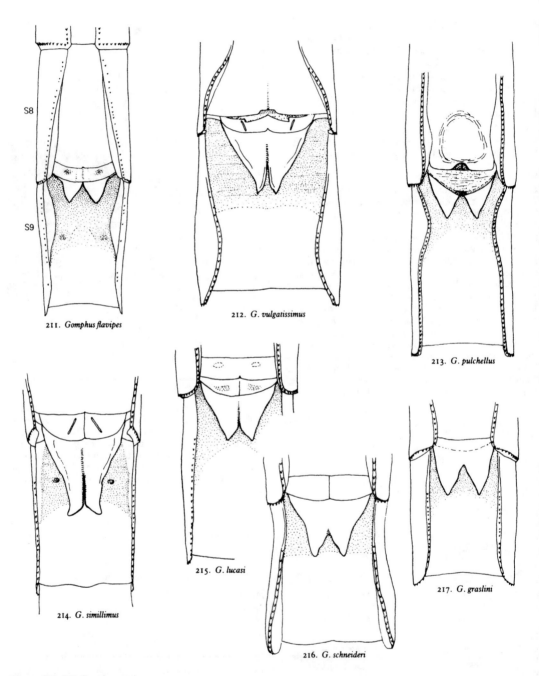

211. *Gomphus flavipes*

S8

S9

212. *G. vulgatissimus*

213. *G. pulchellus*

214. *G. simillimus*

215. *G. lucasi*

216. *G. schneideri*

217. *G. graslini*

Figures 211–217 *Gomphus* species
female abdominal segment S9 and vulvar scales in ventral view

GOMPHUS VULGATISSIMUS (Linnaeus)
Plate 19 (108,109)

Libellula vulgatissima Linnaeus, 1758, *Syst. Nat.* (Edn. 10) 1 : 544.

Type locality: Sweden.

Libellula forcipata Donovan, 1807

E – Club-tailed Dragonfly, **F** – Le Gomphus très commun, **D** – Gemeine Keiljungfer

DESCRIPTION OF ADULT

G. vulgatissimus is the most widespread European *Gomphus* and the type species of the genus. It is also the most extensively black and most robustly-built species.

The back of the head behind the eyes is black with only two small yellow spots, and the labrum is margined with black above and below. The thoracic fasciae are broad (see key, p.121, and fig. 200) but the median dorsal fasciae do not spread laterally at their lower ends to the same extent as in other species. Wings are somewhat broader than in the other species of the genus and the distal side of the forewing discoidal cell normally subtends three cells (usually two in the following species). Costal veins are scarcely lined with yellow. The legs are almost entirely black with yellow restricted to the coxae and a short basal stripe on the inner face of the front femur. The abdomen is stout and apically clubbed in both sexes, S8 and the adjacent halves of S7 and S9 being expanded so that in dorsal view this part of the abdomen is almost (female) or fully (male) as broad as the length of the hindwing pterostigma. The median dorsal abdominal series of yellow marks ceases at S7 and the last three abdominal segments are entirely black mid-dorsally (except as a rare variation). Male superior appendages (fig. 206) are black, cylindrical and without subapical teeth, in lateral view almost squarely and vertically truncated. The female vulvar scale (fig. 212) is two-fifths the length of S9, narrowly incised apically with two subparallel and rather narrow arms each terminating in a small point.

BIOLOGY

Typically associated with slow-flowing, meandering rivers and large streams with muddy beds, occasionally with large lakes, but found away from water in woodland perhaps relatively more frequently than the other species. Perches on bushes as well as on the ground.

FLIGHT PERIOD

Appears early, at the end of April in southerly parts of its range, a month later in Britain, and it is in flight for about five weeks although there are a few records of its capture in August.

DISTRIBUTION (map 56)

Widespread over much of Europe from the English Midlands and southern Scandinavia through France, Italy, Holland, Belgium, Denmark, Germany, Switzerland (now very scarce) and eastern Europe to Russia. Recently recorded from Astorga (Ocharan, 1984) and Rio Cea (Belle, 1985) in northern Spain.

Map 56 – *Gomphus vulgatissimus*

GOMPHUS SIMILLIMUS Sélys   Plate 19 (110,111)

Gomphus simillimus Sélys, 1840, *Monograph. Libell. Europ.*: 85.

Type locality: Aix-en-Provence, France.

Gomphus zebratus Rambur, 1842

F – Le Gomphus similaire, **D** – Gelbe Keiljungfer

DESCRIPTION OF ADULT

G. simillimus is more slender-bodied and narrower winged than *G. vulgatissimus*, with less extensive black markings, although the general colour pattern is similar. The ante-humeral and humeral thoracic fasciae (fasciae 2 and 3) are of similar width and about as broad as the intervening yellow stripe (fig. 201). Fascia 4 terminates near the metathoracic spiracle. Median dorsal yellow marks are present on all abdominal segments. All femora and tibiae are yellow with black stripes, but the tarsi are blackish. These colour characters, in conjunction with the form of male anal appendages and female vulvar scale, distinguish *G. simillimus*. The dark brown superior appendages of the male (fig. 207) are a little flattened, in lateral view more obliquely truncated than those of *G. vulgatissimus*, apically drawn-out into an upturned, sharp point and with a small, ventral, subapical tooth located at the lower end of the truncation. The vulvar scale (fig. 214) is very long, fully half the length of the ninth abdominal segment and extremely narrowly incised so that the two arms are virtually contiguous to near their apices which are divergent.

BIOLOGY

Restricted to rivers.

125

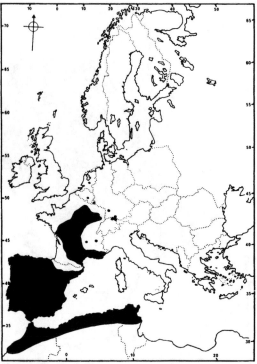

Map 57 – *Gomphus simillimus*

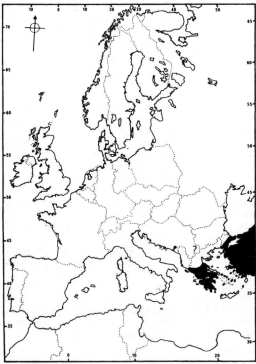

Map 58 – *Gomphus schneideri*

FLIGHT PERIOD
From mid-May to the end of July.

DISTRIBUTION (map 57)
G. simillimus has a restricted west Mediterranean distribution, occurring regularly only in Spain and the southern half of France, with occasional specimens reaching Holland, Belgium, Luxembourg, Switzerland (Haute Rhine) and Germany (Konstanz). Also found in North Africa where the lightly-marked Moroccan population has been separated as subspecies *G. s. maroccanus* Lieftinck, 1966.

GOMPHUS LUCASI Sélys, 1849

A species restricted to North Africa (Morocco, many localities in Algeria and a few in Tunisia). It is closely allied to *G. simillimus* from which it may be distinguished by an anterior yellow margin to the dorsal surface of the vertex (all black in *simillimus*), a narrow postocellar crest produced over the ocelli (thicker and not as produced in *simillimus*), narrower black thoracic fasciae (fig. 204), male superior anal appendages shaped more like those of *G. schneideri* (cf. fig. 208) and a shorter vulvar scale (fig. 215). Lieftinck (1966) provides a detailed key for distinguishing *G. lucasi*, *G. simillimus* and *G. s. maroccanus*.

GOMPHUS SCHNEIDERI Sélys   Plate 19 (112,113)

Gomphus schneiderii Sélys in Sélys & Hagen, 1850, *Revue des Odonates* : 292.
Type locality: Gelemish (Kellemisch), Anatolia, Turkey.
  Gomphus vulgatissimus race *schneiderii* Sélys & Hagen, 1857

G. schneideri was relegated to the status of a race of *G. vulgatissimus* by Sélys and Hagen (1857), but Morton (1915) suggested that it showed closer affinity with *G. simillimus*, a view shared by most recent authors.

DESCRIPTION OF ADULT
G. schneideri is rather poorly known. It is an extensively black insect but less so than *G. vulgatissimus*; S8–10 have, at least in some populations, indications of median dorsal yellow marks, the lower surface of the thorax behind the posterior legs is mostly yellow, the anterior and middle femora each sometimes have a short, basal, yellow line on the upper surface in addition to the basal yellow mark on the inner face present in *G. vulgatissimus*, and the black thoracic fasciae are not quite so broad. Whilst resembling *vulgatissimus* most closely in coloration, *schneideri* has a general body shape approaching that of *simillimus* with a more slender and rather less clubbed abdomen than *vulgatissimus* and slightly narrower wings. The male superior anal appendages (fig. 208) are more obliquely truncated than those of *vulgatissimus*, and are more like those of *simillimus* but have the apical point less upturned and directed more horizontally. The arms of the inferior appendage, in profile, are a little less concave dorsally than in *simillimus*. The vulvar scale (fig. 216) is of much the same relative length as that of *vulgatissimus* but with a broader apical incision.

BIOLOGY
A riverine species.

Mainly in May and June, although recorded as early as 18 April in Turkey (Dumont, 1977c).

DISTRIBUTION (map 58)

A west Asian species recorded from the Caspian region of Russia, Georgia and Turkey, extending west to Greece (Corfu, Peloponnese) from where Buchholz (1954) has described subspecies *G. s. helladicus*, and Yugoslavia (Montenegro).

GOMPHUS PULCHELLUS Sélys
Plate 19 (114,115)

Gomphus pulchellus Sélys, 1840, *Monograph. Libell. Europ.* : 83.

Type locality: Belgium.

Aeschna anguina Charpentier, 1840

F – Le Gomphus gentil, **D** – Westliche Keiljungfer

DESCRIPTION OF ADULT

The most lightly-marked of the *Gomphus* species considered here and easily recognized by the narrowness of the black thoracic fasciae (fig. 202), the antehumeral fascia (fascia 2) being considerably narrower than the yellow band behind it and often more or less incomplete. In contrast, the mesometapleural fascia (fascia 4), although very narrow, is extended dorsally to the top of the thorax whereas in the other species it is short and terminates slightly above the level of the metathoracic spiracle. The thoracic fasciae are brownish grey rather than black, and the yellow colour has a weak greenish tint unlike the pure yellow of mature specimens of the other species. The costal veins are conspicuously yellow and the pterostigma is brown. Abdomen and legs have extensive yellow markings, the yellow colour on the last three abdominal segments being deeper in tone than that on the remainder of the abdomen. The abdomen is slender and not swollen at the apex. *G. pulchellus* differs from *G. simillimus*, the other very yellow European *Gomphus*, in its even narrower thoracic fasciae, the dorsal extension of fascia 4, its yellow tarsi, and especially in the different form of the male anal appendages and female vulvar scale. The male superior appendage (fig. 209) is brownish yellow, blackened only narrowly at the apex, and rather flattened with an oblique, apical truncation which lies almost horizontally with a small, triangular, subapical tooth at its outer end and some irregularities on its lower surface. The general form of the superior appendages is similar to those of *G. graslini* (below) but with much slighter development of the teeth. The vulvar scale (fig. 213) is short, about one-quarter the length of S9, with two widely-separated, triangular lobes, and towards the apex of the eighth abdominal sternite there are two low prominences.

BIOLOGY

Larvae of *G. pulchellus*, unlike those of the other European species of *Gomphus*, are found principally in still water, usually in lakes and large ponds but occasionally in river backwaters. It is abundant in some artificial reservoirs and ponds where the banks are of gently-sloping, sandy soil with scant plant cover, providing appropriate perching places for the adult dragonflies. Long periods are spent by males on the ground; they rarely perch on vegetation.

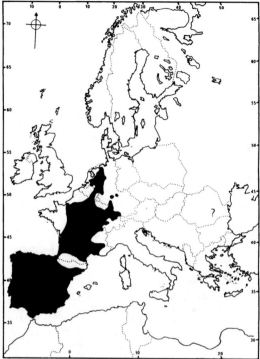

Map 59 – *Gomphus pulchellus*

FLIGHT PERIOD

Mid-May to mid-August.

DISTRIBUTION (map 59)

A west European species like *G. simillimus* but apparently absent from North Africa. Widespread in Portugal, Spain and southern (particularly south-western) France. Local or rare in northern and eastern France, Belgium, Holland, West Germany east to Schweinfurt, and lowland areas of western Switzerland. Rudolph (1980) provides evidence that it has invaded the south-western and north-western parts of the Federal Republic of Germany, the Netherlands and Switzerland during the past century from its probable centre of distribution in France. A single record (Cirdei & Bulimar, 1965) for Romania (near Bucharest) is very doubtful, the easternmost locality probably being the Istrian Peninsula, Yugoslavia (Schneider, 1984).

GOMPHUS GRASLINI Rambur   Plate 19 (116,117)

Gomphus Graslinii Rambur, 1842, *Hist. nat. ins. Neur.* : 158.

Type locality: Forêt de Bercé north of Chateau-du-Loir, Sarthe, France.

DESCRIPTION OF ADULT

A heavily-marked *Gomphus* (fig. 203) but with more extensive yellow markings than *G. vulgatissimus* and *G. schneideri*. The legs, excluding the coxae, are black with yellow stripes only on the femora. Male superior anal appendages (fig. 210) are very characteristic, each being flattened, in dorsal view with the apex obliquely truncated in the horizontal plane with a sharp apical tooth and a large, broadly triangular, lateral tooth subapically on the outer face. There is also a projecting ventral lobe, yellowish in colour (the rest of the appendage is black), just behind the apex on the inner surface. *G. graslini* and *G. pulchellus* are the only European species to possess lateral teeth on the superior male appendages, but those of *graslini* are very much the larger. The female vulvar scale (fig. 217) is about one-third of the length of the ninth abdominal segment and deeply cleft apically. Wings of *G. graslini* are rather broader than those of the preceding species and the pterostigma is somewhat closer to the wing apex from which it is separated by no more than its own length.

BIOLOGY

Exclusively riverine. Males patrol river banks at a height of a metre or so above the water. Rests repeatedly on rocks or stones.

FLIGHT PERIOD

Mid-June to mid-August, appearing rather later in summer than other species of *Gomphus*.

DISTRIBUTION (map 60)

The most local of the west European species of *Gomphus* and found only in central and south-western France west of the Rhone and in the Iberian Peninsula. It is perhaps most numerous in the valleys of Dordogne, Lot and Tarn. There is a single record from Portugal (Cea: McLachlan, 1880) and a few from southern Spain (Overbeek, 1970).

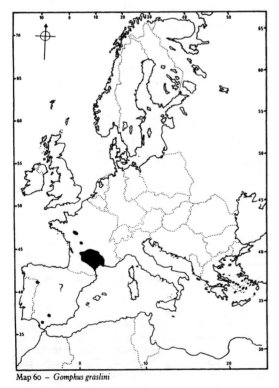

Map 60 – *Gomphus graslini*

PARAGOMPHUS Cowley, 1934
Mesogomphus Förster, 1906 (preoccupied)

This genus is very well represented in Africa and occurs also in India and south-eastern Asia, but just one species penetrates into Europe, and this only in the extreme south. The European species, *P. genei*, is the same as the very common and widespread African species *P. hageni* (Sélys) described from Egypt. As in *Gomphus*, there is no anal field in the hindwing but male *Paragomphus* are immediately distinguishable by the presence of foliaceous ventral extensions of the eighth and ninth abdominal segments and by their long anal appendages. Pterostigmata of fore- and hindwings are approximately equal in length. The basal crossvein between veins Cu and A is further from the arculus in the forewing than in the hindwing.

PARAGOMPHUS GENEI (Sélys) Plate 20(118,119)
Gomphus genei Sélys, 1841, *Rev. zool. Guérin Méneville* : 245.
Type locality: Sicily.
 Onychogomphus hagenii Sélys, 1871b
 Gomphus excelsus Costa, 1884

DESCRIPTION OF ADULT

A small gomphid (body length 37–41mm, the male longer than the female), mainly sandy yellow in colour with only restricted dark markings. The thorax has ill-defined, brownish fasciae on the greenish yellow ground-colour, and the abdomen is yellowish with the intersegmental articulations black and with two pairs of small, brownish marks on S2–7. The legs are mainly yellow with two fine, brownish lines on the tibiae. Male anal appendages are yellowish.

Wings are relatively broad with rather reduced venation; there are, for instance, no more than six crossveins between the node and pterostigma, and in the series of cells between veins R_{4+5} and M, at most only the distal cell is divided. The long pterostigmata are close to the wing apices. The male abdomen is very slender over S3–7 but is laterally expanded in S8 and S9, in addition to carrying foliaceous ventral outgrowths on these segments (fig. 218). The foliations are downwardly-directed extensions of the tergal margins and are as broad as about one-quarter the diameter of the segments that carry them. In the female, these foliations are very much reduced. Male superior appendages (figs 218,219) are very long, in dorsal view parallel, as long as the combined length of S9+10, strongly down-curved and slightly recurved, tapering to a bifid point. The inferior appendage (fig. 219) is slightly more than half the length of a superior, curved in lateral view with the dorsal surface strongly concave, narrowly divided apically for less than half of its own length, the two arms contiguous and each ending in a bulbous head with a subapical, lateral lobe and a small, upwardly-directed, apical tooth. The lobes and teeth are blackish, but the rest of the appendage is yellowish to brownish. The vulvar scale (fig. 220) is quite strongly domed with a short, narrow, median incision. The scale extends over about one-third of the length of the ninth sternum, and the latter has a large, membranous area with a raised rim beneath and beyond the vulvar scale. Female anal appendages are twice the length of the tenth abdominal

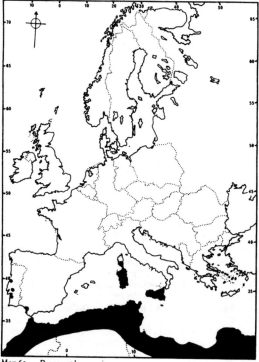

Map 61 – *Paragomphus genei*

segment, attenuated and terminating in sharp, blackish points.

BIOLOGY

Larvae live in slow-flowing streams and nearly stagnant pools with open, sandy bottoms. Observations in southern Spain (Testard, 1975) show *P. genei* to be univoltine with most (85 per cent) adult emergence occurring in early afternoon, an unusual time of day for this.

FLIGHT PERIOD

May to August, with maximum emergence in June in southern Spain (Testard, 1975). In the Middle East, *P. genei* has been taken in April through to mid-September.

DISTRIBUTION (map 61)

In Europe, *P. genei* is confined to the extreme south and has been found only in Corsica, Sardinia (*G. excelsus*) and Sicily (Bucciarelli, 1977), Spain (Testard, 1975) and Portugal (Aguiar & Aguiar, 1985). Benitez Morera (1950) gives four Spanish localities, but his figure of male anal appendages purporting to illustrate *P. genei* are of *Onychogomphus costae*. Found throughout Africa, the Middle East and Asia Minor.

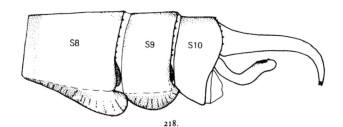

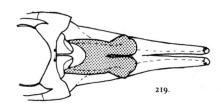

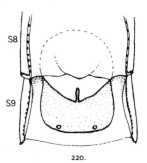

Figures 218–220 *Paragomphus genei*
(218) male terminal abdominal segments in left lateral view
(219) male anal appendages in ventral view, the inferior appendage stippled
(220) female abdominal segment S9 and vulvar scale in ventral view

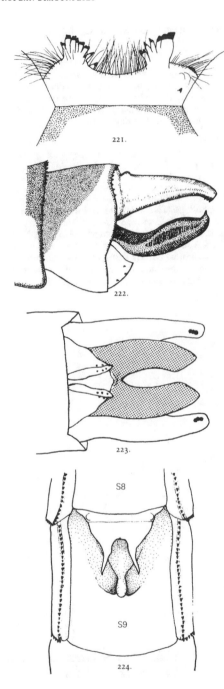

Figures 221–224 *Ophiogomphus cecilia*
(221) female occipital crown between the eyes in posterodorsal view
(222) male terminal abdominal segments in left lateral view
(223) male anal appendages in ventral view, the inferior appendage stippled
(224) female abdominal segment S9 and vulvar scale in ventral view

OPHIOGOMPHUS Sélys, 1854

An anal field in the hindwing comprising three cells, two prominences on the occiput of the female, and relatively short superior appendages in the male, distinguish this genus in the European gomphid fauna. There is only one European species of *Ophiogomphus*, a genus represented also in Asia and especially in North America.

OPHIOGOMPHUS CECILIA (Fourcroy)
Plate 20 (120,121)

Libellula cecilia Fourcroy, 1785, *Ent. Paris.* 2 : 348.
Type locality: near Paris, France.
   Aeschna serpentina Charpentier, 1825
   Aeschna spectabilis Eversmann, 1841

F – Le Gomphus serpentin, **D** – Grüne Keiljungfer

DESCRIPTION OF ADULT

A large gomphid (body length 55–60mm, the male longer than the female) with patterning similar to that of species of *Gomphus*, but with the pale ground-colour of the thorax and two anterior abdominal segments green in mature specimens. Pure yellow coloration is present on the more posterior abdominal segments and on the legs. The median dorsal yellow abdominal marks are broader than in *Gomphus*.

The female occiput carries anteriorly two rather remarkable structures; raised and diverging asymmetrical crests, each crowned by an irregular row of blunt, black teeth (fig. 221). These are contacted by the male appendages in the tandem position. Black thoracic fasciae are narrow and the median dorsal fasciae are not laterally angled in front. Pterostigmata on all wings are of approximately equal length. Male superior abdominal appendages (fig. 222) are slightly longer than S10, bluntly pointed with the apex weakly down-curved and bearing two or three small, black, ventral teeth. There is also a large, rounded and flattened tooth basidorsally on the internal face of the superior appendage. The inferior appendage (fig. 223) is much more narrowly incised than in *Gomphus*, with its two branches slightly convergent so that in dorsal view they lie inside the superior appendages, not directly below them as in *Gomphus*. The black side of the inferior appendage contrasts with the mainly yellowish superior appendages. S7–9 are moderately expanded in the male so that the abdomen appears clubbed, but there are no foliaceous ventral outgrowths. The vulvar scale (fig. 224) is deeply divided with concave sides so that the two points on each side of the apical emargination are slender and weakly divergent.

BIOLOGY

Frequents rivers with sandy beds of similar character to those inhabited by *Gomphus flavipes*. Adults rest on the ground or on vegetation, and it is usually a difficult species to approach closely.

FLIGHT PERIOD

From the end of May, but predominantly in July, and often persisting to the end of September.

DISTRIBUTION (map 62)

O. cecilia is an eastern species that is plentiful in central Asia and quite widespread through Russia and as far west as

Germany and Denmark. It extends north to the Arctic Circle in Finland, and just into Sweden, and there are a few localities in northern Italy (Pisa, and in Lombardy and Piedmont), France (Rivers Rhone and Loire, west to Maine-et-Loire), Holland (probably extinct since 1936) and Switzerland (two localities only, now extinct in the west).

225. *Onychogomphus forcipatus unguiculatus*

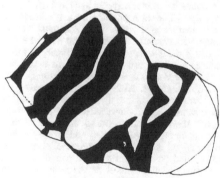

226. *O. uncatus*

227. *O. costae*

Figures 225–227 *Onychogomphus* species synthorax in left lateral view

ONYCHOGOMPHUS Sélys, 1854
Paradigma Buchecker, 1876

The development of an anal field in the hindwing is a character shared with *Ophiogomphus*, but in *Onychogomphus* the field often comprises only two large cells. Males are readily recognizable by their very long and inwardly-curved superior anal appendages, and by their deeply forked inferior appendage which has slender branches. Three species are found in Europe.

Key to species of *Onychogomphus*

1 A sandy yellow species with much-reduced dark markings, the thoracic fasciae brownish and not sharply defined (fig. 227); vertex and occiput yellowish; pterostigmata yellowish brown ...
.... *costae* (p.134)
– Yellow and black species with strongly defined thoracic fasciae; head behind eyes black with two small, yellow spots; vertex with at most a small yellow spot behind ocelli; pterostigmata very dark brown or black 2

2 Thorax with black fasciae more slender (fig. 225); fascia 2 with an indentation near the middle of its anterior margin which is strongly angled inwards at its upper end to narrowly join fascia 1; fascia 3 ventrally not as broad as the yellow stripe between fasciae 2 and 3; fascia 4 not fused with fascia 5, sometimes (subspecies *forcipatus unguiculatus*) broken medially and consisting of a ventral portion which expands dorsally to surround the metathoracic spiracle and an isolated dorsal spot that is occasionally absent (fig. 225). Vertex nearly always with some indication of a yellow spot immediately behind the posterior ocelli. Male superior anal appendage with a subapical tooth (fig. 230, p.133). Female with two small, yellow protuberances on occiput (fig. 228) .... *forcipatus* (p.132)
– Thorax with broader black fasciae (fig. 226); fascia 2 with anterior edge evenly curved and not extending dorsally as far as fascia 1; fascia 3 much broader than the very narrow yellow stripe between fasciae 2 and 3; fascia 4 with both upper and lower segments extended posteriorly to fuse with fascia 5. Vertex behind the ocelli entirely black. Male superior anal appendage without a subapical tooth (fig. 233, p.133). Female occiput without protuberances .... *uncatus* (p.134)

This key is based to a large extent upon colour characters. As these are subject to individual variation, it is recommended that identifications are checked whenever possible by using the sexual characters described under each of the species.

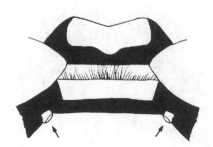

Figure 228 *Onychogomphus forcipatus*
female occiput in dorsal view, with tubercles arrowed

Map 62 – *Ophiogomphus cecilia*

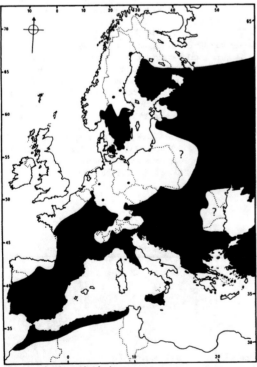

Map 63 – *Onychogomphus forcipatus*

ONYCHOGOMPHUS FORCIPATUS (Linnaeus)
Plate 20 (122,123)

Libellula forcipata Linnaeus, 1758, *Syst. Nat.* (Edn. 10) **1** : 545.

Type locality: Sweden.

    Libellula viridicincta Retzius, 1783
    Aeshna unguiculata Vander Linden, 1820
    Aeschna hamata Charpentier, 1825

F – Le Gomphus à pinces, **D** – Kleine Zangenlibelle

DESCRIPTION OF ADULT

The type species of *Onychogomphus, O. forcipatus* is a medium-sized dragonfly (body length 46–50mm) with a clearly-developed yellow-and-black pattern. Diagnostic features of the coloration of head and thorax are mentioned in the key. Female *O. forcipatus* differ from females of the two other species in having a pair of yellow occipital tubercles, small but conspicuous against the black occiput, just behind the eyes (fig. 228). The hindwing anal field is composed of two or three cells, and the anal triangle of the male is normally three-celled. In both sexes, the legs are black with only the coxae, trochanters and bases of the femora yellow. The abdomen has a median dorsal series of yellow spots, each roughly triangular in shape and partly divided by a transverse black line in the middle. In the male, these spots are present on S2–7, S9 has a medially broken yellow line posteriorly and S10 has a trilobed apical yellow margin. The superior anal appendages are brownish yellow in mature specimens. In the female, the dorsal abdominal yellow spots are broader and may be indicated on S8, S10 is black with a dorsal yellow spot, and the anal appendages are brownish yellow. The male abdomen is distinctly clubbed apically whilst that of the female is rather stout and cylindrical. Male superior appendages (figs 229,230) are sharply angled, the apical third of each being turned downwards and inwards so that it overlaps its fellow, and there is a robust, dorsal, subapical tooth. The inferior appendage is elongated and deeply but very narrowly divided and the two arms are parallel, each ending in an upturned point, and there is a pair of dorsal teeth directed laterally at the level of the base of the apical incision (fig. 230). The tenth abdominal tergite has a pair of sharply-defined longitudinal furrows. The vulvar scale (fig. 231) is about one-third of the length of the ninth segment, deeply but narrowly incised apically so that it consists of two more or less triangular lobes.

    Two forms of *O. forcipatus* are recognized and frequently accorded subspecific rank, but they differ only in colour and intergrade with one another (Lieftinck, 1966). The typical northern form, *O. f. forcipatus*, is more extensively black with thoracic fascia 4 more or less continuous. In the southern form, *O. f. unguiculatus* Vander Linden, fascia 4 is broadly interrupted in the middle (fig. 225).

BIOLOGY

Frequents rivers, and sometimes lakes, with clear water. Adults are occasionally encountered far from water. Sexually mature males perch on stones on the shore-line and react aggressively towards conspecifics, but they do not occupy a fixed territory (Kaiser, 1974c; Miller & Miller, 1985b).

Emerges from the beginning of June and may be found up to the first or second week in September.

DISTRIBUTION (map 63)

Widespread and common over much of Europe, but absent from the British Isles, Denmark and Norway (except the extreme south-east), and very rare in the Netherlands, the northern half of Germany and Poland (last recorded over a century ago (Mielewczyk, 1978)). *O. f. unguiculatus* is found in Iberia, southern France, and peninsular Italy south of the Po (type locality: Bologna), but transition from southern *O. f. unguiculatus* to northern *O. f. forcipatus* is not abrupt. In France, the line between the two forms is drawn from the Loire to Grenoble by Aguesse (1968), but Lieftinck (1966) reports the typical form from as far south as the Dordogne and Lot. Also, *unguiculatus* is recorded from Ticina, Switzerland (De Marmels & Schiess, 1978)

and, in Yugoslavia, *forcipatus* occurs in mountainous regions but *unguiculatus* in Macedonia. Both forms have also been recorded from Greece and Bulgaria, *unguiculatus* always from localities enjoying a Mediterranean climate. Two further very weakly differentiated forms, *O. f. siculus* Sélys in Sélys & Hagen, 1850, and *O. f. meridionalis* (Stein, 1863) occur in Sicily and Greece respectively. *O. f. unguiculatus* is found in North Africa but is less frequent there than *O. uncatus*.

ONYCHOGOMPHUS LEFEBVREI (Rambur, 1842)

This species is closely allied to *O. forcipatus* but is very pale with the black thoracic fasciae much reduced, the tibiae mainly yellow externally and the labrum not bordered with black. It flies in the Middle East and North Africa.

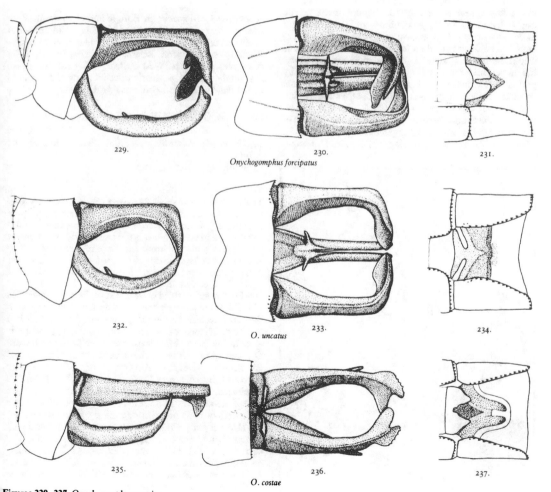

229.

230.
Onychogomphus forcipatus

231.

232.

233.
O. uncatus

234.

235.

236.
O. costae

237.

Figures 229–237 *Onychogomphus* species
(229, 232, 235) male anal appendages in left lateral view
(230, 233, 236) male anal appendages in dorsal view
(231, 234, 237) female vulvar scale and abdominal segment S9 in ventral view

ONYCHOGOMPHUS UNCATUS (Charpentier)
Plate 20 (124,125)

Aeschna uncata Charpentier, 1840, *Libell. europ. descr. depict.* : 123.

Type locality: Montpellier, France.

Gomphus occitanicus Rambur, 1842

F – Le Gomphus à crochets, **D** – Grosse Zangenlibelle

DESCRIPTION OF ADULT

In general appearance, *O. uncatus* resembles *O. forcipatus* but it is a slightly larger insect (body length 50-53mm). Structural differences between *O. uncatus* and *O. forcipatus* include the normally four-celled anal triangle of male *uncatus* (three-celled in most *forcipatus* examined (fig. 198, p.121), although this difference is unlikely to be reliable) and the absence of tubercles on the female occiput in *uncatus*. The thoracic pattern of *uncatus* (fig. 226) is similar to that of the preceding species, but the differences outlined in the key appear to be constant. Tibiae of the female usually have fine, dorsal, yellow lines, and the femora are more extensively yellow than in *forcipatus*: whilst legs are more yellow in *uncatus* than in *forcipatus*, the converse is true of the thorax. In addition, the abdomen of male *uncatus* has larger mid-dorsal yellow spots and one is usually present in a reduced form on S8. The male superior anal appendages are a clearer yellow than in *O. forcipatus*; S10 and terminal appendages in the female are also yellow. Positive identification of males is assured by examination of the male superior abdominal appendages (figs 232, 233). These resemble the corresponding structures of *O. forcipa-tus* in general shape, but they do not overlap apically to the same extent and they lack the subapical tooth which is characteristic of *forcipatus*. The tenth abdominal tergite is without linear furrows (fig. 233, cf. fig. 230). The vulvar scale of the female (fig. 234) is equally diagnostic, being reduced to a pair of inwardly-directed finger-like processes.

BIOLOGY

Larvae live in rivers with a moderate current and clean water. Unlike *O. forcipatus*, *O. uncatus* appears never to be associated with lakes, and on rivers it favours the faster-flowing, rocky reaches whereas *O. forcipatus* is more frequent by deeper, muddy stretches. Rests on sand, gravel or stones, not on plants.

FLIGHT PERIOD

From about the middle of June to early September, but most frequent in July and early August.

DISTRIBUTION (map 64)

A south-western species in Europe and common in Morocco (present also in Algeria and Tunisia). Widespread in Spain and the south-west of France, extending east to Italy. Recently recorded from Belgium (Martens, 1982a), Sicily (Carfi *et al.*, 1980) and West Germany (near Konstanz). An apparently isolated population occurs along the Rhine in Switzerland and possibly Austria (St Quentin, 1959).

ONYCHOGOMPHUS COSTAE Sélys
Plate 20 (126)

Onychogomphus costae Sélys, 1885, *Annls Soc. ent. Belg.* **29** : 146.

Type locality: Oran, Algeria.

DESCRIPTION OF ADULT

The smallest of the three European species of *Onychogomphus* (body length 43–46mm), *O. costae* differs quite considerably from *O. forcipatus* and *O. uncatus* in both coloration and structure. It is a mainly sandy yellow insect with faintly-marked, brownish thoracic fasciae (fig. 227) and an abdominal pattern of reduced brownish markings rather reminiscent of that of *Paragomphus genei*. *O. costae* is, in fact, placed in *Paragomphus* by Davies & Tobin (1985). The apical third of a male superior appendage (figs 235,236) is not strongly bent inwards and downwards as in the two preceding European species, and it terminates in a flattened, bilobed expansion. The inferior appendage is without the lateral basal teeth present in *O. forcipatus* and *O. uncatus*, and it is triangularly incised so that its two arms are strongly divergent. Aguesse (1968) notes a greater degree of divergence in European than in North African specimens. The male has a tuft of long, white hairs at the base of the sternite of S7, the hairs being three or four times the length of hairs in a corresponding position on sternite S6. Males of *forcipatus* and *uncatus* have hairs on sternite S7 that are barely twice as long as those on sternite S6. The vulvar scale (fig. 237) is most like that of *O. forcipatus*, consisting of two triangular lobes, but it is relatively shorter and the apical incision is broader. Lieftinck (1966) provides a full description of the species.

BIOLOGY

Rests on sandy ground in open country not far from streams and rivers in which it breeds.

Map 64 – *Onychogomphus uncatus*

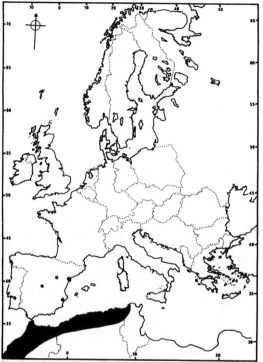

Map 65 – *Onychogomphus costae*

FLIGHT PERIOD

Mainly during May, but captured in Spain in July and August.

DISTRIBUTION (map 65)

In Europe, *O. costae* is confined to Portugal and Spain, where it has been recorded as far north as the River Ebro. However, the range of this species in the Iberian Peninsula is uncertain because of some confusion in the literature between it and *Paragomphus genei*. It is found principally in North Africa (Morocco, Algeria, Tunisia).

LINDENIA De Haan, 1826
Vanderia Kirby, 1890

This genus is easily distinguished from other European Gomphidae by the division of the discoidal cells and by the foliaceous expansions beneath the seventh and eighth abdominal segments in both sexes. There is just one included species.

LINDENIA TETRAPHYLLA (Vander Linden)
Plate 20 (127,128)
Aeshna tetraphylla Vander Linden, 1825, *Monograph. Libell. Europ. Spec.* : 32.
Type locality: Lake Averno, Naples, Italy.
Lindenia quadrifoliata Eversmann, 1854

DESCRIPTION OF ADULT

The largest European gomphid with a body length of up to about 72mm, but slenderly built. It is predominantly sandy yellow with dark brown markings. The thoracic fasciae are rather faintly indicated, much as in *Onychogomphus costae*, and the head is yellow except for the upper occiput and a band across the ocelli which are dark brown. Above the ocelli, the vertex is yellow and bituberculate. The wings of *Lindenia* have moderately large, infuscate membranules (much reduced in other European Gomphidae), and there is some brownish pigmentation at the wing bases. Pterostigmata are yellowish brown and elongated, extending over six or seven crossveins, and the anal loop of the hindwing encloses a large field of three cells. In both sexes, the hind femur carries a ventral row of five or six long, ventral spines, each considerably more than half the depth of the femur and based upon a yellow, conical protuberance; the front and middle femora are equipped with ventral combs of very small spines. The abdomen is yellowish but appears darker in the male owing to numerous minute, black spicules on the dorsal surface from the middle of S3 to the apex of S7. The female is less spinulose. S2 has a pair of submedian, broad, longitudinal, brownish marks and S3–7 are each darkened over about their apical one-sixths, the brown marks being more or less divided on the dorsal mid-line by yellowish ground colour. The last three abdominal segments are dorsally mainly brown. The paired, membranous and foliaceous ventral expansions on S7 and S8 recall those of *Paragomphus genei*, but the latter is a much smaller dragonfly with the foliations on S8 and S9 and fully developed only in the male. The foliations of *L. tetraphylla* are about half the diameter of the segments which bear them. Male superior appendages are straight and pointed, twice as long as S10, and they dwarf the forked inferior appendage which is only about one-third of their length. The vulvar scale of the female is short and forked.

BIOLOGY

A lacustrine species with an apparent preference for large lakes. Maturation occurs far from water. Observations on adult insects are provided by Dumont (1977d), and Schneider (1981) reports a mass migration, an exceptional phenomenon in Gomphidae. Adults are sometimes attracted to lights.

FLIGHT PERIOD

Specimens examined have been captured from April to

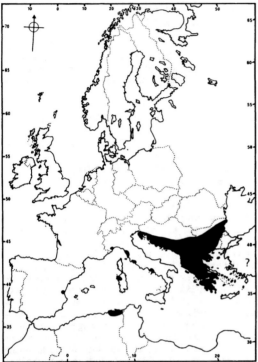

Map 66 – *Lindenia tetraphylla*

August, but most adult emergence probably occurs during June. *L. tetraphylla* is common in Yugoslavia during August (Dumont, 1977d).

DISTRIBUTION (map 66)

In Europe, recorded only from the east coast of Spain, Italy (Tuscany, Venezia Giulia, Campania, Lazio, Sardinia: Galletti, 1978), Yugoslavia (the northernmost locality is Istria) and the Balkans. It is found also in Algeria, and perhaps Tunisia, and is relatively common in parts of the Middle East (Egypt, Israel, Saudi Arabia, Kuwait, Iraq, Iran), the Caucasus and lowland areas of central Asia.

CORDULEGASTRIDAE

Cordulegastridae (often spelt Cordulegasteridae) resemble Aeshnidae in having conspicuous primary antenodal cross-veins, discoidal cells similarly-shaped in all wings and elongated in the long axes of the wings, and in having an ovipositor. The cordulegastrid ovipositor is, however, much larger than that of aeshnids and it is of different structure in the two families. Cordulegastrids differ from aeshnids in their black-and-yellow coloration (no green or blue pigmentation) and in having eyes which only just meet on top of the head (fig. 11b, p.41).

The single European genus, *Cordulegaster*, comprises large, black-and-yellow dragonflies. The male abdomen is narrow and cylindrical centrally but expanded apically and

very deep basally. Males have yellow auricles on the second abdominal segment and distinct anal angles to the hind-wings, and females have very long ovipositors. Males also have a row of blunt tubercles on the ventral edges of the middle and hind tibiae; in females, rows of spines occupy these positions.

Cordulegaster species are not very rapid fliers and they are perhaps the most easily captured of the larger European Anisoptera, a fact suggesting that their black-and-yellow coloration might be aposematic. They fly mostly near small streams of usually clean, running water, and rest on bushes and other vegetation in a vertical posture.

CORDULEGASTER Leach, 1815

Thecaphora Charpentier, 1840
Thecagaster Sélys, 1854

There are fourteen described species of *Cordulegaster*, four in North America and ten in the Palaearctic region. Taxonomy of the Palaearctic species is very confused with many subspecies and geographical races having been described, often from very little material. Although relatively few forms occur in Europe, dispute over their status and nomenclature continues, and much further study is needed before an accurate synopsis of the European fauna can be confidently presented. 'All *Cordulegaster* are evidently very susceptible to the influence of climate and elevation and, like *Calopteryx* among the Zygoptera, prone to a considerable amount of individual colour variation in almost any habitat' (Lieftinck, 1966). The views given here represent the author's understanding of the situation at the present time, but modifications may be necessary in future. Aspects of the problem, which centres mainly on the *C. boltoni* complex or 'formenkreis', are discussed by Morton (1916), Fraser (1929), St Quentin (1952, 1957), Jurzitza (1965), Lieftinck (1966), Dumont (1976), Waterston (1976), Theischinger (1979) and Balestrazzi *et al.* (1983).

The termination 'gaster' of *Cordulegaster* is feminine and specific names should agree in gender; e.g. *bidentatus* should be written *bidentata*, and *pictus* becomes *picta*.

Key to species of *Cordulegaster*

1 Superior male abdominal appendages in lateral view with a small, internal tooth ventrally about one-third of the length of the appendage from its base, the base itself expanded into a large, ventral tooth which may be obscured by the tenth tergite (figs 238,240,242). Inferior appendage transverse to almost square. First abdominal tergite with lateral yellow mark C-shaped or triangular, extending along ventral edge (figs 248–254, p.141). Anal triangle of male usually more than three-celled, often five-celled. Occipital triangle yellow and convex, rising above the inner eye margins. *C. boltoni* complex 2

– Superior appendages of male in lateral view with two submedian teeth on ventral edge (fig. 244). Inferior appendage longer than broad. First abdominal tergite with lateral yellow mark limited to posterior margin. Anal triangle of male three-celled. Occipital triangle sometimes black and flat 5

2(1) Labrum with lower edge yellow (fig. 246, p.138) 3
– Labrum with lower edge black (fig. 247) 4

3(2) Posterior of head behind lower part of eyes almost entirely whitish. Known only from Morocco .... *princeps* (p.139)
– Posterior of head black with a linear yellow mark bordering lower part of eye. Widespread .... *boltoni* (p.138)

4(2) Mesothoracic lateral yellow stripe with hind edge straight. Male superior appendages in dorsal view (fig. 241) touching basally and with apices somewhat dilated and turned inwards beneath so that outer margins appear apically convergent; in lateral view (fig. 240) with inner tooth situated close to base .. .. *picta* (p.139)

– Mesothoracic lateral yellow stripe with hind edge kinked. Male superior appendages in dorsal view (fig. 243) a little separated basally and with pointed apices not turned inwards so that outer margins apically appear subparallel; in lateral view (fig. 242) with inner tooth more distant from base. (Questionably distinct from *picta*) .... *heros* (p.140)

5(1) Occipital triangle black and flat. Frons with pronounced transverse dark mark (fig. 247) .... *bidentata* (p.143)
– Occipital triangle yellow. Frons with at most a weak blackish bar .... *insignis* (p.142)

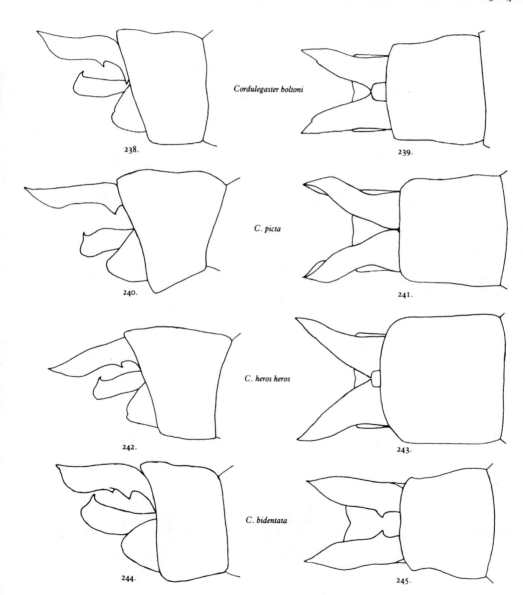

Cordulegaster boltoni

238.   239.

C. picta

240.   241.

C. heros heros

242.   243.

C. bidentata

244.   245.

Figures 238–245 *Cordulegaster* species
male abdominal appendages and segment S10 showing right lateral view (left) and dorsal view (right)

137

CORDULEGASTER BOLTONI (Donovan)
Plate 21 (129,130,132,133)

Libellula boltonii Donovan, 1807, *Nat. Hist. Brit. Ins.* **12** : 97.

Type locality: Yorkshire, England.

Aeshna annulata Latreille, 1805, *nec* Fabricius, 1798 (homonym)
Aeschna lunulata Charpentier, 1825

E – Golden-ringed Dragonfly, **F** – Le Cordulégastre annelé, **D** – Zweigestreifte Quelljungfer

DESCRIPTION OF ADULT

The typical form (*C. boltoni boltoni*) of north-western Europe is a large dragonfly (male about 72–75mm long, female including ovipositor about 80–84mm long). The face is yellow with a central black bar across the lower part of the clypeus extending on to the upper corners of the labrum (fig. 246). The upper edge of this mark usually has two points, and its lower edge is continued narrowly for a short distance down the midline of the labrum. The labrum is margined with black dorsally and partially down the sides, but its lower edge is yellow. There is a dark, transversely rhomboidal mark of variable intensity centrally on the upper part of the frons just below the crest, the vertex is black and the occipital triangle yellow. The eyes are greyish green, bordered behind their lower three-quarters by a narrow, yellow stripe. There are two broad, yellow stripes on the front of the thorax and two on each side. The posterior edge of the lateral mesothoracic stripe is slightly sinuate. Between the broad, lateral thoracic stripes, there is a narrower yellow streak which is broken in its lower part to form one, or occasionally two, isolated yellow spots. The costal veins are extensively yellow, but the remaining venation and pterostigma are black; the membranules are whitish. The male hindwing anal angle is bluntly pointed and the anal triangle usually includes four to seven (rarely as few as three) cells. Yellow abdominal markings (fig. 248, p.141) are relatively small; the largest medial marks are on S3 and S4, but even these are separated from the posterior ends of their segments by almost or fully twice their length. Viewed from the side, the first abdominal segment has a

C-shaped yellow mark running around the posterior and ventral edges of the tergite. Male superior appendages in dorsal view are basally closely-approximated, angled outwards and apically pointed (fig. 239); in lateral view, only a single, small tooth is visible proximal to the centre of the lower edge (fig. 238). The appendages are not quite as long as S10. The ovipositor has a yellow spot at its base.

C. boltoni, the type species of *Cordulegaster*, is polytypic. It varies quite a lot in coloration and the variation is to a large extent geographical. Insects from more southerly regions are generally more extensively yellow. The following key differentiates the various major named forms:

1 Frons with a distinct dark mark below its crest (fig.246). (Yellow marks on the abdomen restricted (fig. 248, p.141), median spots on S3 being separated by at least twice their length from the apex of the segment; yellow marks on S9,10 very small or absent (Plate 21 (129)). Male superior appendages slightly shorter than S10 (fig. 238)) .... *boltoni boltoni*

– Frons without or with but a vestige of transverse dark mark 2

2 Yellow abdominal markings (fig. 251, p.141) smaller than in *b. boltoni*. (Male superior appendages as long as last segment) .... *boltoni trinacriae* Waterston, 1976
(type locality: Biviere di Cesaro, Mt Etna, Sicily)

– Yellow abdominal markings larger than in *b. boltoni* (figs 249,250, p.141) 3

3 Median yellow marks on S7 transversely linear, S9,10 with small yellow marks (Plate 21 (132); fig. 249, p.141) .... *boltoni immaculifrons* Sélys & Hagen, 1850
(type locality: Montpellier, France)

– Median yellow mark on S7 expanded dorsally, S9,10 with relatively large yellow marks (Plate 21 (133); fig. 250, p.141) ... . *boltoni algirica* Morton, 1916
(type locality: Sebdou, Algeria)

BIOLOGY

C. boltoni breeds in streams, brooks and drainage channels, and since a moderate rate of water flow is required by the larvae, it is most common in hill country, ascending to an altitude of about 1500m. The female oviposits in mud or sand whilst in hovering flight, stabbing its long ovipositor into the stream bed where the water is shallow. Males patrol

Figures 246, 247
Heads of
Cordulegaster species
in frontal view to
show extent of black
markings

(*C. b. boltoni* from
Cornwall, England;
C. bidentata from Hautes
Pyrénées, France)

occipital triangle

frons

labrum

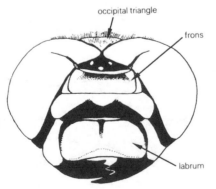

246. *C. boltoni boltoni*

247. *C. bidentata*

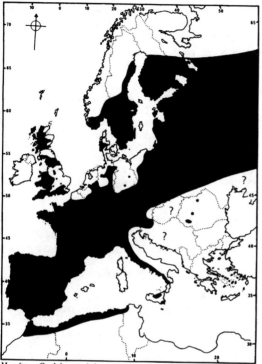

Map 67 – *Cordulegaster boltoni*

stretches of streams, often very small ones, along which they fly, back and forth, just above the water surface. They are aggressive towards other males and will attempt to mate with females visiting the stream, but they do not appear to be tied to particular sites and are therefore not territorial in the strict sense (Kaiser, 1982). Feeding dragonflies may be seen hawking along hedgerows or in woodland glades.

FLIGHT PERIOD

End of May to the beginning of September, but most frequently seen in July.

DISTRIBUTION (map 67)

C. b. boltoni is widespread in Europe from Portugal, northern Spain, France, the British Isles (except Ireland for which there are only a few old and dubious records) and southern Scandinavia to most of central Europe. It is replaced, with considerable intergrading, by *C. b. immaculifrons* on the Mediterranean coast of France, northern Italy (Liguria: Balestrazzi *et al.*, 1983) and in central and southern Spain. *C. b. trinacriae* (referred to *C. picta*? by Bucciarelli, 1977; Balestrazzi *et al.*, 1982, 1983; and Galletti & Pavesi, 1985), is known from Sicily and mainland Italy (Calabria, Lazio). *C. b. algirica* is reported from Tunisia, Algeria, Morocco and from southern Spain (Seville, Andalusia, Malaga (Lieftinck, 1966)).

CORDULEGASTER PRINCEPS Morton, 1916
Plate 21 (134)

This is a second species in the *C. boltoni* complex that is represented in North Africa. Originally thought to be from the Caucasus, Lieftinck (1966) showed its real provenance to be the Middle and High Atlas Mountains in Morocco. The central segments of the abdomen have broad yellow median marks which occupy about half of S4–6, but S9 and S10 have only vestigial yellow marks (fig. 254, p.141), the back of the head is mostly pale, and the superior male appendages are as long as S10. *C. princeps* is typically some 2–3mm longer than *C. boltoni*, but it is of more slender build.

CORDULEGASTER MZYMTAE Bartenef, 1929

This species has been assigned to the *C. boltoni* group (Dumont & Schneider, 1984) although this view has been recently questioned. It was described from the western Caucasus (Georgian SSR) and found subsequently in the Pontic Alps of northern Turkey, west to about 39°E. It is a small, dark species with the labrum broadly black-bordered; male appendages are similar to those of *C. boltoni*, yellow dorsal abdominal spots are very small and absent from S9 and S10, and the male anal triangle is usually three-celled.

CORDULEGASTER PICTA Sélys
Plate 21 (135)

Cordulegaster pictus Sélys, 1854, *Bull. Acad. roy. Belg.* **21** : 87.

Type locality: unknown.

> *Cordulegaster charpentieri* Sélys, 1887, *sensu* Morton, 1916; Fraser, 1929; Waterston, 1976; *nec* Kolenati, 1846

DESCRIPTION OF ADULT

Resembles *C. boltoni* in coloration but the median yellow marks on the central abdominal segments (fig. 252, p.141) are as extensive as in *C. boltoni immaculifrons* whilst other abdominal yellow markings are much as in *C. b. boltoni*, the lateral yellow mesothoracic stripe has a very straight posterior edge, the occipital triangle is partly blackish, the yellowish white mark behind the lower part of the eyes is broader than in *C. boltoni*, the labrum has a narrowly black lower margin and there is usually a blackish mark below the crest of the frons. The second abdominal segment of the male has a broad basal yellow band which is incised medially on its posterior margin, and in the female the yellow ring on the second segment extends forwards mid-dorsally to the base of the segment. The male superior anal appendages of *picta* are distinctive. In dorsal view (fig. 241) they are as long as S10, practically touching at the bases but diverging strongly with their apices rather dilated and turned inwards beneath so that their outer edges appear distinctly sinuate. In lateral view (fig. 240) there is a robust tooth near the base of the superior appendage. The ovipositor has a basal yellow mark as in *C. boltoni*.

More westerly populations of this species may be referable to the subspecies *C. p. intermedia* Sélys & Hagen, 1857, the lectotype designated by Theischinger (1979) originating from north Italy (Livorno). *C. p. intermedia* is larger and darker than *C. p. picta* and it has a pronounced dark mark

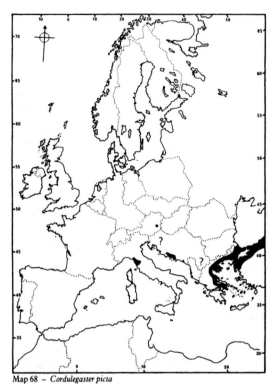

Map 68 – *Cordulegaster picta*

Map 69 – *Cordulegaster heros*

on the upper part of the frons. Balestrazzi *et al.* (1983) consider *intermedia* to be probably the same as *trinacriae* (here treated as a subspecies or form of *C. boltoni*).

BIOLOGY AND FLIGHT PERIOD

Probably similar to *C. boltoni*.

DISTRIBUTION (map 68)

The southern Caucasus, Black Sea coast of Turkey, Greece (Lesbos, Thessaloniki, Pelion Mt, Athens), Yugoslavia (Dalmatia), Italy (Livorno, Florence). The range will be broader if ultimately *C. b. trinacriae* (Sicily) and *C. heros* (Austria) are shown to be conspecific with *C. picta*. Galletti & Pavesi (1985) note the occurrence together in Italy (Lazio) of *boltoni* and *trinacriae*, arguing against their conspecificity and for the assignment of *trinacriae* to *C. picta*.

CORDULEGASTER HEROS Theischinger
Plate 21 (136)

Cordulegaster heros Theischinger, 1979, *Odonatologica* 8 : 29.
Type locality: St Andrä, Niederösterreich, Austria.

DESCRIPTION OF ADULT

This is the largest form in the *C. boltoni* species-complex, indeed the largest of all European Odonata. Males measure 78–84mm and females 93–97mm in overall body length. The general coloration of the body (fig. 253) is similar to that of *C. picta*, but the lateral yellow mesothoracic stripe has a distinct kink on its posterior edge, the dorsal face of the occiput is blackish, all edges of the labrum are black, and the pale marks on the back of the head behind the eyes are broad. The yellow band on the second abdominal segment of the male has its posterior border medially incised as in *C. picta*. The superior appendages of the male are robust, shorter than the last abdominal segment, in dorsal view (fig. 243) basally approximated but less closely so than in *C. picta*, thence strongly diverging to the pointed apices which are not turned inwards beneath; in lateral view (fig. 242) the inner ventral tooth is situated slightly more distad than in *picta*. Further study is needed to establish whether or not these differences indicate a specific separation of *C. heros* and *C. picta*.

Following the description of *C. heros heros*, Theischinger distinguishes the subspecies *C. h. pelionensis* Theischinger, 1979. This has a darker and more sharply-defined mark on the frons below the crest, and rather more slender male appendages, than the nominate form.

BIOLOGY

C. h. pelionensis was found at moderate altitude by a sandy, shallow, very clean stream in company with *C. bidentata*. (*C. picta* has since been found in the same area.)

FLIGHT PERIOD

Late June to early August.

DISTRIBUTION (map 69)

Austria (Niederösterreich, Steiermark), Yugoslavia (Dalmatia, ?Serbia, ?Slavonia, ?Macedonia) and ?Romania (Banat) are cited for *C. h. heros*. *C. h. pelionensis* is known only from the type locality in Greece (Mt Pelion between Makryrrachi and Anelion).

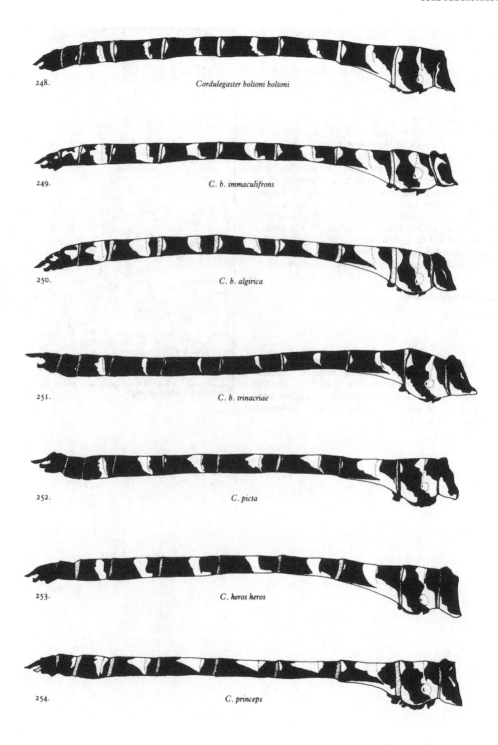

248. *Cordulegaster boltoni boltoni*

249. *C. b. immaculifrons*

250. *C. b. algirica*

251. *C. b. trinacriae*

252. *C. picta*

253. *C. heros heros*

254. *C. princeps*

Figures 248–254 *Cordulegaster boltoni* species complex
male abdomens in right lateral view (Fig. 251 is of the holotype; Fig. 252 is of the neotype of *charpentieri*)

CORDULEGASTER INSIGNIS Schneider

Cordulegaster insignis Schneider, 1845, *Stettin. ent. Ztg.* **6** : 114.
Type locality: Gelemish (Kellemisch), Anatolia, Turkey.
  Aeschna charpentieri Kolenati, 1846 (Dumont, 1976)

DESCRIPTION OF ADULT

Most authors consider *C. insignis* to be allied to *C. bidentata*, but Dumont & Schneider (1984) place it in the *C. boltoni* species complex. A number of subspecies have been described from Asia and one from eastern Europe. The European subspecies, *C. i. montandoni* St Quentin, 1971, was found at Comana, Romania, and it is considered by Dumont (1976) to be the same as *C. i. charpentieri* Kolenati, 1846. It is a darker insect than many of the Asian forms which often have abdomens more yellow than black, but it may be distinguished from *C. bidentata* by the smaller transverse dark bar on the frons, the yellow occipital triangle and a quadrifid apical margin to the yellow band on S7. The male hindwing anal triangle is usually three-celled.

BIOLOGY

Frequents small brooks.

FLIGHT PERIOD

June to August.

DISTRIBUTION (map 70)

C. i. charpentieri ranges from the Caucasus along the Pontic Alps to Bulgaria (Strouma Valley) and Romania (Comana: Dumont, 1976). It has been found on Naxos and may occur also in other parts of Greece. Other subspecies are found in Turkey, Lebanon, Iran, Armenia, Georgia, Turkestan, Uzbekistan and Afghanistan.

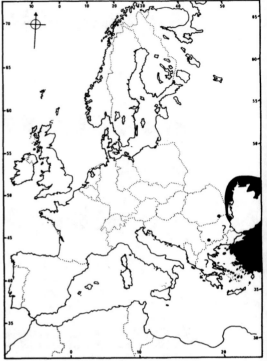

Map 70 – *Cordulegaster insignis*

CORDULEGASTER BIDENTATA Sélys
Plate 21 (131)

Cordulegaster bidentatus Sélys, 1843, *Annls Soc. ent. Fr.* **1** (2nd ser.) : 109.

Type locality: Colonster near Liège, Belgium.

F – Le Cordulégastre bidenté, **D** – Gestreifte Quelljungfer

DESCRIPTION OF ADULT

Although several species and forms in this confusing genus have been mentioned in the foregoing pages, it is only with this species and *C. boltoni* that an entomologist operating in western Europe will be concerned. The yellow markings of *C. bidentata* are more restricted than those of *C. boltoni*.

The occipital triangle is black, the labrum has a brownish ventral border, and there is a strong dark mark on the frons below the crest (fig. 247). The head behind the lower part of the eyes is almost entirely pale. The narrow yellow stripe between the two broad ones on the side of the thorax is obliterated ventrally. Costal veins are less marked with yellow than in *boltoni* and the anal triangle in the male usually comprises only three cells. The first abdominal segment is yellow only on its posterior margin and the ovipositor lacks a basal yellow spot. The superior appendage of the male differs from that of the *boltoni* complex in having a clearly bidentate lower margin (fig. 244). A form from Sicily with more extensive yellow markings than the nominal subspecies is named *C. b. sicilica* Fraser, 1929 (see Galletti & Pavesi, 1985 for a recent account). Although superficially very like species in the *boltoni* complex, Dumont & Schneider (1984) consider *C. bidentata* to be sufficiently different to merit subgeneric separation.

BIOLOGY

Associated with small, shallow streams in hilly or mountainous country, from moderate altitude up to about 1700m. Sélys & Hagen (1850) describe the habitat in Belgium as dry woods in mountains, where *C. bidentata* flies slowly at the edge of tracks in company with the local butterfly, *Euphydryas maturna*.

FLIGHT PERIOD

Appears at the end of June in northern Europe but in May in the south, and is over by mid-August.

DISTRIBUTION (map 71)

Mountainous regions in Spain (Pyrenees), France (Pyrenees, Vosges, Jura, Basses Alpes), Belgium (Ardennes), Germany (e.g. Black Forest, formerly in the Erz Gebirge but not recorded there since 1912), Czechoslovakia, Poland, Hungary, Yugoslavia (Herzegovina), Romania, Greece, and many parts of Switzerland, Austria and Italy (including Sicily, *C. b. sicilica*). Known from several parts of western Asia. *C. bidentata* is absent from Britain, Scandinavia and North Africa.

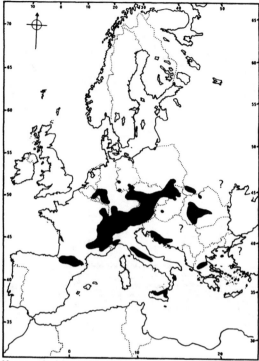

Map 71 – *Cordulegaster bidentata*

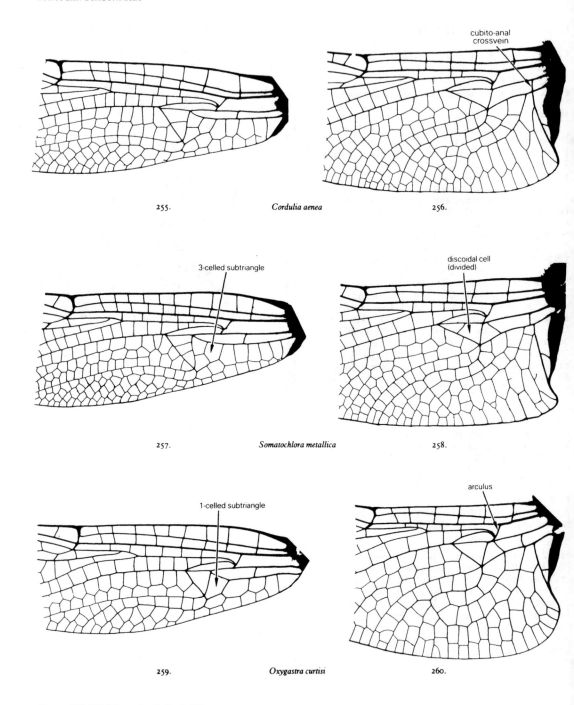

255. *Cordulia aenea* 256.

257. *Somatochlora metallica* 258.

259. *Oxygastra curtisi* 260.

Figures 255–260 Wings of male Corduliidae
bases of left forewings (left) and left hindwings (right)

CORDULIIDAE

Corduliids are medium-sized to large dragonflies sharing some characters with Cordulegastridae and, more especially, Libellulidae. They are metallic green or greenish black (except *Epitheca*), usually marked with yellow and with a dense coat of fine, pale hair on the thorax. The legs are relatively long or very long. Males have angled anal margins to the hindwings (only weakly so in *Epitheca*) and auricles on the second abdominal segment, as in Cordulegastridae, but the ovipositor is short and the triangular discoidal cells are dissimilarly-shaped in fore- and hindwings as in Libellulidae. Also as in the latter family, the hindwing sometimes has a foot-shaped anal loop. There is a small but distinct prominence near the centre of the posterior eye margin (fig. 34, p.53), and the eyes are contiguous on top of the head. The tibiae of Corduliidae are sexually dimorphic having a thin, ventral keel in males only. In *Cordulia*, *Epitheca* and *Oxygastra*, this keel is on all tibiae, but in *Somatochlora* and *Macromia* only fore and hind tibiae are keeled.

Five genera and nine species of Corduliidae are represented in Europe, but only *Somatochlora* includes more than a single species.

Key to genera of Corduliidae

1 Hindwing with large, dark brown, basal marks. Thorax yellowish brown, not metallic. Discoidal cell of forewing normally comprising three cells .... *Epitheca* (p.153)
 – Wings clear or with small amber or yellow basal marks. Thorax shining metallic green. Discoidal cell of forewing undivided or comprising usually no more than two cells ... 2

2(1) Forewing with thirteen to fifteen antenodal crossveins. Length of body about 70mm. Abdomen narrow, black and conspicuously marked with yellow .... *Macromia* (p.155)
 – Forewing with nine or fewer antenodal crossveins. Body length seldom exceeding 50mm. Abdomen relatively broader, greenish to black with yellow marks absent or relatively inconspicuous 3

3(2) Hindwing with a single cubito-anal crossvein and discoidal cell undivided with its basal edge slightly proximal to the arculus (fig. 256). Hindwing strongly tinted yellow-orange basally up to the level of the cubito-anal crossvein and entirely filling the anal cell of the male .... *Cordulia* (p.145)
 – Hindwing with two cubito-anal crossveins (figs 258,260), discoidal cell divided (fig. 258) or undivided (fig. 260) with its basal edge continuous with (fig. 258) or slightly distal to (fig. 260) the arculus (if slightly proximal to the arculus, then abdomen or frons are marked with lateral yellow spots, absent in the alternative). Hindwing basally usually at most weakly yellowish 4

4(3) Discoidal cell of hindwing with basal edge displaced distad of arculus (fig. 260); all discoidal cells usually undivided; forewing subtriangle (adjacent to basal edge of discoidal cell) of one cell (fig. 259). Abdomen with mid-dorsal yellow marks on S1-7 and S10. Frons without lateral yellow spots. Hindwing membranule white .... *Oxygastra* (p.154)
 – Discoidal cell of hindwing with basal edge continuous with or (*S. flavomaculata* and occasionally *S. metallica*) slightly basad to the arculus (fig. 258); discoidal cell of forewing with one (exceptionally two, fig. 257) and of hindwing with one

(none in *S. arctica*) crossvein; forewing subtriangle usually three-celled (fig. 257). Abdomen with S4–10 unmarked or with lateral yellow spots. Frons with lateral yellow spots. Hindwing membranule apically infuscate ....
.... *Somatochlora* (p.148)

CORDULIA Leach, 1815

Chlorosoma Charpentier, 1840

Only two species known, the type species *C. aenea* and a North American species.

CORDULIA AENEA (Linnaeus)   Plate 22 (137)

Libellula aenea Linnaeus, 1758, *partim*, *Syst. Nat.* (Edn. 10) **1** : 544.
Type locality: southern Sweden.
 Cordulia linaenea Fraser, 1956
E – Downy Emerald, **F** – La Cordulie bronzée,
D – Gemeine Smaragdlibelle

The name that should be applied to this species is complicated by the fact that Linnaeus in *Systema Naturae* uses *aenea* for both it and *Somatochlora flavomaculata*. Later (1761), he distinguished the two species and applied *aenea* to *S. flavomaculata* and *aenea* var. β to the present species (Fraser 1956). Thus use of the name *aenea* in the present context does not accord with Linnaeus' intention, but it has been so generally employed in this sense that any change would create considerable confusion.

DESCRIPTION OF ADULT

C. aenea bears a general resemblance to species of *Somatochlora*, but it may easily be separated from them by the wing characters given in the key to genera and by the absence of lateral yellow spots on the frons. The thorax is shining metallic green with a dense coat of buff-coloured hairs, and the legs are black. The membranule of the hindwing is white proximally but dark fuscous in its apical two-thirds. The abdomen is greenish black with bronze reflections and yellow (male) to whitish (female) marks on the second and third segments laterally and ventrally. In the male it is clubbed (*cordulē* (Greek), a club), narrow at S3 then expanding posteriorly to S8. Male anal appendages (fig. 261, p.146) are very characteristic with the inferior appendage deeply divided, each branch carrying a subapical tooth, and the superior appendages apically blunt, slightly divergent, and having only small ventral teeth. The superior appendages are not quite as long as the combined length of S9+10, and only slightly longer than the inferior appendage. The female vulvar scale (fig. 275, p.153) is inconspicuous, short and deeply incised. Body length 48–50mm.

BIOLOGY

Males patrol the margins of tree-sheltered lakes and canals, or of small pools in mountain areas, flying rapidly close to the water surface. Away from water, the species frequents woodland edges and broad rides, often hawking at canopy level. Females oviposit unaccompanied, usually selecting open water some 10–40cm deep near emergent vegetation. The water surface is rapidly tapped with the abdomen and

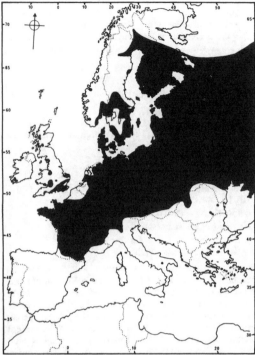

Map 72 – *Cordulia aenea*

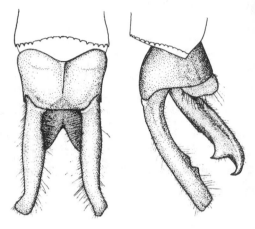

261. *Cordulia aenea*

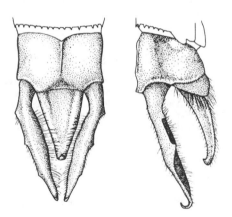

262. *Somatochlora metallica*

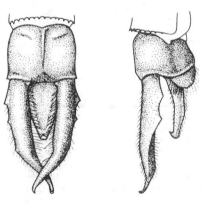

263. *S. alpestris*

at each blow about ten eggs are released which sink to the bottom or become caught on plants. Ubukata (1975) describes the life history and behaviour of the Japanese subspecies.

FLIGHT PERIOD

Mid-May to early August in northern Europe where this is one of the earliest dragonflies. It emerges in April in more southerly parts of its range.

DISTRIBUTION (map 72)

Northern and central Europe eastwards through Turkey (Schneider, 1986b) and Siberia to the Amur and Japan (*C. a. amurensis* Sélys, 1887). Absent from Spain and Portugal, and in Italy restricted to alpine localities in the north and near Potenza in Basilicata (Bucciarelli, 1972). A lowland insect in Britain, where it extends locally as far north as central Scotland and reported recently (Goyvaerts, 1979) from Ireland (Co. Cork), but mainly found in mountainous parts of southern Europe.

Figures 261–269 Male Corduliidae
male abdominal appendages and segment S10 in dorsal view (left) and right lateral view (right) (Fig. 268 shows *Epitheca* in dorsal view only)

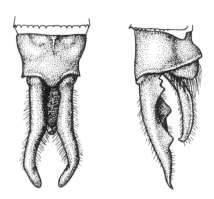

264. *S. arctica*

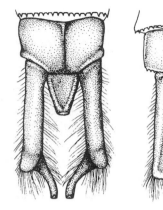

265. *S. sahlbergi*

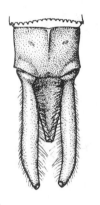

266. *S. flavomaculata*

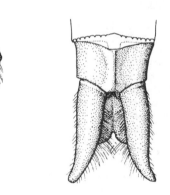

267. *Oxygastra curtisi*

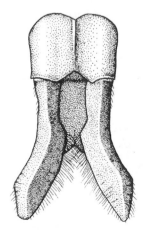

268. *Epitheca bimaculata*

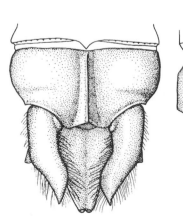

269. *Macromia splendens*

SOMATOCHLORA Sélys, 1871

Medium-sized dragonflies with metallic green heads and thoraxes, and black to greenish black abdomens. The abdomen does not expand posterior to S5. The frons has a yellow spot on each side, connected in *S. metallica* by a yellow bar across the lower margin of the frons. The wings are mainly clear but there is often a yellowish to amber basal mark (paler and less extensive than in *Cordulia*) and, in older females, a general yellowish tinge most pronounced in the costal area. The hindwing membranule is whitish basally and infuscate distally. There may be small yellow marks (larger in females) at the base of the abdomen and, in *S. flavomaculata*, on the sides of the more posterior abdominal segments as well. Male superior appendages are long, pointed, with ventral teeth and, except in *S. arctica*, upcurled apices. The inferior appendage is entire, edentate, slightly concave dorsally, and at most about two-thirds as long as a superior appendage.

There are forty described species which are distributed across the northern hemisphere from North America, Europe and Asia to Japan.

Key to species of *Somatochlora*

The shapes of male superior abdominal appendages and female vulvar scales are diagnostic for every European species of *Somatochlora*: these parts are specifically constant in form and should always be examined to confirm identifications.

1   Forewing with two cubito-anal crossveins .. *alpestris* (p.150)
–   Forewing with one cubito-anal crossvein (fig. 257, p.144) .. 2

2(1) S5–8 with lateral yellow spots .... *flavomaculata* (p.152)
–   S5–8 without lateral yellow spots 3

3(2) Frons with a complete yellow bar along its lower margin connecting two lateral yellow spots .... *metallica* (p.149)
–   Frons with two isolated lateral yellow spots 4

4(3) Hindwing discoidal cell usually undivided .... *arctica* (p.151)
–   Hindwing discoidal cell traversed by a crossvein (cf. fig. 258) .... *sahlbergi* (p.151)

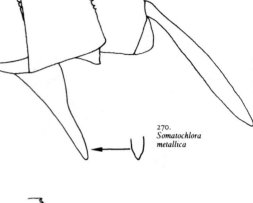

270.
Somatochlora metallica

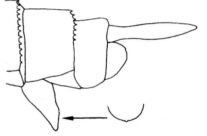

271. *S. alpestris*

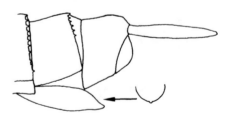

272. *S. arctica*

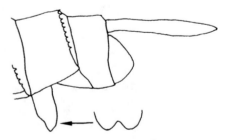

273. *S. flavomaculata*

Figures 270–273 Female *Somatochlora* species apex of abdomen in left lateral view and apex of the vulvar scale in posterior view

SOMATOCHLORA METALLICA
(Vander Linden)   Plate 22 (138,139)

Libellula metallica Vander Linden, 1825, *Monograph.*
  Libell. Europ. Spec. : 18.
Type locality: Brussels, Belgium.
E – Brilliant Emerald, **F** – La Cordulie métallique,
D – Glänzende Smaragdlibelle

DESCRIPTION OF ADULT

S. metallica, type species of *Somatochlora*, is more brilliantly green than its European congeners. It is relatively large (body length usually 48mm or more) with a virtually complete transverse yellow bar on the lower part of the frons connecting the two lateral yellow spots. Small yellowish spots, larger in the female than in the male, are present laterally on the second and third abdominal segments, and the intersegmental division between these segments, as well as spots above the male accessory genitalia, are bright yellow. In the typical form, the thorax laterally is without yellow markings, but in the quite well-defined subspecies *meridionalis* Nielsen, 1935a, there is a pyriform yellow spot on each side. This latter subspecies has clear wing bases (pale amber in the typical form) and a very dark pterostigma.

  Male superior anal appendages (fig. 262) are as long as the combined lengths of S9+10, in dorsal view basally divergent then converging with two ventral teeth, the distal tooth in the posterior half of the appendage. The vulvar scale (fig. 270) is perhaps the most distinctive feature of female *S. metallica*, being very prominent, almost as long as the last two abdominal segments, in profile beak-shaped, acutely pointed, and directed almost vertically downwards.

BIOLOGY

A lowland insect in northern Europe, flying near slow-flowing water courses or the margins of lakes, generally over open water but sometimes hunting in light woodland where it may be found far from water. Further south in its range it ascends to altitudes in excess of 1000m, and in the Alps it may breed in the same montane pools as *S. alpestris*. As in all Corduliidae, females oviposit unaccompanied by males. *S. metallica* lays its eggs in shallow water over mud or in mats of floating vegetation very close to the bank, hovering over the selected place and beating a very rapid tattoo on the surface with its elongated vulvar scale. The last two abdominal segments and the anal appendages are held upwards, at right angles to the rest of the abdomen, during oviposition and the eggs, which emerge from between the vulvar scale and the ventral abdominal surface, have unrestricted access to the water.

FLIGHT PERIOD

From late in May to mid-August, but most plentiful during July.

DISTRIBUTION (maps 73A,73B)

Northern and central Europe to east of the Volga, and south-east to Greece and Asia Minor. Very local in Scotland and northern Scandinavia, not uncommon locally in south-eastern England and widespread in central Europe although mainly at higher altitudes (as the typical subspecies, *S. m. metallica*) in the more southerly parts of its range. *S. m. meridionalis* (type locality: Gerano near Rome, Italy) is found in peninsular Italy (both subspecies recorded

Map 73A – *Somatochlora metallica metallica*

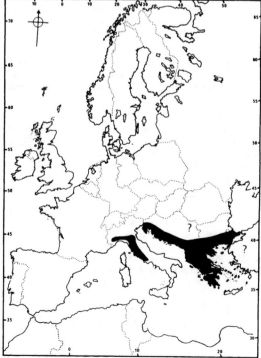

Map 73B – *Somatochlora m. meridionalis*

near Alessandria, Piedmont by Balestrazzi *et al.*, 1977), Yugoslavia, Greece, European Turkey and Asia Minor (Schneider, 1986a).

SOMATOCHLORA ALPESTRIS (Sélys)
Plate 22 (140,141)

Cordulia alpestris Sélys, 1840, *Monograph. Libell. Europ.* : 65.

Type locality: Grosse Scheidegg, Bernese Oberland, Switzerland.

E – Alpine Emerald, **F** – La Cordulie des Alpes,
D – Alpen-Smaragdlibelle

DESCRIPTION OF ADULT

A rather small (body length usually about 45mm), dark greenish species with almost black abdomen and pale coloration limited to a whitish or yellow ring at the apex of the second abdominal segment. Venation towards the wing apex is fine, there being generally eight or nine cells along the margin between the distal ends of veins R_2 and R_3, and no other European species of *Somatochlora* normally has two cubito-anal crossveins in the forewing. I have a specimen of *S. metallica* with two such veins but the more distad of these is just proximal to the discoidal cell whereas in *S. alpestris* it seems always to be proximal to the arculus. The superior abdominal appendages of the male (fig. 263) are rather less attenuated than those of *S. metallica* and, in dorsal view, their outer edges are more strongly double-angled. The distal ventral tooth is in the basal half of the appendage. The vulvar scale of the female (fig. 271) is robust, triangular and broadly rounded, in profile about as long as the dorsal edge of S9 only, with its posterior border at right angles to the long axis of the body.

BIOLOGY

Frequents small lakes, tarns and *Sphagnum* bogs, usually flying close to the water surface near the margins. In the Alps, its altitudinal range is approximately 400–2400m, and it is found mostly above 1200m.

FLIGHT PERIOD

From mid-June to late September, but the majority of records are for July. St Quentin (1938) finds the flight period to vary with altitude: 600-900m, 15 June to 30 July; 900-1600m, 1 July to 15 August; 1600-2100m, 30 July to 15 September.

DISTRIBUTION (map 74)

S. alpestris has a boreo-alpine distribution and is found in northern Europe and in the mountains of central Europe. Outside Europe there are scattered records over a wide area of Siberia (from Yenisey 68°N, Transbaikal) to Japan (Hokkaido). It is widespread in Scandinavia where it extends far north of the Arctic Circle, and there is a solitary and very doubtful record for Scotland (Inverness, 1926 – see Fraser, 1947). In central Europe (Ander, 1950) it occurs from eastern France (Vosges, Dauphine), Germany (Black Forest, Harz Mountains, Thüringer Wald, etc.), throughout the alpine regions of Switzerland, northern Italy and Austria, to the Tatra and Carpathian Mountains in Czechoslovakia, Hungary and Poland. Two localities are cited for Romania (Ienestea, 1956 – see Schneider, 1972).

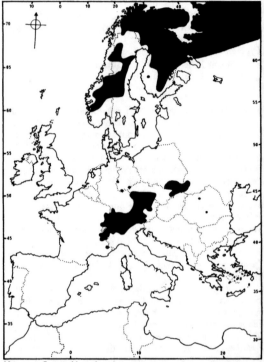

Map 74 – *Somatochlora alpestris*

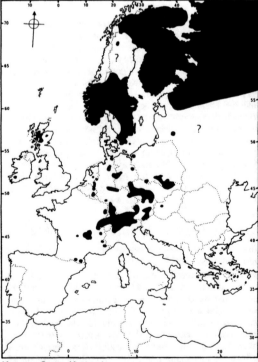

Map 75 – *Somatochlora arctica*

SOMATOCHLORA ARCTICA (Zetterstedt)
Plate 22 (142,143)

Aeschna arctica Zetterstedt, 1840, *Insecta Lapp. descr.* : 1041.

Type locality: near Alte, Finmark, Norway.

Cordulia subalpina Sélys, 1840

E – Northern Emerald, **F** – La Cordulie arctique, **D** – Arktische Smaragdlibelle

DESCRIPTION OF ADULT

Like the preceding, this is a rather small, dark greenish-black species. It has a single cubito-anal crossvein in the forewing, which distinguishes it from *S. alpestris*, and the hindwing discoidal cell is usually undivided, in contrast to the condition in all other European species of the genus in which it is normally crossed by a single vein. The apical wing venation of *S. arctica* is less fine than in either *S. alpestris* or *S. sahlbergi*, with only six or seven cells on the wing margin between the ends of veins R_2 and R_3. Pale coloration on the abdomen is more extensive than in *S. alpestris*, the female having two large, yellow-orange spots dorsally on S3, behind the yellow ring at the apex of the second segment. In addition, there are two small, yellow spots anteriorly on the dorsal surface of S2. These pale marks are also present in males although smaller. In dorsal view, the superior male anal appendages (fig. 264) are calliper-like, parallel basally then curving outwards before converging apically. They are strongly bidentate beneath. The female vulvar scale (fig. 272) is rather longer than the dorsal surface of the ninth abdominal segment, projecting at an angle of only about 30° to the long axis of the abdomen. Its apex is obtusely pointed.

BIOLOGY

Larvae develop in moorland bogs and pools, the females frequently ovipositing in flooded *Sphagnum*. Adults fly low over wet meadows, in open areas in pine forests or on moors with scattered trees. *S. arctica* occurs almost at sea level in Scotland and northern Scandinavia, but it is a montane insect further south in its range. Generally, however, it occurs at lower altitudes than *S. alpestris*. The habitats of *S. arctica* and *S. alpestris* in northern Italy are described by Morton (1928).

FLIGHT PERIOD

End of May to mid-September, but mostly in July. In Norway, from the second week in June to the first week in August (Midttun, 1977).

DISTRIBUTION (map 75)

From Ireland (Killarney), where it is the only *Somatochlora* recorded, Scotland, all Scandinavia, very rare in the Netherlands and Belgium (Ardennes), France (Vosges, Jura, Alpes (Degrange & Seassau, 1970) south-west to the western Massif Central in Haute Vienne (Dommanget, 1984) and Auvergne (Francez, 1985)), widespread in the southern uplands and northern heathlands of Germany and the mountains of central Europe (Switzerland, Austria, northern Italy, Poland, Czechoslovakia, Hungary) eastwards across Siberia to Kamchatka and Japan. A recent report (Boudot *et al.*, 1987) of its occurrence in the eastern Pyrenees (Aude, Ariège) significantly extends its range to the south-west.

SOMATOCHLORA SAHLBERGI Trybõm
Plate 22 (144)

Somatochlora sahlbergi Trybõm, 1889, *Bih. K. svenska VetenskAkad. Handl.* **15** : 7.

Type locality: Yenisey, Siberia, USSR.

Somatochlora theeli Trybõm, 1889
Somatochlora walkeri Kennedy, 1917

DESCRIPTION OF ADULT

S. sahlbergi superficially resembles *S. alpestris*. It has a dark green, metallic thorax with some bronze reflections, and a blackish abdomen. Pale coloration on the frons is restricted to two small, lateral yellow spots, and on the abdomen to the apical ring on the second segment and the ventral surface of S3. The pterostigma is brown, rather paler than in other species, and the body is provided with a dense and relatively long coat of hair. It is a slightly larger insect than *S. alpestris* (body length about 48mm) but resembles it in having a rather stout abdomen. The apical venation of the wings is dense, as in *S. alpestris*, but the forewing has only one cubito-anal vein. The male superior abdominal appendages (fig. 265) are robust and strangely contorted. Each curves outwards in its basal two-thirds, then bends abruptly downwards and inwards, narrowing after the bend and terminating in an upcurled point. There is a single ventral tooth at the base. The inferior appendage is short, hardly one-third the length of a superior appendage. The female vulvar scale is 'somewhat rounded, tapering to a point and notched at the apex. It is hardly longer than half of the ninth segment' (Trybõm, 1889).

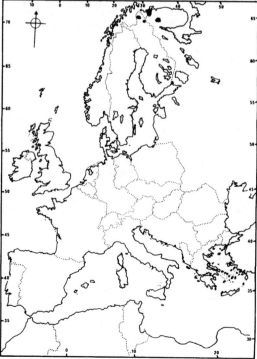

Map 76 – *Somatochlora sahlbergi*

BIOLOGY

Inhabits northern moorlands at the edge of the tundra. Larvae require cold, quite deep, standing water in ponds and bog pools in which aquatic mosses are dominant; they are described by Cannings & Cannings (1985).

FLIGHT PERIOD

From about the third week of June until August.

DISTRIBUTION (map 76)

A circumboreal species recorded from Alaska, the North-west Territories of Canada, northern Siberia, Pechenga district and Lake Imandra in north-west Russia, and the Lake Inari region of Finland (Hämäläinen, 1984c; Sahlén, 1987). All western Palaearctic records are from between 67°40′N and 69°48′N (Valle, 1952), but further east the species reaches more southerly latitudes and there is a possibly isolated population in the flood plain of the rivers Ob and Ket in the Tomsk district of Siberia.

SOMATOCHLORA FLAVOMACULATA
(Vander Linden)   Plate 22 (145,146)

Libellula flavomaculata Vander Linden, 1825, *Monograph. Libell. Europ. Spec.* : 19.
Type locality: Gheel, Belgium.

   Libellula aenea Linnaeus, 1758 (*partim*)

E – Yellow-spotted Emerald, **F** – La Cordulie à taches jaunes, **D** – Gefleckte Smaragdlibelle

DESCRIPTION OF ADULT

A relatively large *Somatochlora*, males usually exceeding 48mm and females 50mm in length. It is easily recognized by the yellow spots on the sides of the central abdominal segments; these spots are larger in females. In addition, there are yellow marks on the mesothorax and metathorax. The apical wing venation is rather coarse with usually seven or fewer cells on the wing margin between the ends of veins R_2 and R_3. The male superior appendages (fig. 266) are long and subparallel with two widely-spaced ventral teeth, the distal tooth distinctly posterior to the apex of the inferior appendage. The vulvar scale of the female (fig. 273) is shorter than S9, directed downwards at an angle of about 80° and with its apex notched.

BIOLOGY

Found near small ponds, marshes, boggy meadows, dykes and ditches, usually at low altitude and often in rich, cultivated land. *S. flavomaculata* seldom flies over open water, frequenting instead reed-beds and rank vegetation in ditches and woodland rides and clearings. Its flight is less rapid than that of other species of *Somatochlora*.

FLIGHT PERIOD

Beginning of June to the end of August.

DISTRIBUTION (map 77)

A mainly central and east European species, ranging from western France (Bordeaux) and Italy (from Tuscany north-wards) to southern Sweden and eastwards into Siberia. Absent from the British Isles, Spain, Portugal and Norway.

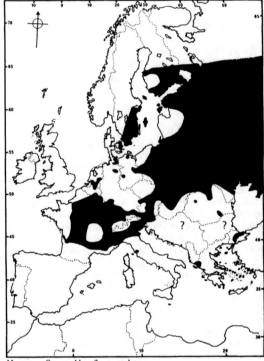

Map 77 – *Somatochlora flavomaculata*

Map 78 – *Epitheca bimaculata*

EPITHECA Charpentier, 1840

A small genus (two species) of non-metallic dragonflies that bear a superficial resemblance to species of *Libellula*. Only one species is found in Europe.

EPITHECA BIMACULATA (Charpentier)
Plate 23 (147,148)

Libellula bimaculata Charpentier, 1825, *Horae entomol.* : 43.
Type locality: Silesia, Poland.

Libellula fuchsiana Eversmann, 1836

F – La Cordulie à deux taches, **D** – Zweifleck

DESCRIPTION OF ADULT

A large dragonfly (body length 60–65mm) of robust build with a broad, rather flattened abdomen which is triangular in section. The yellowish brown thorax is virtually devoid of metallic reflections and coated with buff-coloured hair.

In life, the eyes are greyish blue with little of the green coloration characteristic of the smaller Corduliidae. There is a single cubito-anal crossvein in the forewing, and the discoidal cell is usually three-celled as is the subtriangle; the hindwing discoidal cell is usually two-celled, and the anal field is foot-shaped. Each hindwing has a dark brown, triangular basal mark and the wings of the female are tinged with yellow anteriorly. The membranule of the hindwing is mostly white but with an infuscate apex. *Epitheca* has very long, black legs. S1 and S2 are yellowish, S9 and S10 are black, and S3–8 are broadly black mid-dorsally with lateral yellow to orange-brown spots. The male superior appendages (fig. 268) are longer than S9, divergent, apically rounded and lacking ventral teeth; the inferior appendage is broad, apically incised, and terminating in two large, divergent arms. The female vulvar scale (fig. 274) is exceedingly long, extending to the middle of S10, and it is deeply cleft apically.

BIOLOGY

Males fly rapidly over the surface of lakes far from the bank. The female ejects a large mass of brown eggs (about 2000) whilst resting on vegetation, and these adhere to the end of her abdomen. She then skims over the water surface where there is a good growth of submerged plants and dips the apex of the abdomen below the surface; this causes the eggs to become drawn out in long, gelatinous strands, sometimes half a metre in length, which become entangled amongst submerged vegetation.

FLIGHT PERIOD

Brief, from mid-May to mid-July.

DISTRIBUTION (map 78)

A north Eurasian or Siberian insect with occasional records from the Netherlands (nineteenth century), Belgium (five localities), eastern France (six localities, mostly in the Jura and Ardennes) and northern Italy (e.g. Lake Garda, but no recent records) marking its westerly limits. There are a number of breeding populations in southern Scandinavia and it is widespread but local in central Europe from West Germany (moderately plentiful but declining in Schleswig-Holstein and a few localities in Baden-Württemberg) eastwards. Martens (1982b) reviews the status of *E. bimaculata* in western Europe. It occurs across Siberia as a distinct subspecies, *E. b. sibirica* Sélys, 1887; Sonehara (1967) provides an account of the life history of this form.

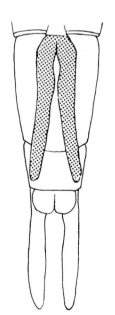

274. *Epitheca bimaculata*

275. *Cordulia aenea*

276. *Oxygastra curtisi*

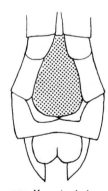

277. *Macromia splendens*

Figures 274–277 Female Corduliidae
apex of abdomen in ventral view with vulvar scale stippled

OXYGASTRA Sélys, 1871

This genus includes just one species. It is a medium-sized, metallic green corduliid that differs from *Cordulia* and *Somatochlora* especially in wing venation and the structure of the anal appendages.

OXYGASTRA CURTISI (Dale)   Plate 23 (149,150)

Cordulia curtisii Dale, 1834, *Mag. nat. Hist.* 7 : 60.
Type locality: Ramsdown, Hampshire, England.

Libellula nitens Fonscolombe, 1837

E – Orange-spotted Emerald, **F** – La Cordulie à corps fin,
D – Gekielte Smaragdlibelle

DESCRIPTION OF ADULT

The body is quite bright metallic green, although in flight it looks dark. It is about 50mm long, with a slender abdomen which is marked with yellow mid-dorsally on all segments except S8 and S9.

The frons is without lateral yellow spots. Wing bases are slightly yellowish and this colour suffuses most of the wings in immature specimens. Discoidal cells of fore- and hind-wings (figs 259,260, p.144) are more similar in shape than in *Cordulia* and *Somatochlora* and they are normally not crossed by veins. The basal edge of the hindwing discoidal cell is distal to the arculus, the forewing subtriangle, immediately basal to the discoidal cell, is single-celled (three-celled in *Cordulia* and *Somatochlora* except in un-common varieties), and the anal field of the hindwing (fig. 260) is not foot-shaped, differing in this respect from other corduliid genera except *Macromia*. The membranules are white and the anal angle of the male hindwing is more rounded than in *Cordulia* and *Somatochlora*. Male superior abdominal appendages (fig. 267) are shorter than the combined length of S9+10, and each has a long, slender, downwardly-directed projection from its inner edge near the base, and also a small, obtuse, apical tooth just distal to the level of the apex of the inferior appendage. The inferior appendage is broad and terminates in two slightly upturned arms. The tenth segment of the male abdomen bears a dorsal, yellow crest (fig. 267). The vulvar scale (fig. 276) is very short and inconspicuous, scarcely extending beyond the base of S9 and not concealing two tubercles at the base of the ninth sternum. Its apex is semicircularly emarginate.

BIOLOGY

Larvae of *Oxygastra*, unlike those of other European Corduliidae except *Macromia*, require running water for their development. They inhabit slowly-flowing and tree-lined rivers and canals with muddy bottoms. Adults feed at the borders of woods or in clearings, often circling for prolonged periods at canopy level. Observations on *O. curtisi* are recorded by Heymer (1964).

FLIGHT PERIOD

June to August.

DISTRIBUTION (map 79)

O. curtisi has a restricted western Mediterranean distribution and is most numerous in Portugal, central and northern Spain, and in southern and western France. It is local in Italy (north-west and Campania) and very rare in Switzerland (Geneva, Ticino), West Germany (Rhine

Map 79 – *Oxygastra curtisi*

Province), Belgium (rediscovered in 1976 on River Ourthe: Dumont, 1977a) and the Netherlands (near Eindhoven, recently rediscovered after not being seen for half a century). Colonies, now extinct, used to exist in Hampshire, Devon and possibly Cornwall in southern England, the last record being from Moors River, Hampshire in 1951. The species has also been found in Morocco (one female 30km south of Rabat: Lieftinck, 1966).

MACROMIA Rambur, 1842

A genus of 112 described species with a wide distribution encompassing North America, Africa, India, China, Japan, Malaysia, Indonesia and Australia. Only one species, however, is found in Europe. *Macromia* differs so much from other corduliids in venation and some other characters (see below) that it is sometimes placed in a separate family, Macromiidae.

MACROMIA SPLENDENS (Pictet)

Plate 23 (151,152)

Cordulia splendens Pictet, 1843, *Magasin Zool. Paris* : 131.
Type locality: River Lez, Montpellier, France.
F – La Cordulie splendide

DESCRIPTION OF ADULT

A striking insect of large size (body length about 70mm). Eyes in life are green. The frons is yellow with a black mark on its upper part extended dorsally into three points. The thorax is dark green and shining with some coppery reflections and it has a pair of broad, yellow stripes on the mesepisterna and another pair on the metepisterna. Legs are black and very long. Several venational characters distinguish *M. splendens* from other European Corduliidae: forewing with thirteen or more antenodal crossveins (at most nine in other genera); anal field of hindwing rounded and encompassing only six to nine cells; each wing has usually three cubito-anal crossveins and there are one to three crossveins dividing the supradiscoidal cells; the discoidal cells are normally undivided and that of the hindwing is placed far distal to the arculus; the veins leaving the arculus are fused for a short distance; and the forewing subtriangle is one-celled. The membranules are white. The abdomen is long and cylindrical, the four posterior segments slightly expanded; it is blackish with faint greenish reflections and marked with yellow in a pattern not unlike that of *Cordulegaster*. In dorsal view there is a broad, yellow bar on S2, paired yellow spots on S3–6 which are confluent on S3 and S4 and which decrease in size posteriorly so that those on S6, at least in the male, may be almost imperceptible, a large, yellow spot at the base of S7 and another, in the male only, on S8. The male auricles are mainly yellowish. Superior anal appendages of the male (fig. 269) are hardly longer than S10, robust, convergent in dorsal view and each has a large latero-ventral tooth. Between this tooth and the apex, there is a ventral field of small tubercles. The inferior appendage is longer than the superior appendages and weakly bilobed at its apex. Male secondary genitalia have well-developed anterior hamules, and in this respect *Macromia* differs from other Corduliidae, and indeed from all other Libelluloidea. In the female, the terminal appendages are very short, no longer than S10, and conical; the vulvar scale (fig. 277) is broad and apically rounded, scarcely projecting but extending to the middle of S10.

BIOLOGY

Breeds in warm and quite deep and broad rivers such as the Lot, Celé and Tarn, and probably also in smaller water courses in the south of Les Landes (Tiberghien, 1985), where the current is slow allowing the deposition of

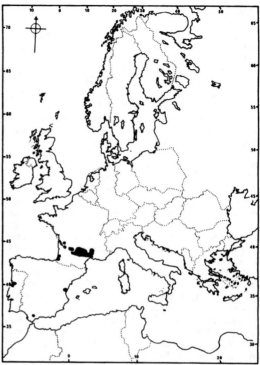

Map 80 – *Macromia splendens*

sediment and the growth of a variety of aquatic plants. Males fly rapidly and for long periods over the water. Specimens have also been observed feeding at the borders of woods at some distance from rivers. Females stay for only a short time over water when ovipositing, 'tapping' the surface four or five times with the end of the abdomen and then disappearing into trees on the river bank (Belle, 1983).

FLIGHT PERIOD

Apparently brief, from mid-June to the end of July.

DISTRIBUTION (map 80)

A very local dragonfly found only in south and south-western France, and in Spain and Portugal. Morton (1925) recounts its rediscovery in France, on the Lot at Cahors. Encountered quite frequently on the Tarn in 1982 (Belle, 1983). The northernmost unquestioned locality, albeit one in which the species has not been rediscovered since its initial nineteenth-century discovery, is Jarnac on the Charente. The Portuguese localities near Coimbra and the Spanish locality in Castellón require confirmation. The only recent Iberian record for *M. splendens* is from Cadiz. Tiberghien (1985) provides a recent distributional survey and adds two sites from the extreme south-west of France.

LIBELLULIDAE

This is the dominant family of Anisoptera in Europe and most other parts of the world, and it includes species that show the most advanced anisopteran characters. Most libellulids are small to medium-sized dragonflies, called 'darters' from their habit of making rapid forays in pursuit of prey or other dragonflies from a perch. The perch is often a prominent stem or twig on which the insect rests with wings drooping and its body more or less horizontal or even, in hot sun, raised upwards. *Zygonyx* is an exception in resting vertically as in aeshnids, cordulegastrids and corduliids. Libellulids are generally, although not always, associated with standing water and sometimes with very small and temporary accumulations of water. Typically, males are more vividly coloured than their females, mature male *Orthetrum* and *Libellula* (except *quadrimaculata*) bearing a dense, bluish pruinescence and male *Sympetrum* (except *danae*), *Trithemis* and *Crocothemis* being bright red. Green and blue pigments are not present in European species and metallic colours are scarce, weakly present only in *Trithemis* and *Zygonyx*.

There are considerable differences between fore- and hindwings (fig. 17, p.45). The forewing discoidal cell is narrow and transverse to the long axis of the wing, situated far distal to the arculus, whilst that of the hindwing is longitudinally elongated and close to (sometimes even below) the arculus. There are no primary antenodals and normally all antenodals, except in some genera the most distal, are aligned with the crossveins between subcosta and radius. These features are shared with Corduliidae, but Libellulidae differ from corduliids in having a more strongly-developed foot-shaped anal loop in the hindwing, in the absence in males of tibial keels, auricles, angled hindwing anal margin and anal triangle, and in being without a well-defined prominence on the posterior eye margin. The labium has a very small median lobe (fig. 31, p.53).

The abdomen in Libellulidae is usually relatively short and often rather broad (very broad in *Libellula* and *Crocothemis*), and it is triangular in cross-section across the middle segments (except *Zygonyx*). The ovipositor is vestigial. Male anal appendages tend to differ little between species, but good diagnostic characters are present instead in the secondary genitalia.

Eleven genera and thirty-four species of Libellulidae are recognized as occurring in Europe.

Key to genera of Libellulidae

1   Hindwing with a conspicuous dark brown to blackish basal mark, sometimes small and sometimes surrounded by yellowish suffusion 2

–   Hindwing without a dark basal mark, or with an extremely small one (female *Zygonyx*), but sometimes with a yellow to amber basal patch 4

2(1)   Forewing with at least eleven antenodal veins and discoidal cell with one to four (usually two or more) crossveins; usually three to five bridge crossveins ..... *Libellula* (p.158)

–   Forewing with at most nine antenodal veins and discoidal cell without or with only one crossvein; one (occasionally two or three) bridge crossveins 3

3(2)   Face white. Forewing (fig. 282) with distal antenodal crossvein complete (i.e. continuing across subcosta to radius); discoidal cell with one crossvein; subtriangle three-celled .... *Leucorrhinia* (p.184)

–   Face blackish. Forewing (fig. 279) with distal antenodal crossvein incomplete and very oblique; discoidal cell open; subtriangle one- or two-celled .... *Diplacodes* (male) (p.169)

4(1)   Forewing with distal antenodal vein complete, extending across subcostal space to vein R_1 5

–   Forewing with distal antenodal vein incomplete, oblique, reaching only to vein Sc 6

5(4)   Forewing (fig. 278) with ten or more antenodal veins; discoidal cell with crossvein; postdiscoidal field (trigonal space) with a row of three cells basally adjacent to discoidal cell. Males extensively blue pruinescent .... *Orthetrum* (p.160)

–   Forewing (fig. 284) with six antenodal veins; discoidal cell without crossvein; postdiscoidal field with two cells adjacent to discoidal cell. Males blackish .... *Selysiothemis* (p.193)

6(4)   Forewing (e.g. fig. 280) with three cells at base of trigonal space adjacent to discoidal cell 7

–   Forewing (fig. 279) with two cells at base of trigonal space adjacent to discoidal cell .... *Diplacodes* (female) (p.169)

7(6)   Discoidal cell of forewing (e.g. fig. 281) with crossvein. Pterostigmal membrane uniformly coloured 8

–   Discoidal cell of forewing (fig. 280) without crossvein. Pterostigma proximally whitish, distally brown .... *Brachythemis* (p.170)

8(7)   Forewing (fig. 17, p.45) with seven (exceptionally eight) antenodal veins .... *Sympetrum* (p.173)

–   Forewing (e.g. fig. 281) with nine or more antenodal veins 9

9(8)   Body length not exceeding 40mm. At least the anterior wing veins red. Fore- and hindwing pterostigmata of about equal length. Hindwing with basal amber patch 10

–   Body length in excess of 45mm. Venation black or yellowish, never red. Pterostigma of forewing slightly longer than that of hindwing. Hindwing basally clear, or with a weak yellowish suffusion, or (*Zygonyx* female) with a very small brown mark 11

10(9) Row of cells between veins IR_3 and Rspl partly double (fig. 283). Pterostigma about 2.6 mm long. Legs mainly black. Abdomen narrow .... *Trithemis* (p.189)

–   Row of cells between veins IR_3 and Rspl single (fig. 281). Pterostigma about 3.5mm long. Legs mainly reddish. Abdomen broad .... *Crocothemis* (p.171)

11(9) Hindwing with two cubito-anal crossveins basal to discoidal cell and with weak basal yellowish suffusion; wing tips usually faintly infuscate. Body yellowish brown .... *Pantala* (p.191)

–   Hindwing with one cubito-anal crossvein basal to discoidal cell; wings hyaline in males, faintly brownish in females which have also a very small, brown, basal mark .... *Zygonyx* (p.190)

Venational features of Libellulidae are important in the separation of genera and, for convenience, some of the more useful characters, as they appear in European species, are shown in Table 3 (p.157).

Table 3 Wing characters of European Libellulidae

	Libellula	Orthetrum	Diplacodes	Brachythemis	Crocothemis	Sympetrum	Leucorrhinia	Trithemis	Selysiothemis	Zygonyx	Pantala
forewing (fw) antenodals*	12–16	10–14	6½–7½	7½	8½–10½	6½–7½	7–8	9½–10½	6	10½–11½	11½–13½
fw subtriangle cells	3–6	3	1–2	1	3	3	3	3	1	3	3–5**
fw discoidal cell crossveins	1–3	1	0	0	1	1	1	1	0	1	1
cells adjacent to discoidal cell distally	3–5	3	2	3	3	3	3	3	2	3	3
bridge crossveins	2–4	1	1	1	1	1	1–2	1	1	1	1
fw trigonal space widens appreciably distad	√	√	√	×	√	×	√	×	×	×	×
cell rows between IR_3 and Rspl (maximum)	2–3	1–2	1	2	1	1	1–2	2	1	2	2
hindwing (hw) cubito-anal crossveins	1–3	1	1	1	1	1	1–2	1	1	1	2
hw discoidal cell crossveins	0–3	0–1	0	0	0	0	0–1	0	0	0	0

* ½ signifies that the distal antenodal is incomplete
** subtriangle in *Pantala* poorly-defined

NOTE. Venational abnormalities are not uncommon and specimens may occasionally be found which depart from the characters indicated. Ranges may be exceeded and some of the single values given are modal rather than absolute.

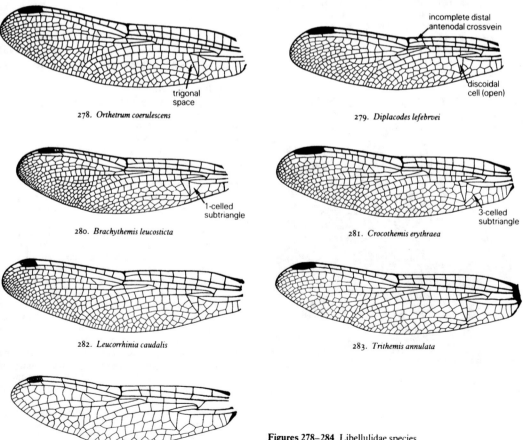

278. *Orthetrum coerulescens*

trigonal space

279. *Diplacodes lefebvrei*

incomplete distal antenodal crossvein

discoidal cell (open)

280. *Brachythemis leucosticta*

1-celled subtriangle

281. *Crocothemis erythraea*

3-celled subtriangle

282. *Leucorrhinia caudalis*

283. *Trithemis annulata*

284. *Selysiothemis nigra*

Figures 278–284 Libellulidae species
Left forewings (Note *Crocothemis* (281) with complete distal antenodal crossvein)

157

LIBELLULA Linnaeus, 1758

Platetrum Newman, 1833
Leptetrum Newman, 1833
Ladona Needham, 1897

Eb. Schmidt (1987) recommends the generic separation of the three European species as *Libellula quadrimaculata*, *Platetrum depressum* and *Ladona fulva*. However, in this work I have retained them in *Libellula*.

Large, dark brown marks at the bases of the hindwings distinguish *Libellula* from all other European genera of Libellulidae except *Leucorrhinia*, and the coloured (not white) face and numerous antenodal crossveins (usually at least thirteen in the forewing) prevent any confusion with the latter genus. *Libellula* are moderately large dragonflies of robust build, with broad and densely hairy bodies and ample wings. The females' wings are very slightly wider than those of the males. Body colour is brown patterned with yellow, black and sometimes a blue pruinosity. Venational features of *Libellula* include distal broadening of the postdiscoidal field in the forewing, pronounced undulation of vein R_3, the presence of two or more bridge crossveins, the separation to their origins of the two sectors of the arculus, and the position of the arculus below the cell between the first and second antenodal crossveins. The last antenodal in the forewing is always complete.

The genus includes about thirty species and is represented in North and South America, Europe and Asia. The three European species of *Libellula* frequent still or slowly-flowing waters. They emerge as adults relatively early in the year and are usually found close to their breeding sites, perching on the tops of reeds and other prominent water-side vegetation. Migrating individuals of two of the species, however, may appear almost anywhere.

Key to species of *Libellula*

1 Node of each wing with a dark mark; forewing lacking brown basal marks. Never with bluish pruinosity ....
.... *quadrimaculata* (p.158)

– No dark nodal marks but forewings with brown basal marks. Mature males develop blue abdominal pruinosity 2

2 Abdomen very broad, S4–6 each about three times as broad as long. Dark brown mark at the base of each wing large, extending far beyond the arculus; membranule whitish; wing-tip of female unmarked. Male with a pair of spines, basally connected, on the ventral surface of the first abdominal segment (fig. 285, p.160) .... *depressa* (p.160)

– Abdomen narrower, the middle segments only about twice as broad as long. Dark brown marks at wing bases smaller, not usually extending beyond the arculus; membranule grey; wing-tip of female with a small, brown mark. Male without ventral spines on first abdominal segment .... *fulva* (p.159)

LIBELLULA QUADRIMACULATA Linnaeus
Plate 24 (153)

Libellula quadrimaculata Linnaeus, 1758, *Syst. Nat.* (Edn. 10) **1** : 543.

Type locality: Sweden.

> *Libellula quadripunctata* Fabricius, 1781
> *Libellula maculata* Harris, 1782
> *Libellula praenubila* Newman, 1833
> *Libellula relicta* Belyshev, 1973

E – Four-spotted Libellula, **F** – La Libellule à quatre taches, **D** – Vierfleck

DESCRIPTION OF ADULT

The sexes of *L. quadrimaculata* are similar, males never developing the blue pruinosity of male *L. depressa* and *L. fulva*.

The thorax is brown without dorsal stripes. Wing markings of *L. quadrimaculata* are rather variable but the dark nodal infuscation on each wing seems always to be present, although sometimes small, and is very characteristic of the species (the specific name refers to this feature). The brown, triangular mark at the base of the hindwing, crossed by yellowish veins as in the two other species, is extended around the arculus into the discoidal cell to limit the orange area between arculus and wing base. The hindwing discoidal cell is darkly pigmented but lacks crossveins in contrast to the two following species. The space between vein R_1 and the costa is yellowish orange in both fore- and hindwings, usually as far as the node, although the intensity and extent of this coloration is variable and diminishes with age. There is no basal brown mark on the forewing, but the cubital space is orange-tinted to the arculus or somewhat beyond. The not uncommon variety, form *praenubila* Newman, may occur throughout the range of the species; in this form there is a dark suffusion below the pterostigma (see Plate 24 (153)), often accompanied

Map 81 – *Libellula quadrimaculata*

by an exaggeration of the yellowish orange pigmentation. The membranule is white. In both sexes, the abdomen is mainly yellow-brown with the last four segments, and the sixth posteriorly, black dorsally. There are yellow lateral markings, narrower than those of *L. depressa*, on S4 to S8 or S9. *L. quadrimaculata* has the terminal abdominal appendages in both sexes almost as long as the last two abdominal segments together. The abdomen is rather elongate, tapering in the male, S4 almost twice as broad as long in the female but narrower in the male. Body length 39–48 mm.

BIOLOGY

Usually found at the margins of shallow ponds and lakes where there is a moderate growth of emergent vegetation, but an area of open water is also required. The species is tolerant of acid conditions and in northern Europe it frequents pools on mosses and moors, in particular those situated in light woodland. Territorial behaviour is described by Warren (1964) and Mokrushov (1982). *L. quadrimaculata* is a well-known migrant; Dumont & Hinnekint (1973) review the incidence of migrations in northern Europe and suggest that these have a ten-year periodicity.

FLIGHT PERIOD

April (in the Mediterranean region) to August or early September, but especially June and the first half of July.

DISTRIBUTION (map 81)

L. quadrimaculata has an Holarctic distribution, being found in North America, almost throughout Europe, and in north and central Asia east to Japan (subspecies *L. q. asahinai* Schmidt, 1957). It extends further north in Europe than its two congeners, but its distribution in the extreme south is rather restricted and it is reported in North Africa only from the Middle Atlas and Rif Mountains of Morocco (Lieftinck, 1966).

LIBELLULA FULVA Müller   Plate 24 (154,155)

Libellula fulva Müller, 1764, *Fauna Insect. fridrichsdalina* : 62.

Type locality: Denmark.

  Libellula fridrichsdalensis Müller, 1764
  Libellula fugax Harris, 1782
  Libellula conspurcata Fabricius, 1798
  Libellula quadrifasciata Donovan, 1807
  Libellula bimaculata Stephens, 1835

E – Scarce Libellula, **F** – La Libellule fauve,
D – Spitzenfleck

DESCRIPTION OF ADULT

The thorax and abdomen in females and immature males are a rich, tawny colour with black triangular marks on the dorsal surfaces of S4–10; the last abdominal segment is almost entirely black. As the male matures, its thorax becomes darker brown and a greyish blue pruinosity develops on the abdomen over all the dorsal surface with the exception of the first two segments, which remain dark brown, and a median black line which expands posteriorly from S8.

The eyes are greyish (brown in the two other *Libellula* species). Each wing has one or two dark brown, basal streaks reaching to about the arculus, and the hindwing

has, in addition, a small, brown triangular area between the membranule and base of the anal loop. Females generally have a pronounced yellowish orange suffusion behind the anterior margin of each wing, and there is a small, brown mark at the wing-tip; these features are absent or only weakly indicated in males. The membranule is grey. In the forewing, the posterior cubital vein is more strongly arcuate in *L. fulva* than in the two other species. The abdomen of *L. fulva* is of similar proportions to that of *L. quadrimaculata*, but the anal appendages are shorter.

BIOLOGY

L. fulva requires for oviposition an open water surface without floating vegetation (Robert, 1958) and it usually breeds in slow-flowing streams, dykes and small rivers, although it is found also in lakes. The colonies are often large.

FLIGHT PERIOD

Rather short-lived as an adult insect with a synchronized emergence, flying chiefly between May and early July with only occasional individuals surviving into August.

DISTRIBUTION (map 82)

A widespread but local species in Europe, from the Mediterranean north to southern England and Finland, and east to Russia. It occurs in the Middle East as *L. f. pontica* Sélys, 1887 (= *conspurcata* Schneider, 1845), but is not found in North Africa.

Map 82 – *Libellula fulva*

LIBELLULA DEPRESSA Linnaeus
Plate 24 (156–158)

Libellula depressa Linnaeus, 1758, *Syst. Nat.* (Edn. 10) **1** : 544.

Type locality: southern Sweden.

E – Broad-bodied Libellula, F – La Libellule déprimée, D – Plattbauch

DESCRIPTION OF ADULT

The very broad, flattened abdomen of *L. depressa* distinguishes it at a glance from its two European congeners.

The thorax is at first yellow-brown with two dorsal, yellow stripes, but it darkens with age and the stripes almost disappear in males. The dark brown patches at the wing bases of *L. depressa* are larger than in the two other species of *Libellula*, reaching the level of the discoidal cell in the forewing and filling it in the hindwing. The spaces proximal to the arculus and for a short distance between vein R_1 and the radial sector are orange-brown in all wings, but this lighter coloration does not extend beyond the limits of the dark brown pigment. The female's abdomen is broader than that of the male, tawny-yellow dorsally with large, clear yellow, lateral marks on S4–7 and a distinct, black, mid-dorsal keel. In the newly-emerged male, the abdomen is similarly coloured to that of the female, but as it ages it becomes quite a bright blue with developing pruinosity, only S1 and S2 remaining brown and S10 blackish. The lateral yellow marks of the male, which are smaller than in the female, become obliterated in pruinescent specimens except for those on the third and fourth abdominal segments. Males turn blue rather late in the season in northern Europe and often mate before pruinescence has developed. Old females sometimes become pruinose. A diagnostic character of the male is a bifid spine (fig. 285) beneath the first abdominal segment. Body length 39–48mm.

BIOLOGY

Oviposits in shallow water at the edges of weedy ponds, and sometimes in very artificial situations such as the mud under lawn sprinklers. Large water-bodies are seldom used as breeding places. *L. depressa*, like some other libellulids, may be attracted to shining surfaces, for example polished car bonnets. It is a migrant and can be found at some distance from water.

FLIGHT PERIOD

May to August in most of Europe, but found as early as 8 April in Spain and recorded as late as 4 September in Austria (Landmann, 1981).

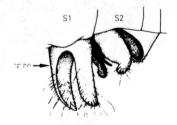

Figure 285 *Libellula depressa* male
anterolateral view of the bifid process (spine) beneath abdominal segment S1 and in front of the accessory genitalia on S2

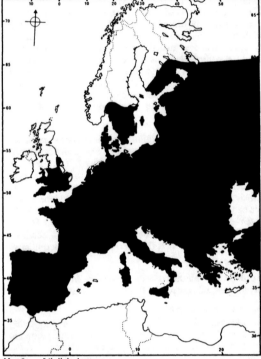

Map 83 – *Libellula depressa*

DISTRIBUTION (map 83)

Almost all Europe to 60°N, but not recorded from Ireland, Scotland or North Africa and scarce in southern Scandinavia, but elsewhere a common species. It is widespread in the Middle East and eastwards extends into western Asia.

ORTHETRUM Newman, 1833

Libella Brauer, 1868
Hydronympha Buchecker, 1876

An Old World genus extending to Australia and comprising about sixty species of medium-sized dragonflies. There are at least ten antenodal crossveins, the most distal one of which is continuous across the subcostal space to vein R_1. These two characters distinguish *Orthetrum* from all other European genera of Libellulidae except *Libellula*, and the absence of a brown basal mark on the hindwing of *Orthetrum* makes its recognition easy. *Orthetrum* species have strong and sustained flight, the bluish pruinose males characteristically being seen patrolling back and forth along the margins of ponds and lakes only a few centimetres above the water surface. Pruinosity is due to a powdery exudate which accumulates on the body surface as the males mature. It may also develop in old females. The exudate is quite easily rubbed off and males, after mating, often have dark marks on the central part of the abdomen where the females' legs have scratched away the deposit to reveal the underlying body surface.

The eight European species are all rather similar and specific determination requires care.

Key to species of *Orthetrum*

1 Pterostigma very dark brown or black. Abdomen of female and immature male patterned on S3–8 with broad, black, lateral lines which curve inwards above the lateral carina to isolate lateral yellow spots 2

– Pterostigma yellowish to brown. Abdomen without broad, black, lateral lines, but with a fine, blackish mid-dorsal carina 3

2(1) Superior anal appendages in both sexes black. Female abdomen broader with S4 almost quadrate, S10 dorsally yellow and brown, and mid-dorsal abdominal carina on central abdominal segments yellow with only the spicules dark .... *cancellatum* (p.163)

– Superior anal appendages in both sexes whitish. Female abdomen narrower with S4 longer than broad, S10 dorsally white, and mid-dorsal carina visible as a more or less complete, fine, black line .... *albistylum* (p.164)

3(1) Abdomen longer than a hindwing, more or less cylindrical, S2 and S3 maximally at least four times as deep as S4. Body length 57–60mm. Row of cells between veins IR$_3$ and Rspl partially double .... *trinacria* (p.161)

– Abdomen shorter than a hindwing, somewhat compressed dorsoventrally and triquetral, S2 and S3 maximally at most twice as deep as S4. Body length not more than 50mm. Sometimes only a single row of cells between veins IR$_3$ and Rspl 4

4(3) Vein R$_1$ yellowish to the level of the node; costa very pale postnodally. Female S8 in profile with ventral edge (lateral carina) straight. Body length usually 46–49mm ....
.... *nitidinerve* (p.165)

– Vein R$_1$ black. Female S8 in profile with ventral edge slightly convex. Body length seldom exceeding 45mm 5

5(4) Side of thorax immediately behind humeral suture with a conspicuous pale stripe. Membranule infuscate ....
.... *chrysostigma* (p.163)

– Side of thorax without a conspicuous pale stripe (unless very teneral). Membranule whitish 6

6(5) Pterostigma 2.5–3.0mm long, not longer than the breadth of the rather broad abdomen. Row of cells between veins IR$_3$ and Rspl extensively (at least four cells) doubled (fig. 287). Male accessory genitalia with anterior lamina in profile much shorter than the conspicuously bilobed hamule (fig. 293, p.165). Male pruinose on abdomen and thorax ....
.... *brunneum* (p.166)

– Pterostigma 2.8–4.0mm long, longer than the breadth of the more slender abdomen. Row of cells between veins IR$_3$ and Rspl mainly not doubled (figs 278, p.157; 286) or rarely with up to five cells doubled. Male accessory genitalia with the anterior lamina in profile extending to at least the apex of the hamule (figs 294,295, p.165). Male sometimes lacking pruinosity on the thorax 7

7(6) Male accessory genitalia (fig. 294, p.165) with anterior lamina broad and apex rounded and slightly swollen. Pterostigma at least 3.0mm and usually 3.5mm or more in length. Wings often weakly infuscate apically. Pruinosity of male usually on abdomen only. Body length usually 40–43mm .... *coerulescens* (p.167)

– Male accessory genitalia (fig. 295, p.165) with anterior lamina triangular in profile, pointed and slightly longer. Pterostigma at most 3.0mm long and usually slightly less. Wings clear. Pruinosity of male usually on both abdomen and thorax. Body length usually 36–40mm *ramburi* (p.168)

This key largely avoids specific characters in the male accessory genitalia and female vulvar scale. Identifications may be checked by referring to the figures and descriptions of these structures.

ORTHETRUM TRINACRIA (Sélys)

Plate 25 (159,160)

Libellula trinacria Sélys, 1841, *Rev. zool. Guérin Méneville* : 244.

Type locality: Catania, Sicily.

> *Libellula bremii* Rambur, 1842
> *Libellula clathrata* Rambur, 1842

DESCRIPTION OF ADULT

This species differs from all others included here in having a very elongated abdomen which, including the anal appendages, exceeds the length of a hindwing by 3–4mm. Total body length is 60mm or just a little less, making this the largest European species of *Orthetrum*. In both sexes the abdomen is narrow and almost cylindrical, basally strongly swollen so that, in profile, the suture between S2 and S3 is at least four times as long as the depth of S4. The pterostigma is about 4mm long and yellowish brown but the membranule is infuscate, a combination of colour characters shared only with *O. chrysostigma* among European *Orthetrum*.

On the sides of the thorax the humeral and metapleural suture are finely but conspicuously black, and there is a black longitudinal stripe centrally on each mesepisternum. In females, a yellow stripe lies internal to each black mesepisternal stripe. There are, in part, two rows of cells between veins IR$_3$ and Rspl, and the venation is black with only the costa and posterior sections of antenodal crossveins yellowish. The legs have rather longer spines than those of other species; this is especially noticeable on the hind tibia where each spine is distinctly more than twice as long as the breadth of the tibia. The abdomen has a broad, mid-dorsal and two lateral black lines. Its ground colour is yellowish or olivaceous in females and immature males, but darker and covered basally by bluish pruinescence (appearing a deeper shade of blue than in other species) in mature males and old females. The anterior lamina of the male accessory genitalia (fig. 288, p.165) has long hairs and spicules on its anterior face and a bilobed apex, and the posterior hamule has a blunt tooth on its posterior face. All parts of the secondary genitalia are blackish. The apical incision of the female vulvar scale (fig. 296, p.167) is very broad but shallow, and with a pair of inwardly pointing projections at the sides. S8 in the female is straight ventrally and the anal appendages are relatively long, distinctly longer than S9 (in other species the anal appendages are at most as long as S9).

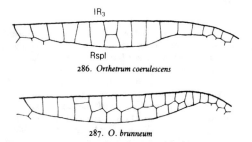

286. *Orthetrum coerulescens*

287. *O. brunneum*

Figures 286, 287 Venation between veins IR$_3$ and Rspl

BIOLOGY

Associated with large bodies of standing water (e.g. reservoirs). Pinhey (1961) comments that its hawking flight resembles that of an aeshnid and, more frequently than other species of *Orthetrum*, it takes smaller dragonflies as prey. *O. trinacria* has the typical libellulid habit of perching on the ground, but it will also rest on tree branches. Males guard ovipositing females.

FLIGHT PERIOD

Mid-June to mid-October in Sardinia, Sicily and Spain; and most abundant during September.

DISTRIBUTION (map 84)

Widespread in Africa and common in the north (although not yet known from Tunisia) extending east to Sudan, Egypt and the Middle East (Palestine, Iraq). Known in Europe only from Sicily, where the type specimen was taken by Ghiliani and where the species has recently been rediscovered, Sardinia (Bucciarelli, 1977) and Spain (Huelva: Belle, 1984; Adra: Conesa Garcia, 1985a).

ORTHETRUM SABINA (Drury, 1770)

(= *ampullacea* Schneider, 1845)

This species is allied to *O. trinacria* with a slender, cylindrical abdomen which is much swollen at the base. The abdomen is much shorter than in *O. trinacria*, however, being only about as long as a hindwing. The thorax is conspicuously marked with two greenish or yellowish white mesepisternal stripes, each of which is bordered on the outside by a blackish stripe (as in *trinacria*). Laterally, the humeral and metapleural sutures are heavily blackened, and brownish stripes cross the mesepimeron, metepisternum and metepimeron, all of these stripes being readily seen against the yellowish ground-colour. The abdomen has a mid-dorsal black line from S3, very broad on S4 and S5, and expanding at the posterior end of S4 and at both ends of S5 and S6. S7–9 are almost entirely black, and S10 has a pair of yellow spots which are sometimes confluent. Pruinosity does not develop on mature specimens of *O. sabina*. The membranule is infuscate.

O. sabina was described from China and it has a very broad range which encompasses semi-desert regions of Africa (north to the Algerian Sahara), the Middle East, Arabia, Turkey, India, Japan, Malaya, the Philippines, New Guinea and Australia (Pinhey, 1961), but it does not penetrate into Europe west of the Caucasus.

ORTHETRUM TAENIOLATUM (Schneider, 1845)

(= *brevistylum* Kirby, 1898)

This is another species whose distribution very nearly extends into Europe. It was described from Rhodes and occurs in the Middle East, northern India, Nepal, Afghanistan, Pakistan and northern Africa south to Nigeria and Ethiopia. It is a small *Orthetrum* with, when immature, pale stripes on the thorax as in *O. sabina*. These are, however, obscured by the pruinosity which covers thorax and abdomen of mature males. There is mainly just a single row of cells between veins IR_3 and Rspl, the membranule is infuscate, the pterostigma is 2.0–2.5mm long, and the narrow abdomen is triquetral and little inflated basally.

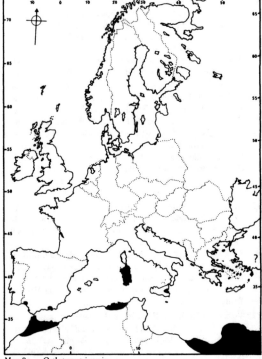

Map 84 – *Orthetrum trinacria*

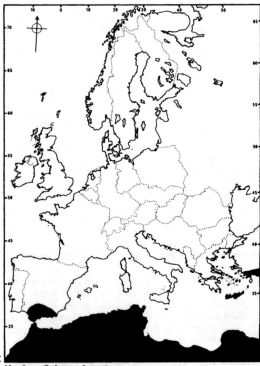

Map 85 – *Orthetrum chrysostigma*

ORTHETRUM CHRYSOSTIGMA (Burmeister)
Plate 25 (161,162)

Libellula chrysostigma Burmeister, 1839, *Handbuch Entomol.* **2** : 857.

Type locality: Tenerife, Canary Islands.

Libellula barbara Sélys, 1849

DESCRIPTION OF ADULT

This species has, like *O. trinacria*, *O. sabina* and *O. taeniolatum*, pale stripes laterally on the thorax and dark membranules. Also, the slender abdomen is basally swollen, although not to the same extent as in *trinacria* and *sabina*. All of these four species are predominantly inhabitants of areas with a hotter climate than that afforded by Europe, and only *O. chrysostigma* maintains a foothold in southern Europe.

The body length of *O. chrysostigma* usually falls within the range 39–43mm. On each side of the synthorax, immediately behind the blackened humeral suture, there is a conspicuous, creamy white stripe which is visible in males until obscured by the pruinosity which eventually covers all of the thorax. There are no other pale thoracic stripes, but the dorsum of the thorax, between the wing bases, is very pale. The ground-colour of the thorax is olivaceous with a pink to mauve tint on the sides. The pterostigma is dusky yellow to golden in colour (hence the specific name) and 2.8–3.3mm long, the membranule is infuscate, and the wing membrane is clear except for very small yellow patches usually present at the base of the hindwing. The row of cells between veins IR_3 and Rspl includes from nil to five doubled cells. The abdomen is narrow and shorter than the hindwing, triquetral, and quite strongly inflated at the base, in the male the suture between S2 and S3 being a little more than twice as long as the depth in profile of S4. The abdomen is yellowish brown becoming pruinose in mature males except on S1 and S2. Male accessory genitalia (fig. 289, p.165) have the anterior lamina long, triangular in profile, with an entire apex; the hamule has a black apical projection. The female vulvar scale (fig. 297, p.167) is apically sinuate with a pair of dark, rounded lobes beneath, and the lateral carina of S8 is slightly convex.

BIOLOGY

Occurs in open country, often in semi-desert localities. It breeds in marshy places, pools, ditches and small streams. For an account of territorial and mating behaviour see Miller (1983); after mating, females are guided and 'nudged' by males towards oviposition sites.

FLIGHT PERIOD

March to the beginning of September in North Africa (Aguesse, 1968); to the middle of December in Andalucia (Ferreras Romero & Puchol Caballero, 1984).

DISTRIBUTION (map 85)

All Africa, widespread in the north, and the Middle East. Recorded from several localities in southern Spain (Benitez Morera, 1950), and from Crete (Schmidt, Eb., 1978a).

ORTHETRUM CANCELLATUM (Linnaeus)
Plate 25 (163,164)

Libellula cancellata Linnaeus, 1758, *Syst. Nat.* (Edn. 10) **1** : 544.

Type locality: Sweden.

Libellula frumenti Müller, 1764
Libellula lineolata Charpentier, 1825

E – Black-lined Orthetrum, **F** – L'Orthétrum réticulé,
D – Grosser Blaupfeil

DESCRIPTION OF ADULT

Quite a large dragonfly (44–49mm in body length) which, together with the following species, differs from others in the genus in having broad, black, festooned lateral stripes on the abdomen. Outside these stripes there are crescentic yellow spots. The median dorsal abdominal carina on S3–10 is yellow but furnished with minute, black spicules so that it appears as a very fine, dark line. The abdominal pattern is obscured by pruinosity in mature males (and sometimes in old females), but S1 and S2 remain brown and the apical segments (after S6 or S7) are black with lateral yellow spots visible at least on S6–8. Pruinosity does not develop to any marked extent on the thorax which is devoid of pale stripes but has black antehumeral stripes (incomplete posteriorly). In addition to the black lateral abdominal stripes, *O. cancellatum* may be distinguished from all other species except *O. albistylum* by the very dark brown to black pterostigmata. The wings are clear, the membranules grey, and there is a mostly double row of cells between veins IR_3 and Rspl. Legs are black with the dorsal surfaces of the femora, especially the front femora, yellow in females and immature males. *O. cancellatum* differs from the closely related *O. albistylum* in having entirely black superior anal appendages in both sexes, S10 in the female is yellowish dorsally and dark brown laterally, and the pterostigmata are rather short, 2.5–3.0mm in length. Male secondary genitalia (fig. 290, p.165) have the anterior lamina apically attenuated, deeply incised, and with long hairs on its anterior face. Stout hairs cover the genital lobe. In the female, the vulvar scale (fig. 298, p.167) is almost rectangularly incised and the ventral edge of tergite S8 is quite straight.

BIOLOGY

Chiefly a lowland species. Larvae develop in ponds and lakes, sometimes in somewhat brackish water. Males fly swiftly and low over the surface of the water close to its margin. Krüner (1977) found male territories to occupy 10–50m lengths of bank, the size varying inversely with male density and the situation determined mainly by the presence of suitable oviposition sites. Copulation and oviposition take place in the male's territory and he guards his mate during oviposition. Territories are held usually for just one day. Territorial males usually perch on stones but in hot weather (>26°C) they use branches.

FLIGHT PERIOD

End of April to the end of August, depending on region, but over much of Europe adults are most abundant in June and July. The average adult life span is twenty-five days (maximum about sixty days) (Krüner, 1977).

Map 86 – *Orthetrum cancellatum*

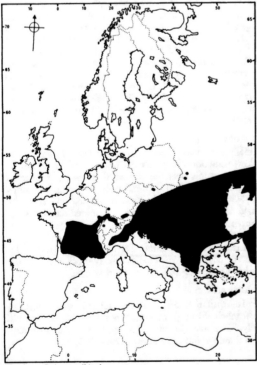

Map 87 – *Orthetrum albistylum*

DISTRIBUTION (map 86)

Widespread and generally common throughout Europe (except northern Britain and northern Scandinavia), North Africa, the Middle East, and in Asia east to Kashmir and Mongolia.

ORTHETRUM ALBISTYLUM (Sélys)
Plate 25 (165,166)

Libellula albistyla Sélys, 1848, *Revue zool.* **11** : 15.
Type locality: Lyon, France.

F – L'Orthétrum à stylets blancs, **D** – Östlicher Blaupfeil

DESCRIPTION OF ADULT

O. albistylum is very similar in size and pattern to *O. cancellatum*, but it may be identified with certainty by the predominantly whitish superior anal appendages. In males, the basal half of the appendage may be blackish, but in females it is entirely white as also is the dorsal surface of the last abdominal segment. Other differences between the two species are the narrow but almost complete dark brown antehumeral stripe of *albistylum*, the slightly narrower and more cylindrical abdomen in *albistylum* which has the dark brown lateral stripes more arcuate in each segment and the median dorsal carina somewhat darker, the greenish to whitish tinge to the yellow ground colour of *albistylum* which is particularly noticeable on the lateral yellow abdominal spots (these appear golden-yellow in *cancellatum*), the longer pterostigma of *albistylum* which usually slightly exceeds 3.0mm in length, and the darker, blackish membranule in *albistylum*. The hindwing discoidal cell is frequently (in about 50 per cent of individuals) crossed by a vein, a circumstance that is exceptional in *cancellatum*, and the veins in the heel of the anal loop are more symmetrically arranged. Male accessory genitalia (fig. 291) have the apex of the anterior lamina less deeply incised compared with *cancellatum*. The ventral edge of S8 is very slightly convex and the apex of the vulvar scale (fig. 299, p.167) is deeply and narrowly incised.

BIOLOGY

Inhabits ponds and lakes, often with *O. cancellatum*; less commonly associated with streams and rivers. Females oviposit near to the water margin, hovering a few centimetres above the water surface down to which they repeatedly dart with down-curved abdomen so that their eggs are washed into the water.

FLIGHT PERIOD

Beginning of June until late in August.

DISTRIBUTION (map 87)

A predominantly eastern Palaearctic insect extending from Japan and China (subspecies *O. a. speciosum* Uhler, 1858) through central Asia to the Middle East and Europe where it almost reaches the Atlantic coast of France (Dordogne, Lot et Garonne). It is not found in northern Europe or the Iberian Peninsula and in western Europe is confined to a relatively narrow band across southern France and northern Italy and it is recorded from the extreme south of Germany (Schmidt, Eb., 1980a) and from Switzerland. Eastwards it becomes progressively more numerous and is plentiful through Austria, Hungary, the Balkans, Romania and southern Russia. In eastern Europe it appears to be extending its range northwards.

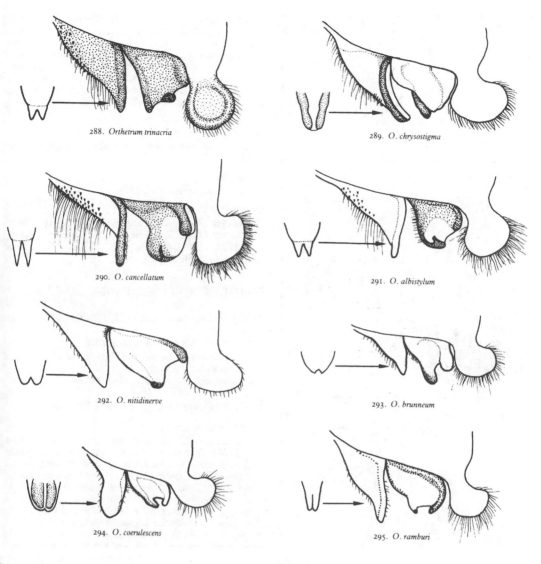

288. *Orthetrum trinacria*

289. *O. chrysostigma*

290. *O. cancellatum*

291. *O. albistylum*

292. *O. nitidinerve*

293. *O. brunneum*

294. *O. coerulescens*

295. *O. ramburi*

Figures 288–295 Male *Orthetrum* species accessory genitalia in left profile with apices of anterior laminae in frontal view

ORTHETRUM NITIDINERVE (Sélys)

Plate 26 (167)

Libellula nitidinervis Sélys, 1841, *Rev. zool. Guérin Méneville* : 243.

Type locality: Girgenti, Sicily.

Libellula baetica Rambur, 1842

DESCRIPTION OF ADULT

This and the three remaining European species in *Orthetrum* are of similar general appearance with yellowish brown body ground-colour and pterostigmata, the body covered by bluish white pruinosity in mature males, and the abdomen lacking the dark lateral stripes of *cancellatum* and *albistylum*. The character which most readily separates *nitidinerve* from its congeners is the yellow colour of the radial vein $(R+M/R_1)$ from base to node. This vein is black in the other species. The costa too is mainly yellow, conspicuously so between node and pterostigma. The row of cells between veins IR_3 and Rspl is mostly doubled and the membranule is white, as in *O. brunneum*, but the pterostigma is longer (3.8–5.0mm) than in the latter species. The hindwing discoidal cell is crossed by a vein in the specimens examined. Bluish pruinosity covers the top and sides of the synthorax, and all of the abdomen except S1, in the mature male. There are no pale thoracic stripes.

165

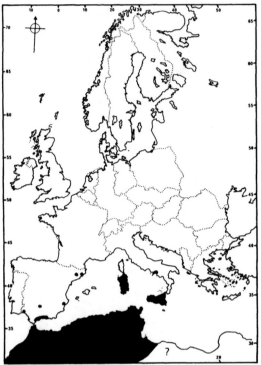

Map 88 – *Orthetrum nitidinerve*

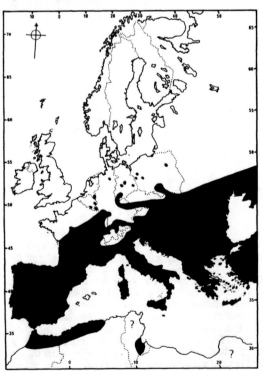

Map 89 – *Orthetrum brunneum*

O. nitidinerve is relatively large (46–50mm long), and the abdomen is moderately broad, triquetral, and not very deep basally. Male accessory genitalia (fig. 292) have the anterior lamina apically incised, though less deeply so than in the three previously described species, and there is only short and sparse pilosity on its anterior face (and on the genital lobe). S8 of the female abdomen is ventrally almost straight and the vulvar scale (fig. 300) is broadly but shallowly excavated.

BIOLOGY

Breeds in standing waters, sometimes at altitudes above 2300m.

FLIGHT PERIOD

April to October (Aguesse, 1968).

DISTRIBUTION (map 88)

A Mediterranean species occuring in North Africa from Morocco to Libya (Tripolitania), and in southern Spain, Italy (Campania: Consiglio, 1952), Sicily and Sardinia (Nielsen, 1941; Bucciarelli, 1977).

ORTHETRUM BRUNNEUM (Fonscolombe)
Plate 26 (168,169)

Libellula brunnea Fonscolombe, 1837, *Annls Soc. ent. Fr.* **6**: 141.
Type locality: Aix-en-Provence, France.
  ?*Libellula sardoa* Rambur, 1842 (type material lost, Sélys in Sélys & Hagen, 1850 : 382)
  Libellula anceps Schneider, 1845, *partim*
  Libellula cycnos Sélys, 1848

F – L'Orthétrum brun, **D** – Südliche Blaupfeil

DESCRIPTION OF ADULT

A medium-sized *Orthetrum*, 41–45mm long, of robust build with a rather broad abdomen which tapers posteriorly only slightly in the female. Pruinosity covers the thorax as well as the abdomen in mature males. The thorax is without clearly-defined pale stripes in the typical form. Ground-colour of the abdomen is dull olive-brown, often with a greyish to mauve tint, and the median dorsal carina is black. The space between veins IR_3 and Rspl is mostly occupied by two rows of cells (figs 287,304), the membranule is white, the radial vein is black, and the pterostigma is short (2.5–3.0mm long). The arculus tends to be more proximal in position than in allied species, arising well before the second antenodal crossvein. Males have very characteristic secondary genitalia (fig. 293) with a short anterior lamina, weakly incised apically, which does not reach the level of the apex of the hamule. The hamule in profile is obviously bilobed. The vulvar scale (fig. 301) of the female is broadly excavated between two rounded lobes, and the ventral edge (lateral carina) of S8 is quite strongly convex (cf. fig. 305, p.168).

The form from Corsica and Sardinia (*cycnos* Sélys (= ?*sardoa* Rambur)) seems to be a distinct subspecies, differing from the typical form mainly in colour: the female synthorax has two, distinct, pale stripes in front and two on each side, and the male legs are entirely black.

BIOLOGY

Usually associated with streams, canals and dykes with a

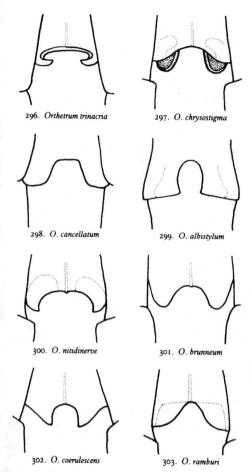

296. *Orthetrum trinacria*

297. *O. chrysostigma*

298. *O. cancellatum*

299. *O. albistylum*

300. *O. nitidinerve*

301. *O. brunneum*

302. *O. coerulescens*

303. *O. ramburi*

Figures 296–303 Female *Orthetrum* species
vulvar scales in ventral view

slow rate of flow, but also found beside ponds and lakes. It is a species which has taken advantage of man-made excavations such as gravel pits. Typically a lowland species, not found above 800m altitude (Robert, 1958). A behavioural study by Heymer (1969) showed that after copulation the male flies around the perched female and then directs her towards the oviposition site in his territory, guarding her against the approaches of other males. Similar behaviour has subsequently been observed in other species of the genus.

FLIGHT PERIOD

June to about the middle of September.

DISTRIBUTION (map 89)

Most common in the Mediterranean region, local in central Europe, and absent from the British Isles and Scandinavia. In the Netherlands it has been recorded from only two localities in the south, the last observation being in 1902 (Geijskes and van Tol, 1983). *O. brunneum* also occurs in North Africa, the Middle East and in Asia east to Kashmir, Gobi and Mongolia.

ORTHETRUM RANSONNETI (Brauer, 1865)

This species is closely allied to *O. brunneum* from which it differs in having a single row of cells between veins IR$_3$ and Rspl. It occurs over much of Africa north to the Sahara (Hoggar, Algeria) and eastwards through the Middle East to Afghanistan.

ORTHETRUM COERULESCENS (Fabricius)
Plate 26 (170,171)

Libellula coerulescens Fabricius, 1798, *Suppl. entomol. Syst.*: 285.

Type locality: Turin, Italy.

 Libellula biguttata Donovan, 1807
 Libellula donovani Leach, 1815
 Libellula opalina Charpentier, 1825
 Libellula olympia Fonscolombe, 1837
 Libellula dubia Rambur, 1842

E – Keeled Orthetrum, **F** – L'Orthétrum bleuissant, **D** – Kleiner Blaupfeil

DESCRIPTION OF ADULT

This, the type species of *Orthetrum*, is small (body length 40–44mm) with a markedly triquetral abdomen which is narrower and rather more tapering than that of either *O. brunneum* or *O. ramburi*. There are two conspicuous, pale yellowish stripes on the front of the synthorax. Pruinosity develops on the abdomen of the male (very occasionally also on the female), but it is slow to spread on to the thorax, thus providing a useful feature for distinguishing *coerulescens* from both *brunneum* and *ramburi* in the field. The wing-tips of *O. coerulescens* are usually slightly infumate and a yellow suffusion may develop on the anterior parts of the female's wings. The row of cells between veins IR$_3$ and Rspl is mainly single (at most five cells are doubled) (figs 278, p.157; 286, p.161). There is some overlap in this character between *coerulescens* and *brunneum* (fig. 304), but the numbers of doubled cells are usually different on the four wings of an individual (there are frequently more on the anterior wings), and when all wings are considered, it

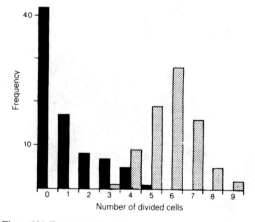

Figure 304 Frequency distribution of numbers of doubled or divided cells between veins IR$_3$ and Rspl in the wings of twenty specimens each of *Orthetrum coerulescens* (solid) and *O. brunneum* (stippled)

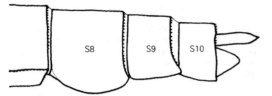

Figure 305 Female *Orthetrum coerulescens*
terminal abdominal segments in left lateral view

provides a reliable means of distinguishing between the two
species. The pterostigma is long (3.6–4.2mm) and bright
yellowish brown. Membranules are white. The abdomen is
yellowish brown with a fine but well-defined black, dorsal
keel. This keel is crossed transversely in the posterior parts
of S3–7 by a small, blackish mark. Pruinosity covers the
entire abdomen of the mature male excepting for its base,
but the dark, transverse marks show through as short,
narrow lines. Male accessory genitalia (fig. 294) are characteristic with the anterior lamina projecting at right angles to
the ventral surface of the abdomen, stout and with its apex
slightly inflated and rounded (i.e. it is not triangular in
profile). The vulvar scale (fig. 302) is apically notched
between broad points, and the lateral carina of S8 of the
female is convex (fig. 305).

BIOLOGY

Found in bogs and marshes and beside small ponds, ditches
and streams with a gentle current. Behavioural and ecological studies have been made by Heymer (1969), Lödl (1978)
and Parr (1983b). Males guard their ovipositing mates.

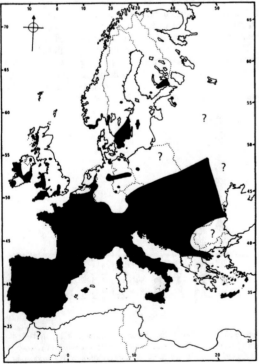

Map 90 – *Orthetrum coerulescens*

FLIGHT PERIOD

From the middle of May to the end of September in
southern parts of the range; late June until mid-August near
the northern limits (southern Finland) (Hämäläinen,
1984a).

DISTRIBUTION (map 90)

Widespread in western and central Europe: Iberian Peninsula, France, Ireland, Wales, the west of Scotland, south
and west England, southern Scandinavia (Finnish distribution mapped by Hämäläinen, 1984a) to Poland, Czechoslovakia and Hungary. The eastern limits to the range of
O. coerulescens are uncertain because of confusion between
it and *O. ramburi*. The latter appears to replace *coerulescens*
in the eastern Mediterranean region, but there is probably
considerable overlap between the ranges of the two species.
Apparently absent from the Middle East and reported from
North Africa (Esmir, Morocco) only by McLachlan
(1889b).

ORTHETRUM RAMBURI (Sélys)   Plate 26 (172)

Libellula ramburii Sélys, 1848, *Revue zool.* **11** : 16.
Type locality: Sardinia.

  Libellula anceps Schneider, 1845, *partim*
  Orthetrum helena Buchholz, 1954

Schneider (1985, *Senckenbergiana biol.* **66** : 97–104) con-
siders *Libellula anceps* Schneider, 1845, a valid senior
synonym of *L. ramburii* Sélys, 1848, and if this is accepted
the species should be known as *O. anceps* (Schneider,
1845).

DESCRIPTION OF ADULT

Closely allied to *O. coerulescens* which it much resembles
and with which it has frequently been confused. It tends to
be slightly smaller than *coerulescens* with a relatively broader
and more depressed abdomen, clear wings, the row of cells
between veins IR_3 and Rspl always (?) undivided, the
pterostigma shorter (at most 3.0mm long), and the pale
stripes on the front of the synthorax less clearly-defined. In
the male, pruinosity develops on the thorax as well as
abdomen, and the accessory genitalia (fig. 295) have an
anterior lamina which in profile is pointed and triangular.
This latter character provides the most certain means of
separating *ramburi* from *coerulescens*. The apex of the female
vulvar scale (fig. 303) is less sharply incised than in
coerulescens.

BIOLOGY

Larvae in standing waters and small streams. Many teneral
adults were observed near Rostov-on-Don in August flying
low over grass in an open pine wood about 200m from a
large but shallow pond.

FLIGHT PERIOD

Probably as *O. coerulescens*.

DISTRIBUTION (map 91)

North Africa, Corsica, Sardinia, Sicily, Malta and the
eastern Mediterranean from Hungary (Bakony), Yugoslavia and Greece to the Middle East, Black Sea and
Caucasus through to northern India.

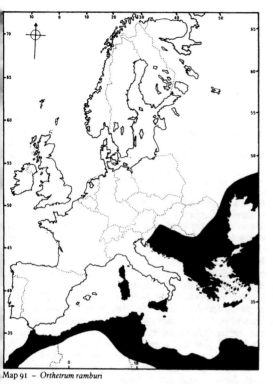

Map 91 – *Orthetrum ramburi*

DIPLACODES Kirby, 1889

Diplacodes is a small genus (nine species) distributed over the warmer parts of the Old World from Africa, India and south-east Asia to Australia. One species, the type species, occurs in Europe. The species are small and characterized by their wing venation. The forewing has the last antenodal crossvein incomplete, not extending below the subcosta, and there are only six or seven complete antenodals preceding it, as in the allied genus *Sympetrum*. *Diplacodes* differs from *Sympetrum*, however, in having only two and not three cells adjacent distally to the discoidal cell (fig. 279, p.157). The discoidal cells are undivided by crossveins. Other venational features are given in Table 3, p.157. The posterior lobe of the pronotum is suberect and crowned with a circlet of long hairs, as in *Orthetrum* and *Sympetrum*.

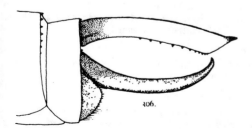

306.

DIPLACODES LEFEBVREI (Rambur)

Plate 26 (173)

Libellula Lefebvrii Rambur, 1842, *Hist. nat. ins. Neur.* : 112.

Type locality: Oasis of Bahryeh (Bahariya), Egypt.

> *Libellula parvula* Rambur, 1842
> *Libellula flavistyla* Rambur, 1842
> *Libellula tetra* Rambur, 1842
> *Libellula concinna* Rambur, 1842
> *Libellula morio* Schneider, 1845
> *Diplacodes ramburii* Kirby, 1890

DESCRIPTION OF ADULT

The several names bestowed upon this one species by Rambur are indicative of its variability. It is a small, delicate dragonfly usually only between 28–32mm long. Females are usually smaller than males. The mature male has a blackish body with only thin pruinosity ventrally on the thorax and anterior abdominal segments. Clypeus, frons and vertex are black with metallic bluish reflections. The legs are entirely black and the thorax is without pale stripes. Wings are hyaline or with their apices variably brownish, and the hindwing has a brownish amber basal patch extending as far as the cubito-anal crossvein. The membranules are grey and the pterostigmata are dark brown, each about 3.0mm long. The abdomen is more or less cylindrical, black, but usually with at least indications of paired, longitudinal, cream-coloured stripes above the lateral carina of S4–8. The superior appendages (fig. 306) are as long as S9, creamy white except for black bases and extreme tips; the inferior anal appendage is black. The female is mainly dark brown, differing from the male in having clypeus and frons creamy and pale stripes on the outer faces of the tibiae and sometimes on the thorax. Wings are hyaline, the coloured patch at the base of the hindwing is yellowish amber and the pterostigmata are brown. The abdomen is stouter than in the male, rather depressed, and with the superior appendages very short, about two-thirds as long as S9. Male accessory genitalia (fig. 307) have a very small anterior lamina which is scarcely visible in profile. The vulvar scale of the female (fig. 308) terminates in two weakly separated, rounded lobes, and it does not project beneath the abdomen.

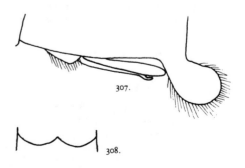

307.

308.

Figures 306–308 *Diplacodes lefebvrei*
(306) male abdominal appendages in left profile
(307) male accessory genitalia in left profile
(308) female vulvar scale in ventral view

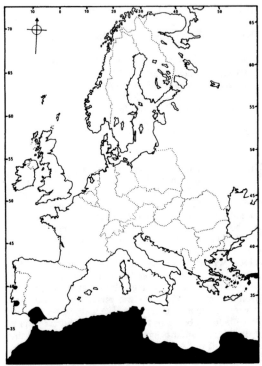

Map 92 – *Diplacodes lefebvrei*

BIOLOGY

Found by stagnant, open waters with little emergent vegetation. It is unusual among libellulids in flying at dusk or later.

FLIGHT PERIOD

May to October in the Mediterranean region (Aguesse, 1968), but earlier, from March, in the Middle East.

DISTRIBUTION (map 92)

Recorded from southern Spain (Almeria, Cadiz, Huelva – breeding: Conesa Garcia, 1985b), Portugal (Compte Sart, 1979) and Greece (Rhodes). Found throughout Africa (including Madagascar, Socotra, Mauritius and the Seychelles), the Middle East, Tadzhikistan and India.

BRACHYTHEMIS Brauer, 1868

Cacergates Kirby, 1889
Zonothrasys Karsch, 1890

Small dragonflies with venation (fig. 280, p.157) similar to that of *Sympetrum* but having the discoidal cells undivided and, in *B. leucosticta*, with a partly double row of cells between veins IR$_3$ and Rspl. The wings are conspicuously marked with yellow to brown patches, at least in mature males. The posterior lobe of the pronotum is not outstanding and not adorned with long hairs. *Brachythemis* is a genus of about six species, and is found in Africa and southern Asia with a single species just penetrating into south-western Europe.

BRACHYTHEMIS LEUCOSTICTA (Burmeister)
Plate 26 (174,175)

Libellula leucosticta Burmeister, 1839, *Handbuch Entomol.* 2 : 849.
Type locality: Egypt.
    Libellula unifasciata Rambur, 1842
    Zonothrasys impartita Karsch, 1890

DESCRIPTION OF ADULT

A small dragonfly, 30–33mm long, with the abdomen moderately broad and triquetral in both sexes. Females and immature males have the clypeus, frons and vertex creamy yellow and the thorax and abdomen greenish yellow conspicuously marked with brown and black. The synthorax has the median carina bordered with brown, the humeral suture is black, two brown bands on the mesepisternum converge to a point posteriorly, the lateral sutures are black and a black line arises from the metapleural suture, near its upper end, and curves forwards and downwards in the direction of the spiracle. The abdomen has a distinct, black, mid-dorsal carina, and there are blackish lateral marks on every segment. Legs are yellowish striped with black. The pterostigma is long (2.9–3.2mm) and creamy in colour with its apical third brown. The wings of young females are hyaline, slightly yellow over a small basal area. As the insect ages, the body and wings darken, the mature male becoming virtually black and old females are brown. The only parts to retain the original yellowish ground-colour are the sides of the frons and the superior abdominal appendages. The mature male has a conspicuous dark brown band across each wing, as in *Sympetrum pedemontanum* but more proximal in position, from the level of the node to two or three (forewing) or one (hindwing) postnodal crossveins before the pterostigma. The wing band increases in size and depth of colour as the male matures, and a similar band, though rarely so dark as in the male, may develop in old females. The brown distal area on the pterostigma may also enlarge with age. Adetunji & Parr (1974) describe maturational colour changes.

The anterior lamina of the male accessory genitalia (fig. 310) is small but more prominent than in *Diplacodes*, and the whitish superior abdominal appendages (fig. 309) appear in profile to be strongly sinuate and dentate ventrally. The vulvar scale (fig. 311) does not project and has two apical, rounded points widely separated by a deep embayment.

BIOLOGY

Pinhey (1961) writes of *B. leucosticta* in eastern Africa as 'gregariously abundant almost wherever there is standing water, whether lake or rain-puddle'. Adetunji & Parr (1974) describe its habitat as large bodies of water, lakes or broad rivers, but occasionally it frequents small, temporary pools. Like *Diplacodes*, it tends to be gregarious and may fly at dusk (it is sometimes attracted to light).

FLIGHT PERIOD

Captured in Spain from the end of June (teneral) to the first half of November (Belle, 1984).

DISTRIBUTION (map 93)

B. leucosticta, like *D. lefebvrei*, is a common African species that encroaches into the south-west of Europe where it has

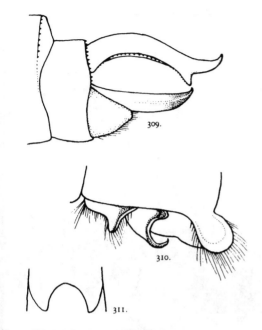

Figures 309–311 *Brachythemis leucosticta*
(309) male abdominal appendages in left profile
(310) male accessory genitalia in left profile
(311) female vulvar scale in ventral view

Map 93 – *Brachythemis leucosticta*

recently extended its range (Belle, 1984). It has been recorded from Portugal (Aguiar & Aguiar, 1983), Spain (Malaga: Compte Sart, 1962, and breeding noted by Conesa Garcia & Garcia Raso, 1982; Cadiz; Miengo in Cantabria, the northernmost known locality: Ocharan, 1983) and recently from Sardinia (Crucitti *et al.*, 1981). It is abundant in the Middle East, Saudi Arabia, and reaches western Turkey and the Caucasus.

BRACHYTHEMIS FUSCOPALLIATA (Sélys, 1887)

This species, which has a very extensive brownish area on the hindwing, is found in Iran, Iraq, Israel, Syria and in Turkey as far north as Adana.

CROCOTHEMIS Brauer, 1868

Rather larger than other European Sympetrinae and somewhat resembling *Orthetrum*, but the abdomen is broad and flattened dorsoventrally, red in the male, and the wing bases are yellow to amber. In the forewing (fig. 281, p.157) there are between nine and eleven antenodal crossveins, the most distal incomplete (the specimen illustrated in fig. 281 is aberrant in this respect), vein R_3 is but weakly sinuate, and the postdiscoidal field widens in its distal half and includes three series of cells proximally. Vein Rspl subtends a single row of cells in both fore- and hindwings.

Crocothemis is an Old World genus of eleven described species, one of which occurs in Europe.

CROCOTHEMIS ERYTHRAEA (Brullé)
Plate 27 (176,177)

Libellula erythraea Brullé, 1832, *Expéd. scient. Morée* **3** : 102.
Type locality: Messini (Nisi), Greece.

> *?Libellula victoria* Fourcroy, 1785
> *?Libellula ferruginea* Vander Linden, 1825
> *Libellula coccinea* Charpentier, 1840
> *Libellula inquinata* Rambur, 1842
> *Orthetrum lorti* Kirby, 1898

F – La Libellule écarlate, D – Feuerlibelle

C. erythraea is very close to the Asiatic species *C. servilia* (Drury, 1770) described from China and recently introduced into Florida, U.S.A. Their synonymy under the earlier name has been suggested (e.g. Dumont, 1977b), but Lohmann (1981) finds good morphological evidence in the structure of the penis for considering the two distinct. Schneider (1985) records the two species flying together in south-east Turkey.

DESCRIPTION OF ADULT

The abdomen of the mature male of *C. erythraea* is the most vividly red of all European dragonflies; the female's abdomen is brownish yellow with lateral yellow marks on S4–8. It is broad and flattened in both sexes. The eyes of the mature male are highly coloured, red above and light purple beneath. The forewings have small, yellow (female) to amber (male) basal areas and the broad hindwings have more extensive areas of the same colour that usually reach the discoidal cell although their size is subject to some variation. Some of the anterior wing veins (costa, radius etc.) are yellow to reddish according to sex, the pterostigma is long and yellowish, the membranule is grey and the legs are mainly yellowish to reddish. Male accessory genitalia (fig. 312) have a small anterior lamina as in all Sympetrinae, but the hamule is prominent and the inner hamule has two denticles at its tip. The female vulvar scale (fig. 313) projects conspicuously, rather like that of *Sympetrum vulgatum*, and its apex is entire. *C. erythraea* is variable in size, the body length usually about 38–39mm but ranging between 33–44mm.

BIOLOGY

Found near shallow, still, eutrophic waters such as small ponds, paddy fields and stagnant drainage channels. It is tolerant of some degree of salinity. A lowland species in Europe but sometimes breeding above 2000m altitude in North Africa. Adults spend much of their time perching on vegetation. Males have a fast, darting flight and hover frequently. Females visiting a pond to lay eggs are repeatedly copulated and they accumulate sperm from a succession of males. Copulation is brief and there is no evidence that male *Crocothemis* remove sperm deposited during a previous mating (Miller, in litt.). Territoriality has been investigated by Falchetti & Utzeri (1974).

FLIGHT PERIOD

The species achieves two adult generations a year in the Mediterranean region where it may be seen between April and mid-November (Aguesse, 1968).

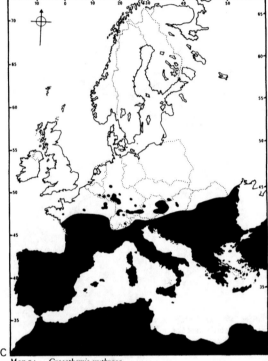

C

Map 94 – *Crocothemis erythraea*

DISTRIBUTION (map 94)

A common and widespread species in the Mediterranean region of Europe, but local in central Europe and extending no further north than the Channel Islands (Le Quesne, 1946), Belgium, Holland (Bakkum, in 1959), Luxembourg and central Germany. It is abundant in the Camargue where hundreds were found as road casualties along a one-hundred metre stretch of road in August, 1981. Widespread in tropical and North Africa and found also from the Middle East, Saudi Arabia, Yemen and Oman to Assam, Pakistan, Tadzhikistan and Afghanistan.

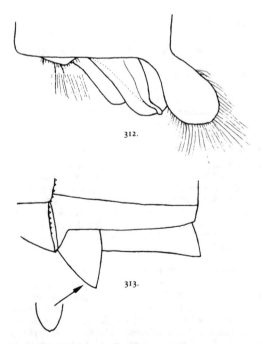

312.

313.

Figures 312, 313 *Crocothemis erythraea*
(312) male accessory genitalia in left profile
(313) female vulvar scale in left profile, and its apex in posterior view

SYMPETRUM Newman, 1833

Diplax Charpentier, 1840

Sympetrum includes eleven European species and altogether about sixty species distributed over most zoogeographical regions except Australasia. The species are small, typically yellowish to olive-brown, but mature males of all but one species (*danae*) assume an orange-red to scarlet coloration of the abdomen. All species have a prominent, posterior prothoracic lobe fringed with long setae. The front of the synthorax is brown with barely a suggestion of pale stripes, and its sides and the front of the head are yellow with black markings. Old females may develop a slight greyish pruinescence on the ventral surfaces of thorax and abdomen. Species of *Sympetrum* may be distinguished on wing characters (Table 3, p.157, and fig. 17, p.45) from other genera of Libellulidae. There is no dark marking at the base of the hindwing (although there may be an orange-yellow area), the postdiscoidal field (distal to the discoidal cell between median and cubital veins) in the forewing does not widen distally, and there are usually seven (sometimes eight) antenodals in the forewing. The last of these is oblique and normally incomplete, terminating at the subcosta and not continuing to vein R_1 like the others, but occasional examples with seven complete antenodals occur, as do individuals with asymmetrical antenodal venation. In the hindwing there are nearly always five (exceptionally six) complete antenodals. Three cells lie along the distal edge of the forewing discoidal cell. These wing characters of *Sympetrum* are shared with only one other European libellulid, *Brachythemis leucosticta*, but this has dark, transverse bands on the wings (like *S. pedemontanum*), a broad, brown and black abdomen, and is without the pilose posterior prothoracic lobe.

Species of *Sympetrum* are typical perchers, using prominent but low vegetation such as grass heads and bracken fronds. Their flight is rapid but usually brief, except on migration. A number of species are strongly migratory and northward flights may take them beyond their regular European breeding ranges. *Sympetrum* breed in shallow, stagnant waters, often exploiting small and temporary pools. Oviposition is frequently (but not always) in tandem, the male lowering his abdomen so that the tip of the female's abdomen is dipped repeatedly into the water.

Key to species of *Sympetrum*

1 Wings with a yellowish brown to reddish brown transverse band from the basal part of the pterostigma to the posterior margin .... *pedemontanum* (p.183)

- Wings without pigmented bands in their apical halves 2

2(1) Hindwing with an extensive basal yellow area which invades one or several cells of the foot-shaped anal loop. In some females the yellow basal area may be rather less extensive, but then the vulvar scale is strongly bilobed (figs 340, 341) 3

- Hindwing yellow at most only at the extreme base. Female vulvar scale not strongly bilobed 4

3(2) Many veins in basal thirds of wings reddish (male) or yellowish (female). Pterostigma pale yellow with contrasting broad black borders. Yellow area of hindwing often not extending into discoidal cell .... *fonscolombei* (p.180)

- Most veins in basal thirds of wings black. Pterostigma yellowish to reddish brown (mature insects) with borders not so strongly contrasting. Yellow area of hindwing reaching into the discoidal cell and sometimes beyond .... *flaveolum* (p.180)

4(2) Legs black (fig. 329) or with at most a narrow yellow stripe on the front femora. Body length usually less than 37mm, wing expanse less than 60mm 5

- Legs, except tarsi, mainly yellowish, or black with conspicuous yellow stripes on femora and tibiae (figs 327,328) (the femoral stripes of *nigrescens* may be more or less obliterated, but this species has dimensions greater than those indicated in the alternative) 7

5(4) Abdomen of mature male black dorsally and of female sordid yellow strongly marked with black laterally. Vulvar scale (fig. 335) very prominent. Thorax laterally with metepisternum blackish with three isolated yellow spots .... *danae* (p.182)

- Abdomen of mature male reddish dorsally and of female with only black stripes laterally. Vulvar scale not prominent. Sides of thorax yellow with narrow, black sutural lines 6

6(5) Cells between radial supplementary vein and posterior margin of forewing numerous (fig. 347), six or seven in each oblique series. Abdomen somewhat flattened, that of male not club-shaped. Male accessory genitalia (fig. 324) with anterior process of hamule much shorter and more slender than the posterior process .... *depressiusculum* (p.182)

- Cells between radial supplementary vein and posterior margin of forewing in series of only four or five (fig. 346). Abdomen more cylindrical, club-shaped in male, the middle segments constricted. Male accessory genitalia (fig. 323) with hamular processes approximately equal in size .... *sanguineum* (p.181)

7(4) Sides of thorax with sutures almost devoid of black markings. Outer surfaces of femora and tibiae yellow to reddish brown, without black markings (fig. 328) 8

- Sides of thorax with conspicuous black sutural stripes. Femora and tibiae with outer surfaces partly black (fig. 327). (Vulvar scale of female prominent; processes of hamule of male robust) 9

8(7) Vulvar scale of female not prominent (fig. 334). Anterior process of hamule of male long and slender (fig. 315) .... *meridionale* (p.178)

- Vulvar scale of female moderately prominent (cf. fig. 330). Anterior process of male hamule short (cf. fig. 324) .... *decoloratum* (p.179)

9(7) Side of thorax with metepisternum dark except for three yellow or reddish spots. Abdomen laterally with more extensive black marks .... *nigrescens* (p.177)

- Side of thorax with metepisternum not darkened, yellow like the rest of the thorax or red in mature males. Abdomen laterally with smaller black marks 10

10(9) Face with the black transverse stripe across the top of the frons terminating at the inner eye margins (fig. 348, p.177). Vulvar scale of female projecting obliquely (fig. 330). Accessory genitalia of male with anterior process of hamule longer than posterior process and with a small terminal hook (fig. 314) .... *striolatum* (p.176)

- Face with black transverse frontal stripe continuing for some distance down inner margins of eyes (fig. 350). Vulvar scale of female projecting almost at right angles from the abdomen (fig. 333). Accessory genitalia of male with anterior process of hamule shorter than posterior process, curved but not apically hooked (fig. 316)...*vulgatum* (p.177)

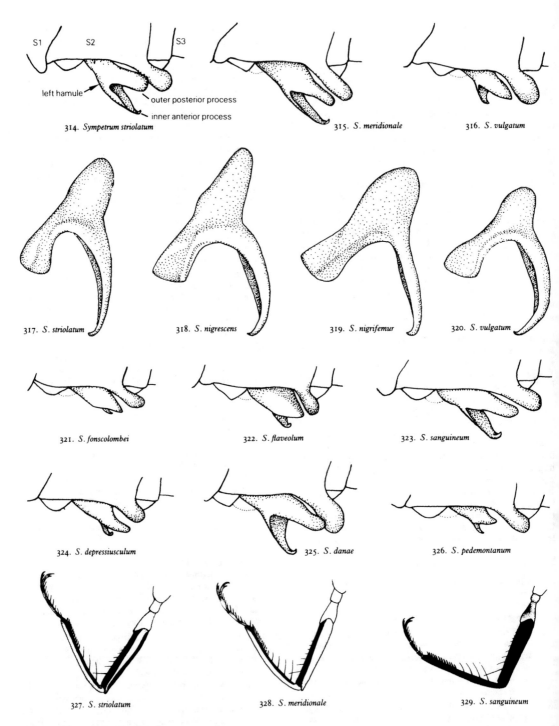

314. *Sympetrum striolatum*

S1 S2 S3

left hamule
outer posterior process
inner anterior process

315. *S. meridionale*

316. *S. vulgatum*

317. *S. striolatum*

318. *S. nigrescens*

319. *S. nigrifemur*

320. *S. vulgatum*

321. *S. fonscolombei*

322. *S. flaveolum*

323. *S. sanguineum*

324. *S. depressiusculum*

325. *S. danae*

326. *S. pedemontanum*

327. *S. striolatum*

328. *S. meridionale*

329. *S. sanguineum*

Figures 314–316, 321–326 Male *Sympetrum* species
accessory genitalia in left lateral view (setae omitted)
Figures 317–320 Male *Sympetrum* species, hamules
Figures 327–329 *Sympetrum* species, legs

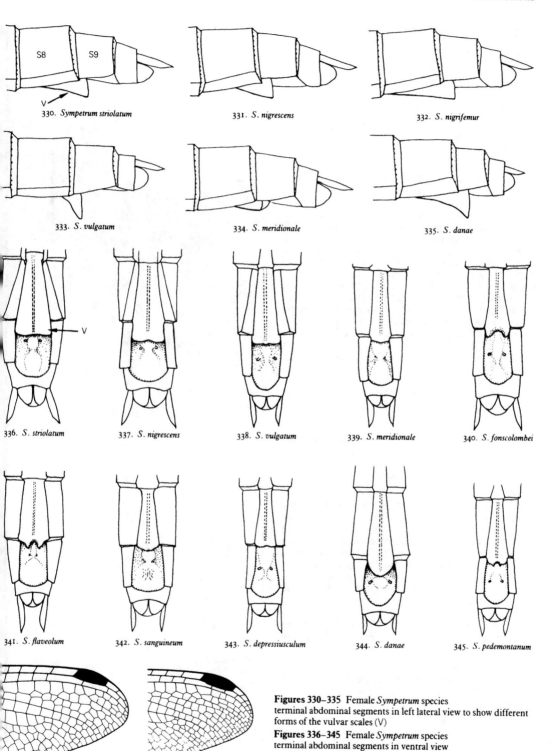

330. *Sympetrum striolatum*

331. *S. nigrescens*

332. *S. nigrifemur*

333. *S. vulgatum*

334. *S. meridionale*

335. *S. danae*

336. *S. striolatum*

337. *S. nigrescens*

338. *S. vulgatum*

339. *S. meridionale*

340. *S. fonscolombei*

341. *S. flaveolum*

342. *S. sanguineum*

343. *S. depressiusculum*

344. *S. danae*

345. *S. pedemontanum*

346. *S. sanguineum*

347. *S. depressiusculum*

Figures 330–335 Female *Sympetrum* species
terminal abdominal segments in left lateral view to show different
forms of the vulvar scales (V)

Figures 336–345 Female *Sympetrum* species
terminal abdominal segments in ventral view

Figures 346, 347 *Sympetrum* species
right forewing apices

SYMPETRUM STRIOLATUM (Charpentier)
Plate 27 (180–182)

Libellula striolata Charpentier, 1840, *Libell. europ. descr.*
  depict. : 78.
Type locality: Silesia, Poland.
  Libellula ruficollis Charpentier, 1840
  Libellula sicula Hagen, 1840
  Libellula macrocephala Sélys, 1841

E – Common Sympetrum, **F** – Le Sympétrum à coté strié,
D – Grosse Heidelibelle

DESCRIPTION OF ADULT

Females and immature males are predominantly yellowish
to olive-brown. The mature male has an orange-red abdomen but the red colour is less intense and perhaps longer in
developing than in many of the other species in the genus.
Small, paired, black points towards the posterior of the
abdominal segments remain surrounded by yellow in the
male. Old females may also develop some red pigmentation
about the abdominal mid-dorsal line. The legs are distinctly
striped with yellow on their external faces (fig. 327), and the
wings have only a very small basal yellow patch and are
mostly clear, although in older individuals a brownish tinge
may spread from the veins in the more apical parts of the
wings. In contrast to most other species, the black line
across the top of the frons does not continue down the eye
margin (fig. 348). The vulvar scale beneath the female's
eighth abdominal segment is moderately prominent, forming an angle of about 30° to the ventral surface of the
abdomen and, in side view, leaving about the middle of
the ventral surface of S8 (fig. 330). The structure of the
accessory genitalia of the male (figs 314,317), especially
the hamule, is diagnostic, as it is for other species in the
genus, but the critical features are small and demand care-
ful examination. One of the larger species of *Sympetrum*,
S. striolatum usually attains a body length of 38–43mm and
a wingspan of 59–63mm, but size is a variable character
and these limits are sometimes exceeded.

BIOLOGY

Larval development takes place in ponds and other stagnant, shallow waters, sometimes slightly brackish, but the
adult, particularly whilst immature, may be found far from
water in woods, parks and gardens. *S. striolatum*, like
several of its congeners, is strongly migratory. Both sexes
perch on prominent twigs and low branches or on stones
and paths, often returning repeatedly to a selected spot.
Females may lay eggs alone or whilst in tandem; in the latter
circumstance the female is lowered to the water surface by
inflexion of the abdomen of the flying male, but in the
former situation it is the female abdomen which is flexed
whilst hovering over water (Moore, 1952b). The life history
of *S. striolatum* is described by Gardner (1950b).

FLIGHT PERIOD

Greatest numbers are to be seen in autumn, but it appears in
mid-May on the Mediterranean coast, and a few persist into
November or, exceptionally, December.

DISTRIBUTION (map 95)

S. striolatum is abundant and widespread, the most frequently encountered species of *Sympetrum* over much of
western Europe. It has been recorded from every European
country except Iceland, and its range extends to North

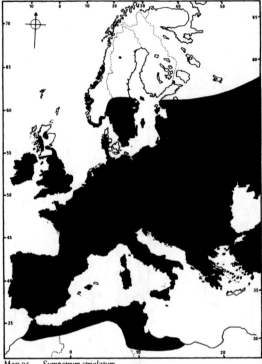

Map 95 – *Sympetrum striolatum*

Map 96 – *Sympetrum nigrescens*

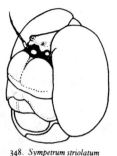

348. *Sympetrum striolatum*

349. *S. nigrescens*

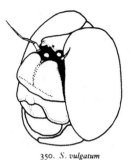

350. *S. vulgatum*

Figures 348–350 *Sympetrum* species, heads in frontolateral view

Africa and, in a number of named forms, right across Asia to Japan. In Europe it is absent only from northern Scandinavia, occurring in Norway mainly in coastal districts as far north as North Trondelag, but scarce and restricted to the south in Sweden. In Finland it is reported only from the Åland Islands, but see under the next species.

SYMPETRUM NIGRESCENS Lucas Plate 27 (183)

Sympetrum nigrescens Lucas, 1912, *Entomologist* 45 : 171.
Type locality: Lochinver, Sutherland, Scotland.

E – Highland Sympetrum

Very closely allied to *S. striolatum* and known as *S. s. nigrifemur* (Sélys) by many British authors until Gardner (1955a) concluded that it is specifically distinct from both *striolatum* and *nigrifemur*. Gardner's interpretation of the status of *nigrescens* is followed here, somewhat tentatively, and the matter is still not firmly resolved.

DESCRIPTION OF ADULT

S. nigrescens is a darker insect than *striolatum* with the metepisternum contrastingly darker than the adjacent parts of the thoracic sides and it has three isolated yellowish to reddish spots. The femora are sometimes almost entirely black, although a pale tibial stripe remains. The black stripe across the top of the frons usually continues downwards for a short distance at the eyes (fig. 349), and the abdomen of females has more extensive lateral black marks than that of *striolatum*. These colour characters are, however, somewhat variable, and in localities where both *nigrescens* and *striolatum* occur, coloration of the two species may intergrade. Hämäläinen (1985a) reports specimens of '*S. striolatum*' with intermediate characters in the Åland Islands, Finland. The most compelling reason to consider *nigrescens* a good species, and not just a melanistic form of *S. striolatum* associated with Atlantic coastal districts of Britain and Scandinavia, is provided by the structure of the genitalia. The anterior process of the hamule (fig. 318) is more curved and it diverges more strongly from the posterior process than in *striolatum*. The female vulvar scale (fig. 331), in lateral view, is shorter and leaves nearer to the apex of S8. *S. nigrescens* is on average slightly smaller than *S. striolatum*.

BIOLOGY

Occurs in light woodland, on moorland and coastal dunes, close to the pools and lochans in which it breeds. *S. nigrescens* does not seem to be migratory.

FLIGHT PERIOD

June to September.

DISTRIBUTION (map 96)

Restricted to coastal parts of Ireland, the Isle of Man, Scotland where it is most frequent near the north and west coasts and on the Inner and Outer Hebrides, south and south-west Norway, and possibly southern Finland (Verhoeven, 1980) although identity of the latter material is in doubt. This is probably a relict distribution (Bartenef, 1919).

SYMPETRUM NIGRIFEMUR (Sélys, 1884)

S. nigrifemur occurs on the islands of Madeira and the Canaries (Gardner, 1963). It is another species in the *striolatum*-group. It was originally described as a race of *Diplax vulgata*, later as a subspecies of *S. striolatum*, but Gardner (1955a) showed it to be specifically distinct. Like *S. nigrescens*, it has extensive black markings but it differs from both *nigrescens* and *striolatum* especially in genitalic characters. The two processes of each hamule (fig. 319) diverge strongly, as in *nigrescens*, but the anterior process is almost straight and basally little expanded; the vulvar scale (fig. 332) in lateral view is robust and leaves near the base of S8. It is a larger species than *S. striolatum* and probably flies throughout the year (records for March, April, August and December).

SYMPETRUM VULGATUM (Linnaeus)
Plate 27 (178,179)

Libellula vulgata Linnaeus, 1758, *Syst. Nat.* (Edn. 10) 1 : 543.

Type locality: Sweden.

Libellula variegata Müller, 1764

E – Vagrant Sympetrum, **F** – Le Sympétrum commun,
D – Gemeine Heidelibelle

DESCRIPTION OF ADULT

This is the type species of *Sympetrum*. It resembles *S. striolatum* but the mature male is deeper red and the paired black points on the abdominal segments are not ringed with yellow. The longitudinal wing veins are basally reddish. In both sexes, the black transverse bar at the top of the frons continues down the inner margins of the eyes (fig. 350). Yellow areas on the sides of the thorax tend to be suffused with olive (females) or red (males). The very prominent female vulvar scale (fig. 333) is clearly visible to the naked eye, and the accessory genitalia of the male (figs 316,320) differ from those of *striolatum* and *nigrescens* especially in having a shorter anterior hamular process. The male abdomen is slightly constricted in the middle, although less club-shaped than that of *sanguineum*. *S. vulgatum* is rather smaller than *striolatum*, usually 35–40mm long with a wingspan of 55–60mm.

BIOLOGY

The habitat of *S. vulgatum* is similar to that of *S. striolatum*, but it is more often found at considerable altitude (e.g. 2500m in the Ötztal Alps). Mayer (1961) gives an account of the species.

FLIGHT PERIOD

Early July to sometimes as late as the beginning of November.

DISTRIBUTION (map 97)

Common in central and north-eastern Europe, extending from Lapland south-west to the Spanish Pyrenees and south to northern Italy. Its range eastwards extends across the Urals to Siberia and China. It does not occur in North Africa. *S. vulgatum* is a rare migrant to south-eastern England where it was last recorded in 1946 (Hammond, 1983).

SYMPETRUM MERIDIONALE (Sélys)
Plate 27 (184)

Libellula meridionalis Sélys, 1841, *Rev. zool. Guérin Méneville* : 245.

Type locality: originally described from insects from Sardinia and Sicily.

Libellula hybrida Rambur, 1842
Libellula nudicollis Hagen in Sélys & Hagen, 1850

F – Le Sympétrum méridional, **D** – Südliche Heidelibelle

DESCRIPTION OF ADULT

The almost total absence of black pigmentation about the sutures of the sides of the thorax distinguishes this *Sympetrum* from all other European species except *S. decoloratum*. The metapleural spiracle is dark and conspicuous against the yellowish ground-colour, and the suture between meso- and metapleuron is indicated by an exceedingly fine, dark line running ventrally from a small, dark point just above and in front of the spiracle. The mature male is bright orange-red and the yellowish brown legs are marked with black only on the tarsi and anterior surfaces of femora and tibiae (fig. 328). The hamule of the male accessory genitalia (fig. 315) has long processes, and the female vulvar scale (figs 334,339), unlike that of the preceding species (to which *S. meridionale* is allied), is scarcely visible in lateral view. *S. meridionale* is of similar size to *S. vulgatum*.

BIOLOGY

Larvae develop in small, shallow ponds or sheltered backwaters of lakes where there is an ample growth of emergent vegetation. It frequently flies far from water, and its flight is rapid.

FLIGHT PERIOD

End of May to mid-October.

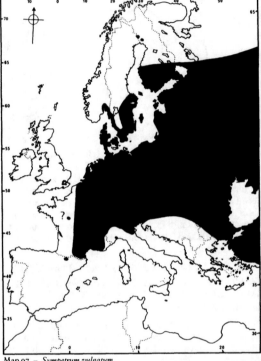

Map 97 – *Sympetrum vulgatum*

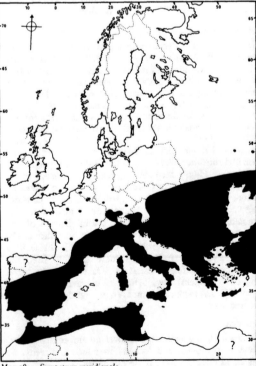

Map 98 – *Sympetrum meridionale*

DISTRIBUTION (map 98)

As its name indicates, this is chiefly a southern species in Europe, recorded from Spain, France, Italy, the Balkans and the Mediterranean islands, but it is also found in central Europe (Belgium, Switzerland, Austria, Czechoslovakia, Hungary, Germany, Poland) and in the Alps it has been observed at an altitude of 3000m. Migrants very occasionally reach the Netherlands (one nineteenth-century record) and the British Isles (one taken in 1847 and one in 1901: Corbet, Longfield & Moore, 1960). Outside Europe, *S. meridionale* occurs in North Africa and eastwards from the Middle East to Kashmir, Manchuria, Mongolia and south-east Siberia.

SYMPETRUM DECOLORATUM (Sélys)

Libellula decolorata Sélys, 1884, *Annls Soc. ent. Belg.* 28 : 35.

Type locality: Asia Minor.

DESCRIPTION OF ADULT

S. decoloratum is an insect of desert country, and like other dragonflies which inhabit this type of terrain, it is very pale in colour. Black marks are reduced to lining the humeral suture at upper and lower ends, a spot ventrally on the metapleural suture, small marks on the sides of the abdomen (but without mid-dorsal black marks in the male on S8 and S9), narrow stripes on flexor surfaces of femora and tibiae, and a small mark at the base of the vertex. The pterostigma is pale yellowish to brown, and there is some amber coloration at the bases of the wings. The female appears yellowish, the mature male is pale red. The female vulvar scale is moderately projecting, much as in *S. striolatum* (figs 330,336), and its apex is weakly bilobed. The structure of the male accessory genitalia is reminiscent of that of *S. depressiusculum* (fig. 324) with a short anterior hamular branch, but the terminal abdominal appendages are distinctive (Bartenef, 1919) having the apex of the inferior appendage situated half-way between the apex and ventral angle of the superior appendage (fig. 351, cf. *striolatum* fig. 352). Body length 34–36mm.

BIOLOGY

S. decoloratum in Algeria congregates on acacias and other bushes in dry, warm valleys (Dumont, 1978).

Map 99 – *Sympetrum decoloratum*

FLIGHT PERIOD

North African material that has been seen was collected between February and May, and between July and September inclusive.

DISTRIBUTION (map 99)

Asia Minor, Caucasus, Iraq, Pakistan and North Africa (Libya, Tunisia, Algeria). Included in the present work on the basis of two specimens from Catalonia, Spain in Sélys' collection, confirmed as *decoloratum* by Ris (1911b). Dumont (1977b) distinguishes the subspecies *S. d. sinaiticum* from Tozeur in western Tunisia, as well as Sinai, and suggests that Iberian material will belong to the same taxon.

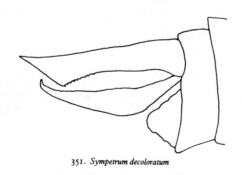

351. *Sympetrum decoloratum*

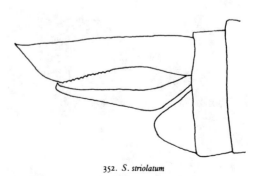

352. *S. striolatum*

Figures 351, 352 *Sympetrum* species
male abdominal appendages in right lateral view

SYMPETRUM FONSCOLOMBEI (Sélys)
Plate 27 (185,186)

Libellula fonscolombii Sélys, 1840, *Monograph. Libell. Europ.* : 49.

Type locality: Aix-en-Provence, France.

Libellula erythroneura Schneider, 1845

E – Red-veined Sympetrum, F – Le Sympétrum à nervures rouges, D – Frühe Heidelibelle

DESCRIPTION OF ADULT

The abdomen of the mature male is bright red and this colour extends to many of the veins in the proximal halves of the wings. In the female, these veins are yellowish. The wings usually have extensive saffron basal patches, exceeded in area only by those of *S. flaveolum*, but the size of these patches is variable. In some populations in south-east Europe they are intense but very small. The pterostigma is pale with conspicuous, broad, black borders. These wing characters alone render *S. fonscolombei* easily recognizable, and the genitalia are also distinctive. The hamule of the male has a very short anterior process (fig. 321), as in *S. flaveolum* but less robust. The vulvar scale of the female (fig. 340) is not prominent, and its distal edge is produced into two widely-separated lobes; the ninth abdominal sternum bears two distinct prominences which are usually more evident than in other species. *S. fonscolombei* is a relatively large species, about 38–40mm in length and with a wingspan of 60–63mm (pterostigma 2.5–3.0mm long).

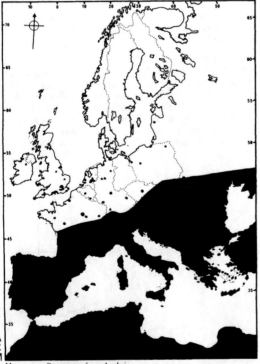

Map 100 – *Sympetrum fonscolombei*

BIOLOGY

Breeds usually in quite large but shallow water bodies, including paddy fields, but as a strongly migratory species it can be found far from water.

FLIGHT PERIOD

Mainly June to November, but recorded from the Camargue as early as 28 March by Aguesse (1968) who finds there are two generations a year in the south of France. Robert (1958) describes migrants arriving in Switzerland in early August to produce adult offspring by mid-September. *S. fonscolombei* and *Crocothemis erythraea* are probably the only European species of Anisoptera sometimes to have two adult generations in one year.

DISTRIBUTION (map 100)

Its European distribution is principally Mediterranean, from Spain to Greece (including Dodecanese) but migrations take the species sporadically north to Belgium, Holland, northern Germany, Poland and the British Isles. Breeding is reported on Jersey, Channel Islands by Le Quesne (1946). In 1911 some individuals reached Scotland. *S. fonscolombei* is found also throughout Africa (including Madeira, the Canary Islands and Azores) and the Middle East to Afghanistan, India, Sri Lanka, Siberia, Mongolia and the Ryukyu Islands.

SYMPETRUM FLAVEOLUM (Linnaeus)
Plate 27 (187,188)

Libellula flaveola Linnaeus, 1758, *Syst. Nat.* (Edn. 10) **1** : 543.

Type locality: Sweden.

Libellula rubra Müller, 1764
Libellula aurea Scopoli, 1772

E – Yellow-winged Sympetrum, F – Le Sympétrum jaune, D – Gefleckte Heidelibelle

DESCRIPTION OF ADULT

The normally very extensive saffron-yellow basal areas on the wings distinguish this species from all other European *Sympetrum* except sometimes *S. fonscolombei*, but from the latter, *S. flaveolum* may be separated by its mainly black venation, red-brown pterostigma, and by its genitalia. The anterior process of the hamule (fig. 322) is small but robust, and the posterior edge of the vulvar scale (fig. 341) bears two quite closely-approximated points; no distinct prominences are visible on the ninth sternum of the female's abdomen. The yellow wing colour is rather variable in extent, especially in females where it normally occupies a rather greater area including a supplementary patch below the node of each wing. In old females there may be a reduction in yellow colour on the wings. The pterostigma is rather short (2.0–2.5mm long), particularly in males where its posterior edge is shorter than the post-pterostigmal section of vein R_1. *S. flaveolum* is usually smaller than *S. fonscolombei*: body length 32–37mm, wing expanse 50–59mm.

BIOLOGY

An insect most often found amongst rushes and other marginal vegetation of ditches, ponds and the still backwaters of lakes and rivers. Females mostly oviposit in tandem amongst vegetation in very shallow water at the edges of water bodies. The flight of *S. flaveolum* is not so fast as that of *fonscolombei* or the species in the *striolatum* group, being rather fluttering and with frequent perching. Nevertheless, it is a migratory species and individuals may appear at places distant from suitable breeding areas.

FLIGHT PERIOD

From the end of June, persisting into November or even December, but over most of Europe it is commonest in August and September.

DISTRIBUTION (map 101)

S. flaveolum may be found over most of southern, central and eastern Europe, but its appearance in Britain and Scandinavia is erratic and subject to the arrival of migrants. Mainly an eastern species, it is very abundant in Siberia and extends to Kamchatka and Japan. It is not found in Africa.

SYMPETRUM SANGUINEUM (Müller)
Plate 28 (189,190)

Libellula sanguinea Müller, 1764, *Fauna Insect. fridrichs-dalina* : 62 (*nec* Burmeister, 1839 : 858).
Type locality: Denmark.

Libellula rufostigma Newman, 1833
Libellula roeselii Curtis, 1838
Libellula nigripes Charpentier, 1840

E – Ruddy Sympetrum, **F** – Le Sympétrum rouge sang, **D** – Blutrote Heidelibelle

DESCRIPTION OF ADULT

This is one of the black-legged species of *Sympetrum*. The outer faces of the femora and tibiae of the fully mature insect (fig. 329) are entirely black, although there may be a small yellow streak internally at the base of the front femur. A club-shaped male abdomen, strongly constricted at S4, is a feature shared by *sanguineum* and *danae*, but the latter species, when mature, has a black abdomen whereas that of *sanguineum* is deep red with relatively large black marks mid-dorsally on S8 and S9. The saffron patch at the base of the hindwing is small but strongly marked. The two processes of the hamule of the male accessory genitalia (fig. 323) are of about equal length, and the female vulvar scale (fig. 342) is neither prominent nor bilobed. Ventrally, the ninth sternum of the female abdomen has very long (about half as long as a superior appendage), perpendicular hairs. Body length typically is about 34–36mm and the wingspan is between 53–59mm.

BIOLOGY

Inhabits weedy ponds and ditches, frequently in woodland, perching on dead twigs or, when the air temperature is low, on bare ground and rocks that radiate heat. *S. sanguineum* may breed in slowly flowing water provided that emergent plants such as *Typha* are present.

FLIGHT PERIOD

From mid-June to mid-October but most numerous in August and September, sometimes lasting until December in frost-free localities.

DISTRIBUTION (map 102)

All of Europe except Iceland and the more northerly parts of the British Isles and Scandinavia. It may also be absent from the Mediterranean islands (except Sicily). It breeds in

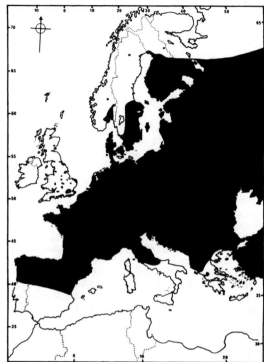

Map 101 – *Sympetrum flaveolum*

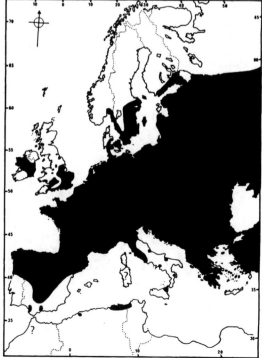

Map 102 – *Sympetrum sanguineum*

southern Britain but its numbers are probably reinforced by migration from the continent. It extends east to western Siberia and is found also in North Africa (Algeria, Morocco).

SYMPETRUM DEPRESSIUSCULUM (Sélys)
Plate 28 (191,192)

Libellula depressiuscula Sélys, 1841, *Rev. zool. Guérin Méneville* : 244.
Type locality: Arona, Piedmont, Italy.

  Libellula genei Rambur, 1842

F – Le Sympétrum à corps déprimé, **D** – Sumpf Heide-libelle

DESCRIPTION OF ADULT

Allied to *S. sanguineum* from which it differs in both sexes in having denser venation in the apical halves of the wings (fig. 347, cf. fig. 346). Between the central part of vein Rspl and the wing margin there are usually six or seven cells in each obliquely running series (four or five in *sanguineum* and all other European species except *pedemontanum*). The hind-wing is relatively broad, much like that of *pedemontanum*; when the costal margin is at right angles to the body, the posterior wing margin reaches to about the apex of S4. The wings are hyaline but, when viewed obliquely, with pronounced golden reflections. The abdomen is somewhat flattened dorso-ventrally (hence the specific name) and, unlike that of *sanguineum* and *danae*, not constricted medially. In the mature male the abdomen dorsally is bright red with yellow side-markings. Between red and yellow on

Map 103 – *Sympetrum depressiusculum*

each side of S4–8 there are narrow, black longitudinal lines. The anterior process of the hamule of the male secondary genitalia (fig. 324) is slender and hooked, much shorter than the posterior process. The ninth abdominal sternum of the female (fig. 343) carries two distinct prominences. In overall dimensions, *S. depressiusculum* is similar to, or very slightly smaller than, *S. sanguineum*.

BIOLOGY

Apparently not a strongly migratory species and found usually quite close to the sedgy ponds and dykes in which it breeds. It is mainly a lowland insect. Aspects of its reproductive behaviour are described by Miller *et al.* (1984).

FLIGHT PERIOD

From the end of June until October or early November.

DISTRIBUTION (map 103)

Eastern Europe to Italy (predominantly in the north), eastern France (common in the Camargue) and the Netherlands, but not in Scandinavia, the British Isles or the Iberian Peninsula. Nowhere common in West Germany, except perhaps in Bavaria, but widespread north to Löningen (Lohmann, 1980). Its reported presence in Sicily and Sardinia requires confirmation. In eastern Europe, *S. depressiusculum* has extended its range northwards during the present century. In Asia it extends east to Manchuria, the east coast of Siberia, and Japan south to the Ryukyu Islands. Three specimens taken at Lake Tonga, Algeria (Martin, 1910) constitute the only record for North Africa.

SYMPETRUM DANAE (Sulzer) Plate 28 (193–195)

Libellula danae Sulzer, 1776, *Ab. Gesch. Ins. Linn. Syst.* : 169.
Type locality: Lac de Joux, Switzerland.

  Libellula scotica Donovan, 1811 (Some uncertainty about the identity of *danae* has led several authors to use Donovan's name, e.g. Fraser, 1954.)
  Libellula nigra Vander Linden, 1825
  Libellula veronensis Charpentier, 1825
  Libellula nigricula Eversmann, 1836

E – Black Sympetrum, **F** – Le Sympétrum noir,
D – Schwarze Heidelibelle

DESCRIPTION OF ADULT

Another small, black-legged species with broad hindwings, the posterior margins of which extend beyond the apex of S4 when the wings are at right angles to the body. The mature male differs from all other European species of *Sympetrum* in having a predominantly black, not red, abdomen. The abdomen is club-shaped, as in *S. sanguineum*. Females and immature males are also extensively black, and the metepisternum is very dark with three isolated yellow spots (similar to the pattern in *S. nigrescens*). Three yellow marks on the ventral surface of the female thorax behind the posterior coxae form a distinctive fleur-de-lis, and the dorsal surface of the synthorax bears a black triangle with its base behind the pronotum. Basal yellow wing marks are small but usually distinct. The black pterostigma is short (2.0mm or rather less in length), especially in males, shorter than the post-pterostigmal section of vein R_1 and at most three times as long as broad.

Male accessory genitalia (fig. 325) are large and robust with the two processes of the hamule almost equal in size, as in *sanguineum* but stouter. The female vulvar scale (figs 335,344) is large and prominent. Body length usually in the range 29–34mm; wingspan 45–54mm.

BIOLOGY

A species of peat-moss and moorland, sometimes in very open woodland, breeding in ponds, bog holes and drainage ditches. It seldom flies over open water. In more northerly parts of its range it occurs at sea-level, but further south it is confined to hill country and in the Alps it ascends to a height of 2000m. Populations often attain a relatively high density. *S. danae* does not often take long flights and is seldom found very far from its breeding places, but migrations have been recorded off the Irish coast (Gardner, 1955b) and in the USSR. For biological observations on *S. danae* (in Japan) see Sonehara (1965). Waringer (1983) describes larval development.

FLIGHT PERIOD

End of June until October.

DISTRIBUTION (map 104)

S. danae is a circumboreal species. It occurs over much of northern Europe (Britain, Holland, Belgium, Scandinavia, Germany, Poland, Russia), but it is confined to mountains further south (southern France, Switzerland, Austria, northern Italy). As different subspecies it extends across North America and Asia (from just south of the Arctic Circle) to Japan.

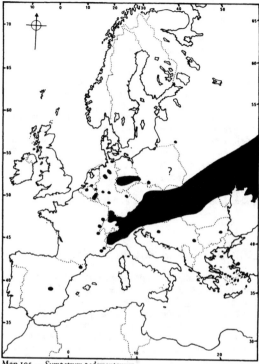

Map 105 – *Sympetrum pedemontanum*

SYMPETRUM PEDEMONTANUM (Allioni)
Plate 28 (196)

Libellula pedemontana Müller in Allioni, 1766, *Mél. phil. math. Soc. r. Turin* 3 : 194.

Type locality: Turin, Italy.

Libellula harpedone Sulzer, 1776
Libellula sibirica Gmelin, 1788

F – Le Sympétrum du Piémont, D – Gebänderte Heide-libelle

DESCRIPTION OF ADULT

S. pedemontanum is allied to the three preceding species and, like them, is small and has black legs and broad hindwings. It is immediately recognizable, however, by the transverse, brownish band behind the pterostigma on each wing, a feature in other European dragonflies seen only in *Brachythemis leucosticta*. The venation of the wings is brownish, fine and apically dense with six or seven cells in series between vein Rspl and the margin (as in *S. depressiusculum*, fig. 347). The pterostigma is elongated, bright red in mature males and yellowish brown in females. The male has a red, broad and unconstricted abdomen, and the accessory genitalia (fig. 326) are not unlike those of *S. depressiusculum* but with the anterior process of the hamule less curved and the posterior process relatively shorter. The female vulvar scale (fig. 345) has a small, apical incision. Body dimensions are similar to those of *S. danae*.

BIOLOGY

Frequents marshes, swamps and pools with dense reed beds, most often in hill country. It has a flitting flight, like

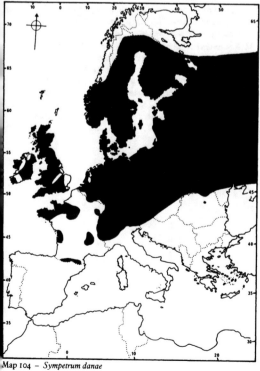

Map 104 – *Sympetrum danae*

that of *S. danae*, and rests frequently on the tips of rushes and sedges.

FLIGHT PERIOD

From mid-July to the end of October.

DISTRIBUTION (map 105)

Central and southern Europe west to Spain (two reported localities). In mountains in northern Italy as far south as the valley of the River Po, Switzerland, Austria, Germany north to Lüneburg Heath and Schleswig-Holstein, spreading northwards to occupy almost all of West Germany after its initial discovery there in 1906, Hungary, Czechoslovakia, Romania and east through Russia and Asia to Japan (*S. p. elatum* (Sélys, 1872)). Rare in Yugoslavia, Bulgaria, Belgium, France and the Netherlands (discovered in 1982 in North Brabant and Limburg: Huijs & Peters, 1984). Absent from the British Isles and Scandinavia.

LEUCORRHINIA Brittinger, 1850

An Holarctic genus of about fifteen species, five of which occur in Europe. They are small to medium-sized dragonflies, easily recognizable by their mostly white faces and dark brown basal marks on the hindwings. The pterostigma is very short, only about 1.5 times to barely twice as long as broad in males, slightly longer in females. Distal to the pterostigma, the costa, vein R_1, and crossveins between the two including the one defining the distal edge of the pterostigma, are whitish when viewed by reflected light. There are seven or eight complete forewing antenodal crossveins (fig. 282, p.157). The posterior lobe of the pronotum is upstanding, slightly bilobed, and provided with a crown of long hair. Relatively long and dense hair clothes the entire thorax, and the abdomen also is hairy. The body is dark but brightly marked with red or yellow in some species, partly pruinescent in others. Antehumeral thoracic stripes are poorly-defined. In all species, the sides of the thorax are black with a yellowish, horizontal, subalar stripe, a yellowish spot above the insertion of each leg, and a large yellow mark on the metepimeron. The legs are black.

All species are moorland or heathland insects, and in southern Europe they occur only in upland areas. They breed in acid, *Sphagnum* pools and fly in late spring and early summer. Females oviposit unaccompanied by males.

Key to species of *Leucorrhinia*

1   Anal appendages black. Abdomen with large red or yellow marks on S7 as well as on some or all of the more anterior segments, never pruinescent. Forewing nearly always with a short, brown mark between the bases of veins CuP and A_1 2

–   Anal appendages whitish. Abdomen of female with pale markings only to S6 on which they are small; mature male with S3+4 strongly pruinescent. Forewing without any proximal brown marking 4

2(1)   Male with a large, triangular yellow spot on the dorsum of S7, contrasting in colour with brownish spots on the more anterior segments, and abdomen expanded at S5 and S6. Female with vulvar scale (fig. 363) terminating in two, closely-approximated, triangular teeth which extend about one-quarter the length of sternum S9 .... *pectoralis* (p.189)

–   Male abdomen with dorsal spots unicolorous red and abdomen not medially expanded. Female vulvar scale of different form and shorter 3

3(2)   Body length usually less than 37mm. Antenodal section of costa dark brown. Pale dorsal abdominal marks smaller. Male hamule with anterior process in profile strongly projecting and only weakly curved, and with a smaller forward-curving posterior projection; anterior lamina very short (fig. 356). Female vulvar scale (fig. 361) terminating in two processes which are about one-fifth as long as sternum S9 .... *dubia* (p.187)

–   Body length usually between 39–41mm. Antenodal section of costa yellowish, edged with black. Pale dorsal abdominal marks larger. Male hamule with anterior process in profile hooked posteriorly, and with a large, rounded, posterior lobe but no projection; anterior lamina large and almost reaching to apex of hamule (fig. 357). Female vulvar scale (fig. 362) apically without processes .... *rubicunda* (p.188)

4(1)   Abdomen in both sexes strongly clubbed, broadest at S7 and S8. Male with pterostigma white on upper surface, dark brown on lower surface. Hindwing with two cubito-anal crossveins. Labium entirely black .... *caudalis* (p.186)

–   Abdomen not clubbed. Both sexes with pterostigmata brown on both surfaces. Hindwing with a single cubito-anal crossvein. Labium with a large, whitish spot laterally below each eye (fig. 353) .... *albifrons* (p.186)

Figure 353
Leucorrhinia albifrons
head in left lateral view
with the pale spot on the
lateral lobe of the labium
arrowed. This area is
entirely black in other
European species of
Leucorrhinia

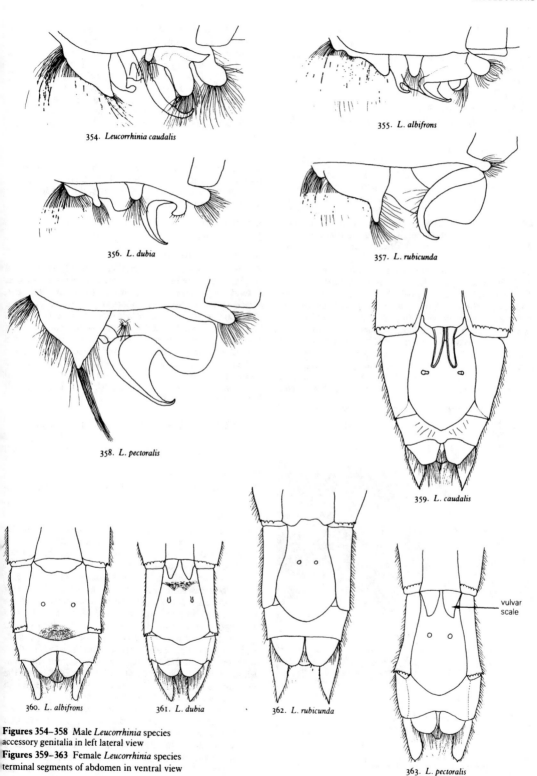

354. *Leucorrhinia caudalis*

355. *L. albifrons*

356. *L. dubia*

357. *L. rubicunda*

358. *L. pectoralis*

359. *L. caudalis*

360. *L. albifrons*

361. *L. dubia*

362. *L. rubicunda*

vulvar scale

363. *L. pectoralis*

Figures 354–358 Male *Leucorrhinia* species accessory genitalia in left lateral view

Figures 359–363 Female *Leucorrhinia* species terminal segments of abdomen in ventral view

LEUCORRHINIA CAUDALIS (Charpentier)
Plate 28 (199,200)

Libellula caudalis Charpentier, 1840, *Libell. europ. descr. depict.* : 89.

Type locality: Brzeg, Silesia, Poland.

Libellula fallax Eversmann, 1841
Libellula ornata Brittinger, 1845

F – La Leucorrhine à large queue, **D** – Zierliche Moosjungfer

DESCRIPTION OF ADULT

The club-shaped abdomen, distinctly expanding after S5, is very characteristic of *L. caudalis*. The club shape is accentuated in males because of their narrower fourth abdominal segment. Mature males have a black abdomen with grey pruinosity on S3–5, and the anal appendages are whitish terminating in very small, black points in both sexes. The female abdomen has yellow, mid-dorsal marks on S2–6 (much reduced on S6). Basal brown wing marks are restricted to the hindwings (absent altogether as a rare aberration), in females broadly surrounded by a yellow cloud. In some females, there is a brown suffusion beneath the pterostigmata of all wings. The pterostigma of the mature male is white on the upper surface but brown on the under surface of the wing (fig. 282, p.157); that of the female is dark brown on both surfaces. Two hindwing cubito-anal crossveins, the proximal one within the basal brown wing mark, seem always to be present, a character seen only occasionally in *L. pectoralis* and very rarely if at all in the other European species. Male accessory genitalia (fig. 354) have a large anterior lamina clothed with long setae,

the anterior hamule is shorter than the anterior lamina and terminates in a knob, the posterior hamule is large and posteriorly hooked over a hair-crowned lobe which has a backwardly-pointing basal tooth overlying the stem of the genital lobe. The apex of the female vulvar scale (fig. 359) terminates in two subparallel processes which extend over about one-third of the length of the ninth sternite. *L. caudalis* is a rather small dragonfly, usually 36–38mm long, but its apically swollen abdomen tends to make it appear larger.

BIOLOGY

Associated with mesotrophic to eutrophic, acid, moorland ponds with sedges and floating plants such as *Nymphaea* and *Potamogeton*. Pajunen (1964b) describes some aspects of behaviour in *L. caudalis*.

FLIGHT PERIOD

Mid-May to early in July.

DISTRIBUTION (map 106)

A west Siberian species extending to mainly eastern and central Europe with many recorded localities in Russia, Poland and Germany. Known also from Hungary, Czechoslovakia, northern Austria, Denmark, Holland, Belgium and eastern France where it occurs west to Indre (Aguesse, 1968). It used to occur in several places in Switzerland, particularly in the east (Robert, 1958), but it is now on the verge of extinction and found only in canton Aargau. Widespread in the southern half of Finland, and southern Sweden, and known also from the south of Norway, but absent from the British Isles, Iberia and all Europe south of the Alps.

LEUCORRHINIA ALBIFRONS (Burmeister)
Plate 28 (197,198)

Libellula albifrons Burmeister, 1839, *Handbuch. Entomol.* 2 : 851.

Type locality: Berlin, Germany.

F – La Leucorrhine à front blanc, **D** – Östliche Moosjungfer

DESCRIPTION OF ADULT

This, the type species of the genus *Leucorrhinia*, resembles *L. caudalis* in that the abdomen of the mature male is dorsally black with greyish pruinescence on the anterior segments, most strongly developed on S3, and the anal appendages are whitish. It is 37–39mm in body length, and the abdomen is fusiform, narrowed at the third segment and broadening slightly posteriorly. The abdomen of the newly-emerged male is marked with yellow as follows: a small spot on each side of S1, a mid-dorsal kidney-shaped spot on S2, a transverse basal band and a small, round, median spot on S3. The female abdomen has a larger yellow mark on S3, and S3–6 have mid-dorsal, longitudinal, yellow streaks. The wings are relatively broad and the pterostigma of the male differs from that of male *caudalis* in being brown and not white above. A feature that distinguishes *albifrons* from the other four European species of the genus is a large, white spot on each lateral lobe of the labium (fig. 353). Male accessory genitalia (fig. 355) have a short anterior lamina, the anterior hamule terminates in a posteriorly-curved process, and the posterior hamule has a weakly-projecting anterior process and a rounded posterior

Map 106 – *Leucorrhinia caudalis*

process. The tooth overlying the base of the genital lobe is shorter than that of *caudalis*. The female vulvar scale (fig. 360) is short and weakly bilobed.

BIOLOGY

Similar to that of *caudalis*, but inhabiting more oligotrophic waters (with floating plants). Reported (Sélys & Hagen, 1850) to fly in company with *L. pectoralis* and *L. rubicunda* 'in woodland clearings, not very rapidly but in an erratic manner'.

FLIGHT PERIOD

Where the two species are sympatric, *L. albifrons* appears about a week later than *L. caudalis* and it persists until the beginning of August.

DISTRIBUTION (map 107)

Like *L. caudalis*, *L. albifrons* has a Euro-Siberian range, but it is scarcer in western Europe and extends westwards only as far as the extreme east of France (Moselle, Vosges, Jura, Alps) with the exception of an apparently isolated population in the Brenne, Indre (Dommanget, 1984). It is not recorded from Belgium, and Geijskes & van Tol (1983) report the finding of only six adults and three larvae in the Netherlands, the last in 1964. Several localities are cited for Germany, but Schmidt, Eb. (1981) states that it is now almost extinct in the Federal Republic. In Scandinavia the distribution of *L. albifrons* is very similar to that of *L. caudalis*, but it is not recorded from Denmark. Also known from Switzerland (now only Aagau, four localities), Austria, Hungary, Czechoslovakia, Poland and Russia eastwards to the Urals.

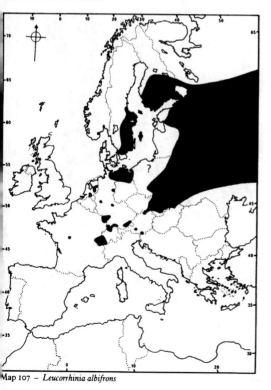

Map 107 – *Leucorrhinia albifrons*

LEUCORRHINIA DUBIA (Vander Linden)
Plate 29 (201,202)

Libellula dubia Vander Linden, 1825, *Monograph. Libell. Europ. Spec.* : 16 (*nec* Rambur, 1842 : 67).
Type locality: Gheel, Belgium.
  Libellula melanostigma Eversmann, 1836
  Libellula leucorrhinus Charpentier, 1840

E – White-faced Dragonfly, **F** – La Leucorrhine douteuse, **D** – Kleine Moosjungfer

DESCRIPTION OF ADULT

This and the two following species differ from the two preceding in having more extensive pale markings on the abdomen, no pruinosity, and blackish anal appendages. *L. dubia* is marginally the smallest of the European species of *Leucorrhinia*, measuring only 33–36mm in length. The head is relatively narrow, black except for the white face, brown occipital triangle and a very small pale spot on each side behind the eyes. The thorax is black with pale antehumeral stripes, reddish brown in the male and yellowish in the female, and bright markings between the wings, red in the male and yellow in the female. The abdomen of the mature male is narrow and parallel-sided, moderately inflated in front of the slight constriction on S3, and dorsally marked with red as follows: a pair of small, lateral spots on S1, the articulation between S1 and S2, most of S2 which is black only antero-laterally, the anterior half of S3, short and very narrow mid-dorsal lines on S4 and S5, a narrow mid-dorsal mark anteriorly on S6 and a similar but broader mark, pointed posteriorly, on S7. The red mark on S7 occupies about half the length of the segment. The three posterior abdominal segments are dorsally black but with their articulations red. The female abdomen is slightly stouter than that of the male and with primrose yellow replacing the red coloration. The dorsal mark on S2 is relatively smaller than that in the male, but those on S4 and S5 are larger, not much smaller than those on S6 and S7. Basal brown marks on the wings are well developed, the forewing being dark brown at the base between veins Sc and R$_1$ and behind vein CuP to the anal margin. This pattern is a miniature replica of that on the hindwings and more extensive than in the other four species of the genus. Male accessory genitalia (fig. 356) have a very short anterior lamina, scarcely half as long as the prominently projecting hamule. The female vulvar scale (fig. 361) ends in two subtriangular lobes which extend over the basal one-fifth of sternum S9.

BIOLOGY

Oligotrophic to mesotrophic acid pools on moors and mosses with *Sphagnum* and often *Eriophorum*. Flies low and rapidly, the male with a dodging flight described by Lucas (1900) as sparrow-like. Settles on the ground or on low vegetation, and is not infrequently ensnared by sundew. Population ecology and sexual behaviour are described by Pajunen (1962, 1964a) and territoriality by Warren (1964). The life cycle generally extends over three years (Norling, 1976, 1984b: Sweden).

FLIGHT PERIOD

May to the beginning of August but with the peak in June in most localities. In northern Finland, it does not emerge until late June (Hämäläinen, 1984c).

DISTRIBUTION (map 108)

L. dubia is the most numerous of the European species of *Leucorrhinia*, at least in western Europe. It extends east across Siberia as *L. d. orientalis* Belyshev, reaching north of the Arctic Circle. The nominotypical subspecies is widespread in northern and central Europe, and is the sole representative of the genus in the British Isles where it occurs in localized colonies in Scotland and England. It extends south to the Pyrenees (both French and Spanish sides) and through the Alps to northern Italy, but is absent from the Balkans and in eastern Europe reaches south only as far as northern Romania (a single record cited by Cirdei & Bulimar (1965)).

LEUCORRHINIA RUBICUNDA (Linnaeus)
Plate 29 (203,204)

Libellula rubicunda Linnaeus, 1758, *Syst. Nat.* (Edn. 10) **1** : 543.

Type locality: southern Sweden.

Libellula infuscata Eversmann, 1836

F – La Leucorrhine rubiconde, **D** – Nordische Moosjungfer

DESCRIPTION OF ADULT

Similar in appearance to *L. dubia* but a larger insect, usually 38–42mm in body length. The abdominal pattern of red (male) or yellow (female) marks resembles that of *dubia* but is more extensive, the dorsal spot on S7 occupying more than half of the length of the segment in the male and almost all of the segment in the female. The pterostigma of the male is reddish, brighter than in *dubia*, and the basal brown wing patches are smaller, noticeably so on the forewing where they are reduced to a very short line between veins CuP and A$_1$. Male accessory genitalia (fig. 357) have a large anterior lamina, projecting almost to the level of the apex of the strongly curved hamule which has a broad posterior lobe. The female vulvar scale (fig. 362) has two widely-separated and very short terminal lobes.

BIOLOGY

Frequents mesotrophic *Sphagnum* bogs on moorland and in forests. Larvae may require quite deep water. Males bask in sun on stones and tree stumps. Territorial behaviour of *L. rubicunda* is described by Pajunen (1966a).

FLIGHT PERIOD

Mainly during May and June, but there are a few records for April and some for July. It is the earliest dragonfly to appear in northern Finland (mid-June) but the flight period is over by late July (Hämäläinen, 1984c).

DISTRIBUTION (map 109)

Similar to that of *L. dubia*, although more strictly boreal and less widespread. It reaches neither the British Isles nor the Pyrenees, and is most plentiful in northern and central parts of eastern Europe. It extends south to elevated regions in the south of Germany, but has not been recorded from Switzerland since the last century. There are a few records for eastern France (Lorraine, Paris Basin and Doubs: Aguesse, 1968), and it is widespread in Scandinavia including Denmark, northern Germany, Poland and Russia. In the Soviet Union, it extends east as far as Lake Baikal and north to the Arctic Circle.

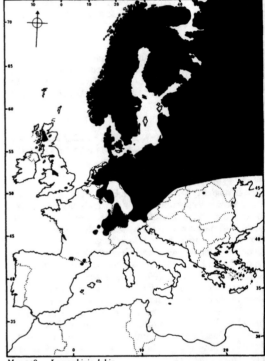

Map 108 – *Leucorrhinia dubia*

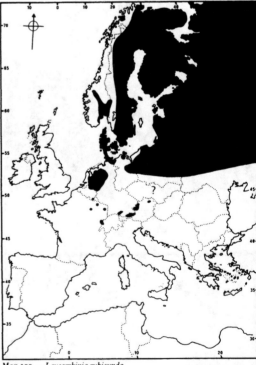

Map 109 – *Leucorrhinia rubicunda*

LEUCORRHINIA PECTORALIS (Charpentier)
Plate 29 (205,206)

Libellula pectoralis Charpentier, 1825, *Horae entomol.* : 46.
Type locality: Silesia, Poland.

F – La Leucorrhine à gros thorax, **D** – Grosse Moosjungfer

DESCRIPTION OF ADULT

Similar in body length to *L. rubicunda*, but a more robust insect with broader head and thorax. The abdomen also is stouter, and it is unusually shaped in both sexes being expanded at S6. This expansion is most pronounced in males. Pale markings on the abdomen form a similar pattern to that seen in the two preceding species but they are more extensive, each dorsal mark occupying all or almost all of its segment's length. The coloration of the male abdomen is distinctive. The pale marks on S1–6 are yellowish to reddish brown, but that on S7 is clear yellow. The pterostigma of the male is very dark brown. Usually there is a single cubito-anal crossvein in the hindwing, but occasionally two are present. Basal brown wing markings are similar to those of *L. rubicunda* and are very small on the forewings. Male accessory genitalia (fig. 358) are of similar proportions to those of *L. rubicunda*, but the anterior lamina is unique in being surmounted by a pencil of very long, stout setae. The female vulvar scale (fig. 363) terminates in two triangular points which extend over about one-quarter of the ninth sternum.

BIOLOGY

Found near small lakes and tarns with mesotrophic to eutrophic, acid water, and also small canals with a dense

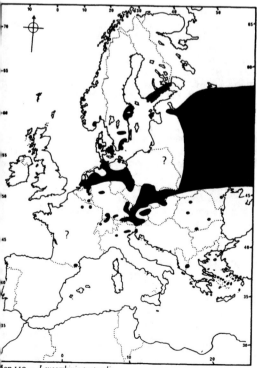

Map 110 – *Leucorrhinia pectoralis*

growth of *Typha* and *Carex* (Balestrazzi & Bucciarelli, 1971a). Kiauta (1964) describes aspects of the behaviour of *L. pectoralis*.

FLIGHT PERIOD

From the end of April until early in July.

DISTRIBUTION (map 110)

L. pectoralis is a Euro-Siberian insect with a rather more southerly range than those of its European congeners. In Scandinavia it is found only in the south, and it extends from the Baltic east to about longitude 90°. Southwards in Europe it reaches to the Balkans, Turkey, northern Italy (four localities: Balestrazzi & Bucciarelli, 1971a; Bucciarelli, 1976) and the French Pyrenees in scattered colonies. It is absent from the British Isles, although Lucas (1900) refers to a solitary specimen taken in 1859 near Sheerness, Kent.

TRITHEMIS Brauer, 1868

Trithemis is represented by about forty species in the Palaearctic, Oriental and, especially, Ethiopian regions, and the one species that is known to breed in Europe is principally African in distribution. Pinhey (1970) has monographed the genus. They are small dragonflies, about the size of *Sympetrum*, which they resemble in venation although with a greater number of forewing antenodal crossveins and a doubling of some cells between veins IR₃ and Rspl. The posterior lobe of the pronotum is not outstanding and does not have a circlet of long hair, thus resembling the form in *Selysiothemis*, *Zygonyx* and *Pantala* which follow.

TRITHEMIS ANNULATA (Palisot de Beauvois)
Plate 29 (207)

Libellula annulata Palisot de Beauvois, 1807, *Insectes recueillis Afr. et Amér.* : 69.
Type locality: Nigeria.

Libellula rubrinervis Sélys, 1841
Libellula obsoleta Rambur, 1842
Libellula haematina Rambur, 1842
Tramea erythraea Brauer, 1867
Trithemis scorteccii Nielsen, 1935b

DESCRIPTION OF ADULT

The mature male is an attractive and distinctive insect with a rose-red abdomen thinly suffused with blue-violet pruinosity. Dorsally, there are black median marks on the three terminal abdominal segments. The thorax is reddish brown, slightly pruinose between the wing bases, and the frons and vertex are metallic violet. Females are mainly yellowish olive in colour with a narrow mid-dorsal black abdominal band broadening on S8–10. The forewings are clear, or indistinctly yellowish basally in females, but the hindwings in both sexes have large, basal, amber patches (paler in females) which almost reach to the arculus. Hindwings are rather strongly narrowed apically. Venation is reddish (males) to yellowish (females), and the pterostigma is reddish brown, about four times as long as broad and 2.5–2.8mm in length. There are nine and a half (occasionally ten and a half) antenodal crossveins in the forewing (fig. 283, p.157), a few of the cells between veins

189

IR₃ and Rspl are doubled, and the membranules are lightly infuscated. The legs are predominantly black, but they may be yellowish about their bases, particularly in females. In males, the superior anal appendages are more than twice as long as S10, and the accessory genitalia are as in figure 364. In females, the vulvar scale does not project and the anal appendages are slightly longer than S10. Body length 34–36mm.

BIOLOGY

Generally associated with standing water although it also occurs by slowly-flowing stretches of rivers.

FLIGHT PERIOD

In Europe, adults have been observed from April to November.

DISTRIBUTION (map 111)

Very common all over Africa, and also found in the Middle East, Arabia, western Asia and the extreme south of Europe. It is recorded from south-western Iberia north as far as central Portugal (Aguiar & Aguiar, 1983, 1985), Sardinia, Sicily (type locality for *rubrinervis* Sélys and *haematina* Rambur), coastal parts of Italy as far north as Tuscany, and from southern Greece. Apparently numerous on Cyprus and Rhodes. Confirmed as a European breeding species by Ferreras Romero (1981a) and Belle (1984) who found, respectively, larvae in three rivers in the extreme south-west of Spain and exuviae in abundance at a reservoir in Huelva, Spain. The many new Iberian records suggest a recent expansion of range northwards from North Africa (Belle, 1984).

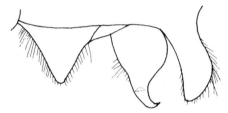

Figure 364 *Trithemis annulata*
male accessory genitalia in left lateral view

The ranges of some other species of *Trithemis* extend almost to the frontiers of Europe:

TRITHEMIS ARTERIOSA (Burmeister, 1839)

Another abundant African species, *T. arteriosa* is found in the Canary Islands, along much of the Mediterranean littoral of North Africa, in the Middle East to Iran, Arabia, and on Rhodes. It is slightly smaller than *T. annulata*, of more slender build, and the abdomen of the mature male is red, lacking the violet pruinosity of *annulata*. There are small but distinct amber marks at the base of the forewings, as well as large, bright amber patches on the hindwings.

TRITHEMIS FESTIVA (Rambur, 1842)

This species is known from Iraq, Rhodes and Cyprus (*cyprica* Sélys, 1887), and east to south-east Turkmenia and the Himalayas.

TRITHEMIS KIRBYI Sélys, 1891

This is an Indian species, but its subspecies *T. k. ardens* Gerstäcker, 1891, occurs widely in Africa and is recorded from several places in Morocco and Algeria. The mature male has a brilliant red abdomen, and the basal amber wing patches are large on both pairs of wings, those on the hindwing extending beyond the discoidal cell. In the female, the hindwing has an isolated yellow patch beyond the anal loop.

I have observed *T. annulata*, *T. arteriosa* and *T. kirbyi ardens* on the same stretch of a small river north of Agadir, Morocco. *T. k. ardens* kept to rocky, open situations where the current was moderate, and males frequently perched on boulders in the river. *T. annulata* and *T. arteriosa* flew together about a pool, bordered by vegetation, where the current was slow. Males of both of these species perched on *Juncus*. Combats between conspecific males were frequent, but no interspecific activity was seen.

ZYGONYX Hagen, 1867
Pseudomacromia Kirby, 1889

Zygonyx species are large libellulids which bear a superficial resemblance to corduliids. Their bodies are slightly metallic and they rest in a more or less vertical attitude with the abdomen directed downwards. Adults fly near or over waterfalls and rapids. In wing venation, they resemble *Trithemis* but have a greater number of antenodals in the forewing (Table 3, p.157).

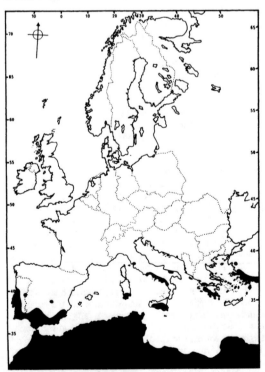

Map 111 – *Trithemis annulata*

There are about twenty species occurring in the Palaearctic, Oriental and Ethiopian regions, but only one species is found in Europe. The termination 'onyx' (claw) of the generic name is masculine and our European species should be known as *Z. torridus*, not *torrida*.

ZYGONYX TORRIDUS (Kirby)   Plate 29 (210)

Pseudomacromia torrida Kirby, 1889, *Trans. zool. Soc. Lond.* **12** : 299.
Type locality: Sierra Leone.
Pseudomacromia atlantica Martin, 1900

DESCRIPTION OF ADULT

The body length of *Z. torridus* is about 50mm. It is a dark-coloured dragonfly, the thorax blackish with light pruinescence (males) to dark brown (females), with antehumeral stripes poorly defined and the yellow lateral stripes discontinuous. The abdomen is long, narrow, parallel-sided and not much swollen basally. It has a similar pattern in both sexes being black with weak metallic reflections, and having a pair of large, lateral yellow spots on S2–9 inclusive, and a very narrow, mid-dorsal, longitudinal, yellow stripe which incorporates the dorsal carina on each of the same segments. There is a narrow, yellow band around the anterior margins of S2 and S3. Anal appendages are black and rather short, the superior appendages of the male being about as long as S9. The legs are black with yellow markings restricted to coxae, trochanters and bases of the front femora. Wings are long, clear in males and slightly brownish in females, with white membranules and

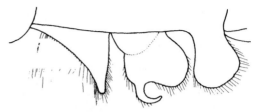

Figure 365 *Zygonyx torridus*
male accessory genitalia in left lateral view

narrow, blackish pterostigmata. Females have a small, brown patch adjacent to the posterior half of the hindwing membranule, but not extending laterad beyond the first series of cells. Male accessory genitalia are illustrated in figure 365; the anterior lamina is densely clothed with setae. The female vulvar scale does not project ventrally.

BIOLOGY

An insect of fast-running brooks and small rivers, often flying in the vicinity of rapids and waterfalls. Oviposition occurs in tandem and with the female settled, which is exceptional for an exophytically ovipositing odonate.

FLIGHT PERIOD

Captures in Europe have been between April and August.

DISTRIBUTION (map 112)

Widespread in Africa (although not recorded from much of North Africa), the Middle East, Arabia and east to India. Its status as a European species rests on records from Alicante and Valencia in the south of Spain (Benitez Morera, 1950).

PANTALA Hagen, 1861

Pantala is a genus of only two species, but nevertheless it has an almost world-wide distribution. The diagnostic venational characters are given in Table 3, p.157; the large number of forewing antenodal crossveins of which the last is incomplete, and the two cubito-anal crossveins in the very broad hindwings (one in the forewing), characterize the genus. The type species, *P. flavescens*, is occasionally reported in Europe.

PANTALA FLAVESCENS (Fabricius)
Plate 29 (208)

Libellula flavescens Fabricius, 1798, *Suppl. entomol. Syst.* : 285.
Type locality: India.
Libellula viridula Palisot de Beauvois, 1805
Libellula analis Burmeister, 1839
Libellula terminalis Burmeister, 1839
E (U.S.A.) – Globe Skimmer

DESCRIPTION OF ADULT

A large dragonfly, 49–52mm in length and predominantly tawny to yellowish brown. The thorax is tawny above and

Map 112 – *Zygonyx torridus*

yellowish laterally without dark stripes. Antehumeral stripes are not defined. The abdomen is moderately broad, yellowish, with a dark, mid-dorsal band which is very narrow at the base of the abdomen but widens posteriorly to end in subtriangular black spots on S8–10. On S3–7 the mid-dorsal band is extended laterally in the middle of each segment as a thin, transverse line, and in the apical quarter of each segment as a broad, transverse bar. Posterior to the central transverse line, the mid-dorsal band is paler, and from the posterior transverse bar forwards to the anterior of the segment there is a whitish mark. The colour pattern of the body is similar in males and females. Legs are black with a yellow, longitudinal stripe externally on the front and middle femora and tibiae; coxae and trochanters are yellowish. The wings are mainly clear, but males may develop a small, light brown patch at the extreme apex of each wing, larger on the hindwings. The anal area of the hindwing is suffused yellowish amber, more especially so in females. The pterostigma is reddish brown, that of a forewing (2.6–2.8mm) longer than that of a hindwing (2.2–2.4mm). Membranules are white. Anal appendages in both sexes are about as long as the combined length of S9 and S10. Male accessory genitalia (fig. 366) have the hamule directed posteriorly, its apex beside the genital lobe, and the genital lobe is more sparsely setose than is usual in Libellulidae. The female vulvar scale projects slightly.

BIOLOGY

Breeds in small pools, often of a temporary nature, in paddy fields, and also in slowly-flowing water. Females oviposit in flight by touching the water surface with the tip of the abdomen, as in most other libellulids. It may be deceived into ovipositing on smooth, shining road surfaces. Larval development is very rapid, as it is in other odonates such as *Hemianax* and *Brachythemis* which may use temporary water bodies. Adults spend much time in flight and feed in large groups. Slope-soaring as an energy-saving behaviour is discussed by Gibo (1981).

FLIGHT PERIOD

Most likely to be seen in Europe during the late summer or autumn. Emergence and life history are described by Byers (1941).

DISTRIBUTION (map 113)

This global tropical migrant, occurring in the Americas, Africa, parts of Asia (including Turkey) and Australasia (twice recorded in New Zealand) is a renowned migrant and has often been observed at sea hundreds of miles from land. It occurs mostly within the intertropical convergence zone (Waterston, 1985), but has also been recorded at a height of 6300m in east Nepal which must be about the altitudinal record for a dragonfly. *Pantala flavescens* appears erratically in Europe and has been found in Spain and southern France, but it is unlikely to breed except perhaps in the eastern Balkans. A specimen was taken in England at Horning, Essex, in 1823 by Sparshall (Lucas, 1900, but see Sélys & Hagen, 1850 (p. 260)) and given the manuscript name of *Libellula sparshalli* by Dale. In North Africa, *Pantala* occurs in Egypt but not north of the Hoggar Mountains in Algeria (Dumont, 1978).

Map 113 – *Pantala flavescens*

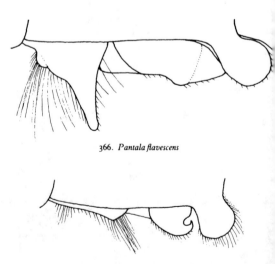

366. *Pantala flavescens*

367. *Selysiothemis nigra*

Figures 366, 367
male accessory genitalia in left lateral view

SELYSIOTHEMIS Ris, 1897

Selysiothemis, together with *Urothemis* and two other genera, are considered by some authorities to be sufficiently distinct from libellulids to justify their inclusion in the separate family Macrodiplactidae.

Selysiothemis is monotypic. It has characteristically sparse and faint venation with only two rows of cells distal to the discoidal cell in the forewing (fig. 284, p.157), and the discoidal cells are without crossveins. These features are shared with *Diplacodes* from which *Selysiothemis* may at once be distinguished by the complete distal (sixth) fore-wing antenodal crossvein.

SELYSIOTHEMIS NIGRA (Vander Linden)
Plate 29 (209)

Libellula nigra Vander Linden, 1825, *Monograph. Libell. Europ. Spec.* : 16.

Type locality: Terracina, Campania, Italy.

Urothemis advena Sélys, 1878

DESCRIPTION OF ADULT

The sole species in the genus, *S. nigra*, is a small dragonfly, 31–34mm long, with the venation very thin and not easy to see. The wings are clear and shining, and the short pterostigma, only about 1.5mm long, is pale yellowish but bordered anteriorly and posteriorly by thick, black edges to the veins so that it resembles an = sign and is the most noticeable feature of the wing (fig. 284). The venation is sparse. The body is yellowish in females, blackish in males. The female thorax is mostly pale but narrowly brown on each side of the median carina and with blackening of the dorsal section of the humeral suture, the ventral part of the mesometapleural suture, all of the metapleural suture, and the sutures surrounding the mesinfraepisternum and metinfraepisternum. Legs are mainly black in males, yellowish only on the coxae and about the femoral bases, but in females they are more extensively pale. The female abdomen is yellowish with S1 and the basal half of S2 broadly brown, the posterior half of S2 has a dark brown, mid-dorsal, triangular mark with its apex basal, S3–7 have a narrow, blackish, mid-dorsal stripe which is absent for a short distance posteriorly on each segment and slightly expanded both anteriorly and posteriorly, and S8–10 each have a subtriangular, black, mid-dorsal mark. Superior anal appendages are short, hardly longer than S10 in the female and about as long as S9 in the male. Male accessory genitalia (fig. 367) have a very short anterior lamina, but its low apex is crowned with long setae. The female vulvar scale does not project ventrally.

BIOLOGY

Breeds in standing water. Oviposition is in tandem, the female releasing eggs as she strikes the water surface with the tip of her abdomen. Adults hover frequently, a metre or so above the ground. Fraser (1936) and Schneider (1981) mention migratory flights of *S. nigra* in the Persian Gulf and Jordan respectively. Adults are sometimes attracted to light.

FLIGHT PERIOD

From June to September in the Balearic Islands (Compte Sart, 1963); from the first half of May until the second half

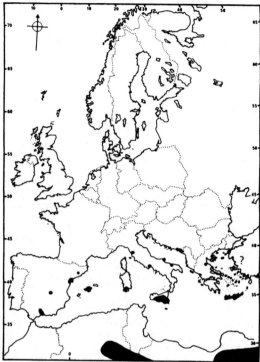

Map 114 – *Selysiothemis nigra*

of August in Andalucia (Ferreras Romero & Puchol Caballero, 1984).

DISTRIBUTION (map 114)

Common in the Middle East and parts of North Africa away from the Mediterranean coast, and ranging east to Sind and Kashmir. There are European records from Bulgaria, Greece, Yugoslavia, Italy (Utzeri & Belfiore, 1976) north to Emilia-Romagna and including Sardinia and Sicily, Malta, and Spain (Ferreras Romero, 1983). Compte Sart (1960) summarizes the then known European distribution and (1963) reports the location of several colonies on Majorca.

UROTHEMIS EDWARDSI (Sélys, 1849)

This species, found on the boundary of the region covered by this work, was described from Lac Hubeira in Algeria and is known from Morocco, Mauretania, Mali, Niger, Sudan as well as southern Arabia and Africa south of the Sahara, and there is a relict subspecies *U. e. hulae* Dumont, 1975, in Palestine. In venation, it shares with *Selysiothemis* the characters of undivided discoidal cells and two rows of cells in the trigonal space distal to the discoidal cell of the forewing. There are, however, seven and not six forewing antenodals (all complete), the subtriangle of the forewing is three-celled, and the hindwing has a large, brown, basal patch (sometimes reduced in males). The pterostigma is about 4mm long and the abdomen has a broad, black, mid-dorsal band. Males develop bluish pruinosity.

VII. KEYS TO THE FINAL-INSTAR LARVAE OF EUROPEAN ODONATA

Fully-grown dragonfly larvae leave the water in which they have developed and climb supports, in many cases the stems of waterside plants, to which they cling. The skins, or exuviae, of these larvae are left attached to the supports after the adult insects have emerged, and they may persist for several months. By searching for and identifying larval exuviae or final-instar larvae (larvae in which the wing sheaths reach to or exceed the posterior margin of the third abdominal segment), it is possible to ascertain which dragonflies are breeding in a locality at a time when adult insects are no longer flying or cannot be found. It is emphasized, however, that the keys are applicable only to the final larval instar and should not be used to attempt to name younger stages.

It is possible, also, to determine sex from the larval integument, the gonapophyses appearing at an early stage. In female Zygoptera, Aeshnidae and Cordulegastridae, the gonapophyses (ovipositor rudiments) are conspicuous. They arise from the ventral surfaces of S8 and S9 and reach posteriorly over S9+10 (Zygoptera) (fig. 368) or S9 alone (Anisoptera) (fig. 369). Male gonapophyses on the ventral surface of S9 are small and inconspicuous by comparison (figs 371,372). In all Anisoptera, male larvae may be distinguished from females by the presence of a male projection at the base of the epiproct (fig. 370), a clearly delimited, elevated area which is absent in females.

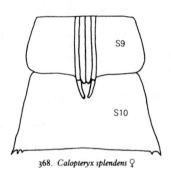

368. *Calopteryx splendens* ♀

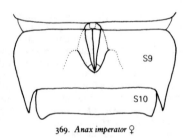

369. *Anax imperator* ♀

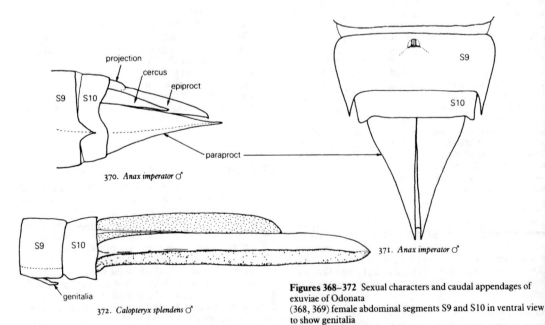

370. *Anax imperator* ♂

371. *Anax imperator* ♂

372. *Calopteryx splendens* ♂

Figures 368–372 Sexual characters and caudal appendages of exuviae of Odonata
(368, 369) female abdominal segments S9 and S10 in ventral view to show genitalia
(370–372) male abdominal segments S9 and S10 in left lateral view (370, 372) and ventral view (371)

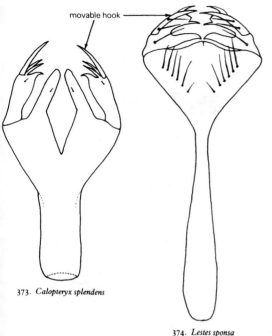

373. *Calopteryx splendens*

374. *Lestes sponsa*

375. *Platycnemis pennipes*

376. *Coenagrion puella*

Figures 373–376 Labia of Zygoptera

Larvae of most European Odonata are recognizable to genus, but identification to species can be very difficult and uncertain. The following keys are intended primarily for generic identification. Specific characters of many larvae, particularly those in the larger genera, have not yet been satisfactorily worked out and there are many contradictory statements in the literature. Much research remains to be done before a fully satisfactory diagnostic key to the larvae of all European species can be presented. It must be emphasized that what follows will undoubtedly be eventually found to contain inaccuracies, but it is hoped that, if the limitations of the keys are borne in mind, their availability will lead to their improvement.

In addition to a number of earlier works, of which special mention must be made of Rousseau (1909) and Lucas (1930), more recent keys to larvae are provided by Gardner (1954b) and Gardner in Hammond (1983) for British species, by Conci and Nielsen (1956) and Carchini (1983) for Italian species, and by Franke (1979) for central European species. In compiling the following keys, I have leant heavily on these studies.

Terminology for the larval labium follows that defined by Corbet (1953); in particular, the largest part of the body of the labium to which the labial palps articulate, is termed the prementum. Abdominal segments are referred to as S1, S2..... S10, as for adults. Body length is measured from the front of the head (antennae excluded and labium retracted) to the apices of the caudal lamellae (Zygoptera) or paraprocts (Anisoptera).

Key to suborders and families

1   Body slender, head much broader than abdomen except in Calopterygidae (figs 4a, p.17; 395,404,405, pp.198–200). Abdomen with three caudal lamellae (gills) attached posteriorly, the median or all three leaf-like (lamellate) (e.g. figs 372; 379–382, p.196). Swimming is only by lateral movements of the abdomen. ZYGOPTERA 2

–   Body more robust, head at most only slightly broader, usually narrower, than abdomen (e.g. figs 4c,d, p.17; 435–437, p.203). Extremity of abdomen never with caudal lamellae but with five short, spinous appendages (fig. 370). Swimming is usually by sudden expulsion of water from the rectum. ANISOPTERA 6

2(1)   Abdomen with lateral gills .... Euphaeidae

–   Abdomen without lateral gills 3

3(2)   Basal antennal segment at least as long as the remaining segments together (fig. 377). Prementum of labium with deep, lozenge-shaped apical cleft (fig. 373). Lateral caudal lamellae triquetral, median lamella lamellate (fig. 372) .... Calopterygidae (p.197)

–   Basal antennal segment short (fig. 378). Prementum entire or (Lestidae) with only a very short apical cleft (figs 374–376). All caudal lamellae lamellate 4

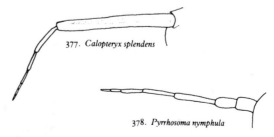

377. *Calopteryx splendens*

378. *Pyrrhosoma nymphula*

Figures 377, 378 Antennae of Zygoptera larvae

4(3) Labial palp with movable (outer) hook furnished with setae and with distal margin deeply cleft to separate a large inner lobe (end hook); prementum with very short, slit-like apical incision (fig. 374). Caudal lamellae apically rounded (except *Lestes barbarus*) and not divided into two regions (fig. 379). Apex of front tibia with bidentate setae ..... Lestidae (p.198)

– Labial palp without setae on movable hook, and with distal margin much less deeply cleft so that end hook is small; apical margin of prementum entire (figs 375,376). Caudal lamellae often apically pointed and often divided into pre- and post-nodal sections (figs 380–382). Apex of front tibia with tridentate setae 5

5(4) Caudal lamella with a long, apical attenuation and undivided (fig. 380). Prementum with no more than four long setae arranged in an almost straight, transverse line (fig. 375) ....
.... Platycnemididae (p.199)

– Caudal lamella apically rounded or with short point, often divided transversely into two parts (figs 381,382). Prementum usually with more than four long setae (except *Ceriagrion*) which are arranged in two oblique series with the mesial ones more basal in position than the outer ones (fig. 376) ......
.... Coenagrionidae (p.200)

6(1) Antenna with four segments, robust (fig. 384). Tarsi of front and middle legs two-segmented. (Labium flat, body dorso-ventrally flattened) .... Gomphidae (p.204)

– Antenna with seven (occasionally six) segments, filiform (fig. 383). All tarsi three-segmented 7

7(6) Labium flat; labial palp with movable hook longer than breadth of palp and distal margin without crenations (fig. 385). Body elongated with hind leg not nearly reaching to apex of abdomen (fig. 4c, p.17) .... Aeshnidae (p.202)

– Labium spoon-shaped; labial palp with movable hook shorter than breadth of palp and distal margin crenated or serrated (figs 386,387) 8

8(7) Body elongated with tarsal claw of hind leg not reaching to apex of abdomen (figs 458,459, p.206). Apex of prementum bifid; labial palp with distal margin bearing deep and irregular dentations (fig. 386) ...... Cordulegastridae (p.206)

– Body short, squat and dorso-ventrally flattened with hind leg reaching beyond apex of abdomen (fig. 4d, p.17). Apex of prementum not bifid; labial palp with distal margin more or less evenly crenated (fig. 387) 9

379. *Lestes sponsa*

380. *Platycnemis pennipes*

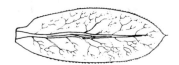

381. *Coenagrion mercuriale*

382. *Erythromma najas*

Figures 379–383 Middle caudal lamellae of Zygoptera larvae

383. *Anax imperator*

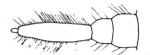

384. *Gomphus vulgatissimus*

Figures 383, 384 Exuvial antennae of Anisoptera larvae

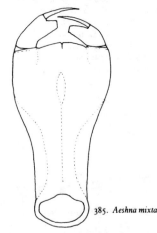

385. *Aeshna mixta*

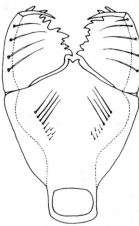

386. *Cordulegaster boltoni*

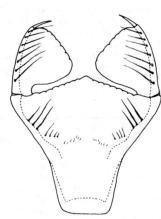

387. *Orthetrum cancellatum*

Figures 385–387 Labia of Anisoptera

9(8) Cerci more than half as long as the paraprocts (fig. 388). Distal margin of labial palp with deeper crenations .... Corduliidae (p.206)

– Cerci less than half as long as the paraprocts (fig. 389). Distal margin of labial palp with shallower crenations .... Libellulidae (p.208)

388. *Somatochlora metallica*

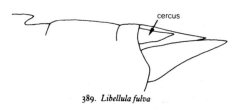

389. *Libellula fulva*

Figures 388, 389 Caudal appendages in Corduliidae and Libellulidae

Keys to genera and some species

CALOPTERYGIDAE

The long, slender larvae of *Calopteryx* are found amongst the roots and stems of aquatic plants, or on the mud of stream beds. Legs may be held close to the sides of the body, and the caudal lamellae are also usually kept closely apposed, so that the brownish larvae look very much like pieces of dead stem or 'stick-like water-bugs' (Lucas, 1900).

1 Median caudal lamella at least three-quarters of the length of a lateral lamella (fig. 4a, p.17). Premental cleft narrower, more than four times as long as broad (fig. 390). Head in lateral view with a strong occipital tooth behind eyes (fig. 392) .... *Calopteryx virgo* (p.54)

– Median caudal lamella at most two-thirds of the length of a lateral lamella (fig. 372). Premental cleft broader, about three times as long as maximum breadth (fig. 391). Head with a weak and inconspicuous occipital tooth (figs 393,394) 2

2 Body length (inclusive of caudal lamellae) 31–37mm; length of first antennal segment 3.7–3.8mm .... *C. splendens* (p.55)

– Body length 26–28mm; length of first antennal segment 3.1–3.6mm (Aguesse, 1968) .... *C. haemorrhoidalis* (p.56)

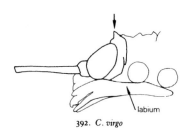

392. *C. virgo*

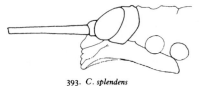

393. *C. splendens*

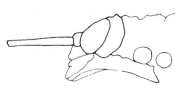

394. *C. haemorrhoidalis*

Figures 392–394 Basal antennal segment, head and thorax of *Calopteryx* species in lateral view, the labium and front and middle leg insertions indicated diagrammatically. The occipital tooth in *C. virgo* is arrowed

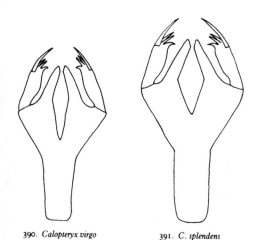

390. *Calopteryx virgo*

391. *C. splendens*

Figures 390, 391 Labia of Calopterygidae

LESTIDAE

Slender larvae with elongated caudal lamellae (fig. 395). The species are distinguished to a large extent on labial setation, but this is subject to variation: the formulae given in the key represent the commonest conditions.

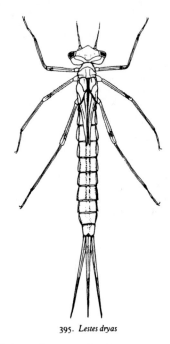

395. *Lestes dryas*

Figure 395 Larva of Zygoptera, Lestidae

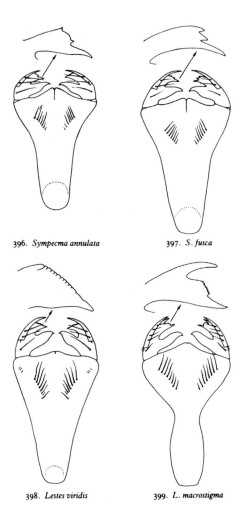

396. *Sympecma annulata*   397. *S. fusca*

398. *Lestes viridis*   399. *L. macrostigma*

1   Prementum subtriangular, not stalked (figs 396–398) 2
–   Prementum clavate, constricted basally to form a stalk (figs 399–403). *Lestes* (*Lestes*) 4

2(1)   Labial palp with apical margin between movable hook and end hook finely crenated (fig. 398) ....
.... *Lestes* (*Chalcolestes*) *viridis* (p. 61)
–   Labial palp with apical margin dentate (figs 396,397). *Sympecma* 3

3(2)   Outermost tooth of apex of labial palp, adjacent to movable hook, small and directed anteriorly; apex of prementum strongly produced forwards medially (fig. 396) ....
.... *Sympecma annulata* (p.66)
–   Outermost tooth of labial palp larger and approximately parallel to adjacent teeth; apex of prementum not strongly produced mesially (fig. 397) .... *S. fusca* (p.65)

4(1)   Movable hook of labial palp with two (rarely three) setae; stalk of prementum much attenuated and, when folded under head, reaching hind coxae (figs 400,401) 5
–   Movable hook of labial palp with three or four setae; stalk of prementum less elongated and, in repose, not reaching bases of hind coxae (figs 399,402,403) 6

Figures 396–403 Labia of Lestidae with apical margin of palps (exclusive of movable end-hooks) enlarged

5(4)   Movable hook of labial palp with two setae; two oblique series of six premental setae (fig. 400) .... *Lestes sponsa* (p.63)
–   Movable hook of labial palp with two or three setae; two oblique series of seven premental setae (fig. 401) (Larval instars described by Gardner, 1952) .... *L. dryas* (p.64) (see also MacNeill (1951a, 1951b) and Carchini (1983) for distinguishing larvae of *sponsa* and *dryas*)

6(4)   Prementum relatively short with a broad stalk; movable hook of labial palp with four setae; two oblique series of eight premental setae (fig. 399) .... *L. macrostigma* (p.62)
–   Prementum with stalk longer and narrower; movable hook of palp with three (rarely four) setae; two oblique series of six or seven premental setae (figs 402,403) 7

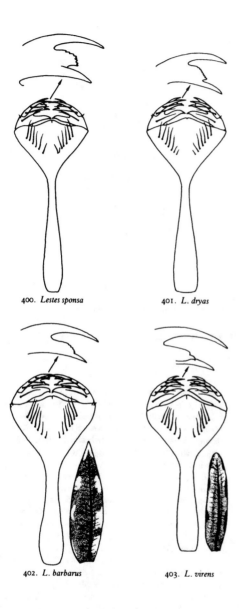

400. *Lestes sponsa*

401. *L. dryas*

402. *L. barbarus*

403. *L. virens*

PLATYCNEMIDIDAE

Distinguishing characters for larvae of the three European species of *Platycnemis* (fig. 404) have not been found.

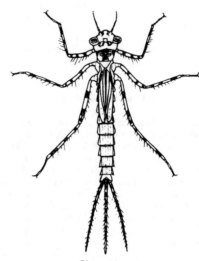

404. *Platycnemis pennipes*

Figure 404 Larva of Zygoptera, Platycnemididae

7(6) Caudal lamella apically pointed (fig. 402). Prementum with two oblique series of seven setae; movable hook of palp with three or four setae (fig. 402) .... *L. barbarus* (p.61)
– Caudal lamella apically rounded (fig. 403). Prementum with two oblique series of six setae; movable hook of palp with three setae (fig. 403) .... *L. virens* (p.62)

COENAGRIONIDAE

Some authors make a primary distinction between species with six and species with seven antennal segments, but the number of antennal segments is not easy to determine on exuviae, and even in fresh material the division between the two apical segments is difficult to see. In the following key I follow Gardner (1954b) and place most reliance on the form of the caudal lamellae, notwithstanding the intraspecific variation of these structures. Exuvial lamellae may be examined by soaking in water and arranging flat on a piece of clear perspex, glass or white card. Note that median and lateral lamellae have different arrangements of marginal setae, the median lamella having a longer and often more closely-spaced row of stout setae on its dorsal surface whereas the lateral lamellae have more stout setae ventrally.

1 Caudal lamella undivided, without nodes, and without sharp segregation of stout and fine marginal setae, but with characteristic dark X-shaped design (fig. 415). (Antenna seven-segmented (fig. 378); body relatively broad with rather rectangular head (fig. 405); labium as in fig. 406) (all larval instars described by Gardner & MacNeill (1950)) ....
.... *Pyrrhosoma nymphula* (p.72)

– Caudal lamella nodate or subnodate, with short, spinous antenodal setae and long, fine postnodal setae, the two setal types being segregated or occurring together over only a short transition zone; lamella not patterned as above
.... 2

2(1) Caudal lamella (figs 416,419,420) with stout marginal setae reaching to or beyond the mid-point of only one margin; subnodate, nodal line if present sharply angled or sinuately very oblique; apex of lamella pointed. (Antenna with seven segments) 3

– Caudal lamella (figs 381,382,417,422) with stout marginal setae extending about mid-way or more along both margins; nodate or subnodate, nodal line if present approximately at right angles to the long axis of the lamella, straight or slightly sinuate; apex of lamella pointed or rounded. (Antenna with six or seven segments)
.... 6

3(2) Prementum with one or two (rarely three) long setae in each series with scattered short setae between them (fig. 414). Head in dorsal view with occiput angulate. Caudal lamella (fig. 420) short and broad, subnodate, with dark spots around postnodal margin
.... *Ceriagrion tenellum* (p.97)

– Prementum with four or more long setae in each series (fig. 407). Head in dorsal view with occiput rounded behind eyes. Caudal lamella more elongated and not marked as above 4

4(3) Caudal lamella (fig. 416) long, slender and pointed, about four times as long as broad. *Ischnura* (only *elegans* and *pumilio* have been adequately described) 5

– Caudal lamella (fig. 419) shorter, about three times as long as broad with distal margins strongly convex to the short, blunt, apical point. (Fraser (1950) describes all(?) larval instars) .... *Coenagrion scitulum* (p.80)

5(4) Labial palp with six or seven (rarely five) setae (fig. 407) .... *Ischnura elegans* (p.92), *I. genei* (p.94) (Conci & Nielsen (1956) distinguish these two species by the form of the apex of the labial palp between movable hook and end hook. In *genei* the margin is seven-toothed,

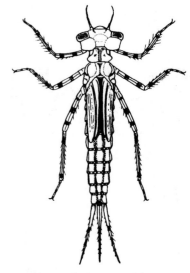

405. *Pyrrhosoma nymphula*

Figure 405 Larva of Zygoptera, Coenagrionidae

but in *elegans* only three or four teeth are discernible)

– Labial palp with five setae. (Fraser (1949a) and Gardner (1951c) describe the larval instars) .... *I. pumilio* (p.92) (*I. graellsi* is described by Ferreras Romero (1981b) as having six palpal setae in the male and five in the female)

6(2) Caudal lamella subnodate, nodal line absent or hardly visible (figs 381,422) 7

– Caudal lamella usually nodate with nodal line distinct (figs 382,417) .... 9 (Specimens in which the nodal line is not distinct, but which do not run from couplet 7, are probably *Coenagrion*)

7(6) Caudal lamella with sharp, elongated point (fig. 421) and irregularly spotted. Overall length of final-instar larva only 18mm. Antenna with six segments ....
.... *Nehalennia speciosa* (p.95)

– Caudal lamella with short apical point, unpigmented or with narrow, transverse, dark bands. Body length 20mm or more. Antenna with six or seven segments 8

8(7) Labial palp with a very small spine on outer margin adjacent to the distal seta, and with six or seven setae (fig. 408). Caudal lamella (fig. 422) usually with one to three narrow, transverse, dark bands. Antenna with six segments .... *Enallagma cyathigerum* (p.89)

– Labial palp without such a spine adjacent to distal seta, and with five setae (fig. 413). Caudal lamella (fig. 381) short and unpigmented. Antenna with seven segments. (All larval instars described by Corbet, 1955b) ....
.... *Coenagrion mercuriale* (p.80)

9(6) Caudal lamella with apex bluntly rounded, postnodal section with three, broad, transverse dark bands (fig. 382). Antenna with six segments. (Labial palp with six setae; prementum with two series of four setae (fig. 409)) ....
.... *Erythromma najas* (p.73)

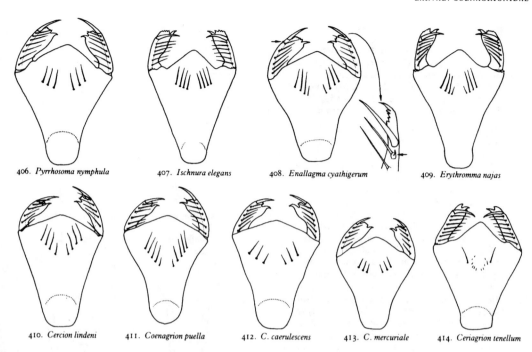

406. *Pyrrhosoma nymphula*   407. *Ischnura elegans*   408. *Enallagma cyathigerum*   409. *Erythromma najas*

410. *Cercion lindeni*   411. *Coenagrion puella*   412. *C. caerulescens*   413. *C. mercuriale*   414. *Ceriagrion tenellum*

Figures 406–414 Labia of Coenagrionidae
(short arrows on figure 408 indicate the small spines on the palps)

(Conci & Nielsen (1956) describe *E. viridulum* as having unpigmented and slightly pointed caudal lamellae, seven antennal segments but the last one rudimentary, and labial palp with five setae)

– Caudal lamella with apex bluntly rounded or pointed, postnodal section not pigmented (figs 417,418). Antenna with six or seven segments. *Coenagrion* and *Cercion* (larvae of several European species have been omitted from this key because they are too poorly known) 10
The remaining part of this key follows Gardner (1954b) and Franke (1979) and is provisional only. Carchini (1983) provides a very useful alternative key.

10(9) Antenna with six segments ....
.... *Coenagrion armatum* (p.83), *C. hastulatum* (see Gardner, 1954a) (p.82), *C. lunulatum* (p.82)
– Antenna with seven segments 11

11(10) Caudal lamella with one margin more strongly curved than the other, nodal line slightly distal to the middle and slightly oblique 12
– Caudal lamella with margins almost symmetrically curved and with nodal line about at right angles to the long axis (fig. 417) 13

12(11) Prementum (fig. 412) with three setae in each oblique series .... *Coenagrion caerulescens* (p.81)
– Prementum (fig. 410) with four setae in each oblique series .... *Cercion lindeni* (p.88)

13(11) Caudal lamella (fig. 418) most often with bluntly pointed apex .... *Coenagrion puella* (p.86)
– Caudal lamella (fig. 417) most often with rounded apex ....
.... *C. pulchellum* (p.87)
(This separation of *puella* and *pulchellum* follows Gardner (1954b) and seems to be unreliable.)

415. *Pyrrhosoma nymphula*   416. *Ischnura elegans*   417. *Coenagrion pulchellum*   418. *C. puella*

419. *C. scitulum*   420. *Ceriagrion tenellum*   421. *Nehalennia speciosa*   422. *Enallagma cyathigerum*

Figures 415–422 Median caudal lamellae of Coenagrionidae in left lateral view (outlines only of *Coenagrion puella* and *Nehalennia speciosa*)

AESHNIDAE

The form of the prothoracic supracoxal armature is frequently used as a specific character, but differences are slight and difficult to appreciate. Consequently, it has not been employed here. Schmidt, Er. (1936a) monographs central European aeshnid larvae. The following key does not include *Aeshna serrata* or *Caliaeschna microstigma*.

1   Head in dorsal view (fig. 423) with posterior corners of occiput almost right-angled. (Prementum (fig. 427) approximately twice as long as broad. S5–9 with strong lateral spines (fig. 435)) .... *Boyeria irene* (p.119)
–   Head in dorsal view (figs 424–426) with posterior corners of occiput rounded 2

2(1) Eyes small, in dorsal view (fig. 424) much shorter than length of lateral occipital margin. (Prementum (fig. 428) short, slightly longer than broad. S7–9 with lateral spines, S9 with a mid-dorsal spine on its posterior margin (fig. 436)) .... *Brachytron pratense* (p.117)
–   Eyes large, in dorsal view (figs 425,426) as long as or considerably longer than length of lateral occipital margin 3

3(2) Head in dorsal view (fig. 425) roundish, eyes flattened dorsally with posterior margins almost straight, approximately parallel to posterior margin of occiput. Lateral spines on S7–9 (fig. 437). *Anax, Hemianax* 4
–   Head in dorsal view (fig. 426) more or less pentagonal, eyes not especially flattened dorsally with posterior margins concave. Lateral spines on S6–9, but those on S6 sometimes very small (fig. 4c, p.17). *Aeshna* 6

4(3) Male projection at base of epiproct not clearly differentiated and apically acute. Length of exuvia less than 45mm (exuvia described by De Marmels, 1975) ....
.... *Hemianax ephippiger* (p.116)
–   Male projection at base of epiproct clearly delimited and apically truncate. Length of exuvia more than 46mm. *Anax* 5

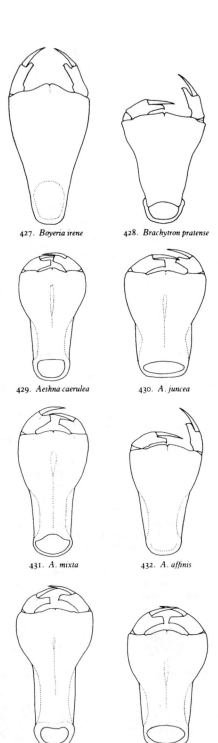

427. *Boyeria irene*   428. *Brachytron pratense*

429. *Aeshna caerulea*   430. *A. juncea*

431. *A. mixta*   432. *A. affinis*

433. *A. cyanea*   434. *A. grandis*

Figures 427–434 Labia of Aeshnidae

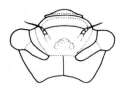
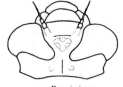

423. *Boyeria irene*   424. *Brachytron pratense*

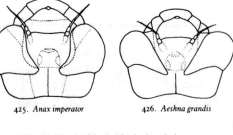

425. *Anax imperator*   426. *Aeshna grandis*

Figures 423–426 Heads of Aeshnidae in dorsal view

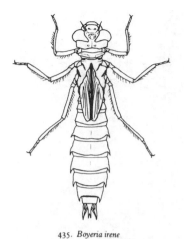

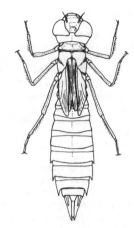

435. *Boyeria irene* 436. *Brachytron pratense* 437. *Anax imperator*

Figures 435–437 Larvae of Aeshnidae

5(4) Male projection at base of epiproct (fig. 438a) as long as broad, half the length of a cercus. (Larval instars described by Corbet, 1955a) .... *Anax imperator* (p.114)

– Male projection (fig. 438b) only about half as long as broad, about one-third the length of a cercus .... *A. parthenope* (p.115)

6(3) Cercus more than half as long as a paraproct (fig. 440) ... 7

– Cercus at most about half as long as a paraproct (figs 439,441) 8

7(6) Cercus fully two-thirds as long as a paraproct (fig. 440). Prementum strongly constricted basally, the base only half the width of the apex .... *Aeshna isosceles* (p.111)

– Cercus about 0.6 times as long as a paraproct. Prementum not strongly constricted basally ...... *A. subarctica* (p.105) (Differences between the very similar larvae of *subarctica* and *juncea* are described by Clausen (1984). In addition to the relatively longer cercus of *subarctica*, the two species may be distinguished by the respective lengths of the front femora, never more than 4mm in *subarctica* but always longer than 4mm in *juncea*.)

8(6) Body length at most 38mm 9

– Body length greater than 38mm 11

9(8) S9 with lateral spines short, reaching over only about one-third of the margin of S10 (fig. 439) and lateral spines on S6 vestigial. Distal margin of epiproct trifid. Labium as in figure 429 .... *A. caerulea* (p.103) (The only European species of *Aeshna* with six and not seven segments in antennae (Fraser, 1955).)

– S9 with lateral spines longer and those on S6 developed. Distal margin of epiproct almost straight 10

10(9) S9 with lateral spines reaching almost to the apex of S10. Prementum (fig. 431) almost twice as long (excluding palps) as maximum breadth. (Larval instars described by Gardner, 1950c) .... *A. mixta* (p.107)

– S9 with lateral spines reaching over about two-thirds of the margin of S10. Prementum (fig. 432) about 1.5 times as long as maximum breadth .... *A. affinis* (p.108)

11(8) Cercus about a third as long as a paraproct 12

– Cercus about half as long as a paraproct (fig. 441) 13

12(11) Lateral spines on S4–9. Larva 50mm or more in length..... .... *A. crenata* (p.106)

– Lateral spines on S6–9. Larva 40–46mm in length. (Labium as in figure 434) .... *A. grandis* (p.111)

13(11) Prementum (fig. 433) more than twice as long as maximum breadth. Distal margin of epiproct (fig. 441) bifid. (S9 with lateral spine reaching slightly beyond a third of the length of S10 (fig. 441) and S6 with lateral spines developed) .... *A. cyanea* (p.108)

– Prementum (fig. 430) less than twice as long as broad. Distal margin of epiproct (fig. 442) trifid 14

14(13) Lateral spine on S9 reaching over about a third of the length of S10, and lateral spine on S6 vestigial (fig. 4c, p.17) .... *A. juncea* (p.104)

– Lateral spine on S9 extending over about half of the length of S10, and lateral spine on S6 developed .... *A. viridis* (p.110)

Figures 438–442 Details of abdominal apices of male Aeshnidae, in dorsal view

Shaded areas of *Anax* species (483a,b) show differently-shaped male projections. Apices of epiprocts of *Aeshna cynaea* (441) and *A. juncea* (442) are drawn to larger scale

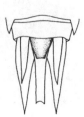

438a. *Anax imperator* 438b. *A. parthenope*

439. *Aeshna caerulea* 440. *A. isosceles*

442. *A. juncea*

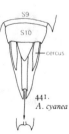

441. *A. cyanea*

GOMPHIDAE

Larvae are squat and dorsoventrally flattened (figs 443–445), often very hairy with particles adhering to their setae. They are slow-moving and burrow in sand, silt or mud. Schmidt, Er. (1936b) describes European species.

1   Tarsi of all legs with two segments. Abdomen ovoid in dorsal view (fig. 443), length 42–45mm, breadth 12–13mm. (Tibiae without an outwardly-directed apical spine. Mid-dorsal spines on S3–9, lateral spines on S7–9. Prementum (fig. 446) squarish. Third antennal segment scarcely broader than second (fig. 452)) ....
.... *Lindenia tetraphylla* (p.135)

–   Tarsi of front and middle legs with two segments, hind tarsi with three segments. Abdomen fusiform (figs 444,445), length 23–30mm. (Front and middle tibiae often with an outwardly-directed apical spine) 2

2(1)   Third antennal segment elongated, at least four times as long as broad and not or scarcely broader than second segment (figs 453,455) 3

–   Third antennal segment inflated, only about three times as long as broad and distinctly broader than second segment (figs 454,456,457) 9

3(2)   Inner margin of labial palp with end hook (distal tooth) directed mesially (figs 447,448). Fourth antennal segment nipple-like, much shorter than breadth of third segment (fig. 455). *Gomphus* (see also Cowley, 1933) 4

–   Inner margin of labial palp straight with end hook directed anteriorly, parallel to movable hook (fig. 450). Fourth antennal segment about as long as breadth of third segment (fig. 453) .... *Paragomphus genei* (p.128)

4(3)   S9 in ventral view at least as long as broad and S10 about half as long as broad. Body length 32–35mm. Abdomen sparsely hairy. Labial palp with inner margin proximal to end hook with four or five teeth (fig. 447). Lateral spines on S6 .... *Gomphus flavipes* (p.123)

–   S9 in ventral view broader than long. Body shorter. Abdomen more hairy. Other characters not all as above 5

5(4)   S10 in ventral view only slightly broader than long. (Lateral spines on S7–9 only. Inner margin of labial palp with seven to nine teeth) .... *G. pulchellus* (p.127)

–   S10 in ventral view about twice as broad as long 6

6(5)   Anterior margin of prementum concave. (Lateral spines only on S7–9. Inner margin of labial palp with six teeth) ...
.... *G. graslini* (p.127)

–   Anterior margin of prementum almost straight 7

7(6)   S9 in ventral view fully twice as broad as long. Abdomen very hairy (fig. 444). (Lateral spines on S6–9. Inner margin of labial palp with nine to eleven teeth) ....
.... *G. vulgatissimus* (p.125)

–   S9 in ventral view less than twice as broad as long. Abdomen rather less hairy 8

8(7)   Inner margin of labial palp not strongly curved and end hook not prominent. (Lateral spines absent from S6. Inner margin of labial palp with five or six teeth) ....
.... *G. simillimus* (p.125)

–   Inner margin of labial palp concave and end hook prominent (Aguesse, 1968) .... *G. lucasi* (p.126)

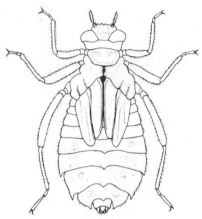

443. *Lindenia tetraphylla*

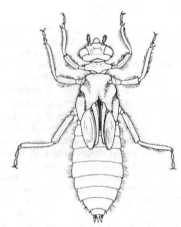

444. *Gomphus vulgatissimus*

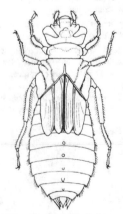

445. *Onychogomphus forcipatus*

Figures 443–445 Larvae of Gomphidae

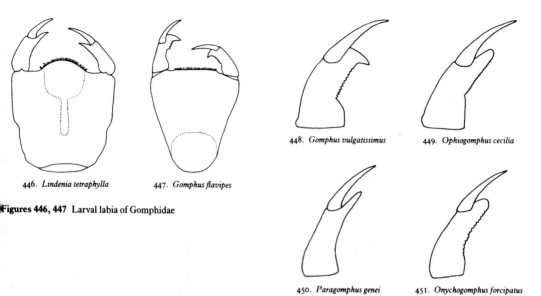

446. *Lindenia tetraphylla*

447. *Gomphus flavipes*

Figures 446, 447 Larval labia of Gomphidae

448. *Gomphus vulgatissimus*

449. *Ophiogomphus cecilia*

450. *Paragomphus genei*

451. *Onychogomphus forcipatus*

Figures 448–451 Left labial palps of Gomphidae

9(2) S9 in ventral view almost twice as long as S10. Large, at least 29mm long. (Lateral spines on S7–9) .... *Ophiogomphus cecilia* (p.130)

–   S9 and S10 in ventral view of approximately equal length. Body length less than 26mm. *Onychogomphus* 10

10(9) Mid-dorsal abdominal spines prominent only on S2+3 (Aguesse, 1968) .... *Onychogomphus costae* (p.134)

–   Mid-dorsal abdominal spines prominent on all but the last one or two segments (fig. 445) 11

11(10) Lateral spines on S6–9; mid-dorsal abdominal spines moderately developed. Outer margin of third antennal segment (fig. 456) moderately convex .... *O. forcipatus* (p.132)

–   Lateral spines on S8+9; mid-dorsal abdominal spines strongly developed. Outer margin of third antennal segment (fig. 457) very convex .... *O. uncatus* (p.134)

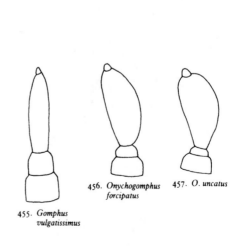

452. *Lindenia tetraphylla*

453. *Paragomphus genei*

454. *Ophiogomphus cecilia*

455. *Gomphus vulgatissimus*

456. *Onychogomphus forcipatus*

457. *O. uncatus*

Figures 452–457 Larval antennae (setae omitted) of Gomphidae

CORDULEGASTRIDAE

The general aspect of a cordulegastrid larva (figs 458,459) is somewhat intermediate between that of a typical aeshnid and gomphid. The body is relatively elongated and hairy. The head shape, however, is much closer to that of certain libellulids, especially *Libellula* and *Orthetrum*. The labium (fig. 386) is concave, as in the Libellulidae, and is diagnostic with its bifid apex and broad, coarsely serrated palps. Cordulegastrid larvae live in the soft bottoms of streams, buried in mud or sand with only part of their heads and anal appendages raised clear of the substrate.

The two most widely-distributed European species, *Cordulegaster boltoni* and *C. bidentata*, can be easily distinguished by the presence in *C. boltoni* (fig. 458) and absence in *C. bidentata* (fig. 459) of short lateral spines on S8 and S9. Balestrazzi *et al.* (1982) describe the exuvia of *C. picta* (?)*trinacriae* (here treated as a subspecies of *C. boltoni*) as having lateral spines on S8+9 larger and more divergent than in nominotypical *boltoni*. The exuvia of *C. heros* is also typical of the *boltoni*-complex, but it has relatively shorter and stouter paraprocts and epiproct than *boltoni* (Theischinger, 1979, photograph).

CORDULIIDAE

The larvae are broad and dorso-ventrally flattened with very long legs (figs 460,461). Ris (1911a) describes central European species.

1 S9 with long lateral spines reaching to the apices of the paraprocts (figs 461,468). Occiput with two pubescent tubercles. Length 30mm. (S2–9 with large mid-dorsal spines) .... *Epitheca bimaculata* (p.153)
– S9 with lateral spines shorter or absent. Without occipital tubercles. Length (except *Macromia*) no more than 25mm .. 2

2(1) Dorsal surface of abdomen with flattened, pectinate setae (fig. 462). (Distal edge of labial palp (fig. 470) usually with seven rounded teeth. Lateral spines on S8+9 (fig. 463); mid-dorsal spines represented by small tubercles) .... *Oxygastra curtisi* (p.154)
– Dorsal surface of abdomen with simple hair-like or spiniform setae 3

3(2) Distal margin of labial palp with five or six well-defined, rounded teeth, each armed with six to ten short setae. A pronounced rostrum is present between the antennae. Large, 28–32mm long. (Strong mid-dorsal spines on S2–9) .. .. *Macromia splendens* (p.155)
– Distal margin of labial palp with eight to nine rounded teeth, each sometimes crowned with fewer setae (figs 469,471). No rostrum between antennae. Smaller, 25mm long or less 4

4(3) S3 or S4 to S8 or S9 with mid-dorsal spines (figs 466,467) ... 5
– Abdominal segments lacking mid-dorsal spines (figs 464,465). (Body very hairy) 7

5(4) Mid-dorsal spine on S9 small or rudimentary, and those on S4–8 relatively small (fig. 466). Abdomen in dorsal view (fig. 460) apically rather truncated. (Lateral spines on S8+9. Usually eight rounded teeth on distal edge of labial palp (fig. 471)) .... *Cordulia aenea* (p.145)

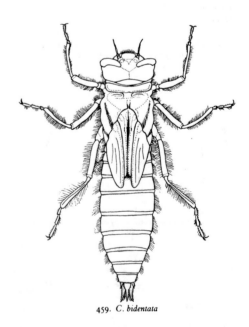

458. *Cordulegaster boltoni*   459. *C. bidentata*

Figures 458, 459 Larvae of *Cordulegaster* species

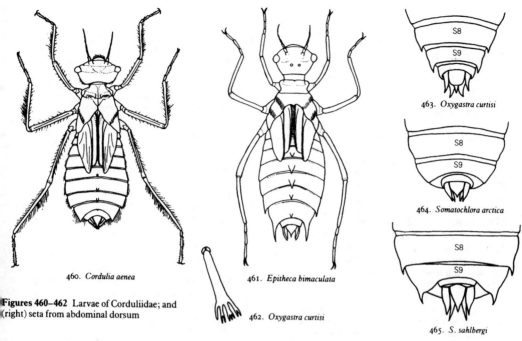

460. *Cordulia aenea*

461. *Epitheca bimaculata*

462. *Oxygastra curtisi*

Figures 460–462 Larvae of Corduliidae; and (right) seta from abdominal dorsum

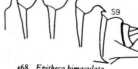

463. *Oxygastra curtisi*

464. *Somatochlora arctica*

465. *S. sahlbergi*

Figures 463–468 Abdominal apices of Corduliidae (setae omitted)
(above) (463–465) in dorsal view; (left) (466–468) in left lateral view

– Mid-dorsal spine on S9 well-developed and those on S4–8 large, pointed and curved posteriorly (fig. 467). Abdomen in dorsal view fusiform 6

6(5) Lateral spines on S9 long, but those on S8 small or absent (Ris, 1911a) .... *Somatochlora flavomaculata* (p.152)
– Lateral spines well-developed on S8+9 (fig. 467) ....
.... *S. metallica* (p.149)

7(4) Lateral spines prominent on S8+9 (fig. 465) (Cannings & Cannings, 1985) .... *S. sahlbergi* (p.151)
– Lateral abdominal spines absent (fig. 464) 8

8(7) Paraproct about twice as long as S10 .... *S. arctica* (p.151)
– Paraproct about three times as long as S10 ....
.... *S. alpestris* (p.150)

466. *Cordulia aenea*

467. *Somatochlora metallica*

468. *Epitheca bimaculata*

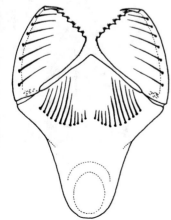

469. *Somatochlora metallica*

470. *Oxygastra curtisi*

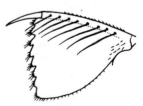

471. *Cordulia aenea*

Figures 469–471 Details of larval labia of Corduliidae showing (left) labium and (above) right labial palps

207

LIBELLULIDAE

Larvae (figs 4d, p.17, and 472–474) resemble those of Corduliidae, having squat, flattened bodies and a deeply spoon-shaped labium (figs 476,477). The following species are not included in the key: *Orthetrum trinacria*, *O. chrysostigma*, *O. ramburi*, *Diplacodes lefebvrei* and *Sympetrum decoloratum*.

1   Head in dorsal view (figs 472,475) with occiput long, at least twice as long as the small and prominent eyes, and with lateral margins subparallel behind the eyes. Body very hairy, legs shorter and stouter 2

–   Head in dorsal view (figs 473,474) with occiput no more than twice as long as the relatively longer eyes, and with lateral margins strongly convergent. Body less hairy, legs longer and more slender 9

2(1)   S8 with mid-dorsal spine. *Libellula* 3

–   S8 without mid-dorsal spine. *Orthetrum* 5

3(2)   Abdomen with long mid-dorsal spines on S4–9 (fig. 478). Prementum with setae arranged as in figure 480; labial palp with four or five setae .... *Libellula fulva* (p.159)

–   Abdomen with shorter mid-dorsal spines on S4–8 (fig. 479). Prementum with setae differently arranged; labial palp with seven or more setae 4

4(3)   Prementum (fig. 481) with nine to eleven long setae and three to five short mesial setae. Head in frontal view (fig. 476) with eyes dorsally protuberant and labial palp with seven or eight well-defined crenations. Abdomen with alternate dark and pale dorsal longitudinal bands. (Gardner (1953b) describes all the larval instars) ....
.... *L. depressa* (p.160)

–   Prementum (fig. 482) with six to nine long setae and four to seven short mesial setae. Head in frontal view (fig. 477) with eyes not so protuberant dorsally, and labial palp with eight or nine shallow crenations. Abdomen uniformly coloured .... *L. quadrimaculata* (p.158)

5(2)   Labial palp with three to five setae 6

–   Labial palp with six to eight setae 7

6(5)   Prementum (fig. 485) with distal margin rounded and on each side two long setae and four to seven short setae with a mesial field of several transverse rows of very small spiniform setae; labial palp with three (rarely four or five) setae .... *Orthetrum coerulescens* (p.167)

–   Prementum (fig. 484) with distal margin angled and on each side three long setae and twelve to sixteen short setae in a single (partly double mesially) transverse row (no spiniform setae); labial palp with five (rarely four) setae ...
.... *O. albistylum* (p.164)

7(5)   Labial palp with six setae (fig. 486). (Two large setae on each side of prementum). (Kumar, 1971 describes all larval instars) .... *O. brunneum* (p.166)

–   Labial palp with eight (rarely seven) setae (fig. 483) 8

8(7)   Abdomen with mid-dorsal spines on S3–6 and small lateral spines on S8+9 .... *O. cancellatum* (p.163)

–   Abdomen without mid-dorsal spines and with lateral spines (very small) only on S8 (Nielsen, 1955) ....
.... *O. nitidinerve* (p.165)

9(1)   Abdomen without mid-dorsal spines 10

–   Abdomen with mid-dorsal spines on at least two segments
.... 12

10(9)   S8+9 with well-developed lateral spines (those on S9 as long as the segment). Labial palp with fourteen to sixteen setae. Body length at least 25mm ....
.... *Pantala flavescens* (p.191)

–   S8+9 with lateral spines short or absent. Labial palp with ten to thirteen setae. Body length less than 20mm 11

11(10) Lateral spines very short on S9 and rudimentary on S8 (fig. 496). Labial palp with twelve to fourteen setae. (Gardner, 1951a describes all larval instars) ....
.... *Sympetrum fonscolombei* (p.180)

–   Short lateral spines on S8+9. Labial palp with ten or eleven setae .... *Crocothemis erythraea* (p.171)

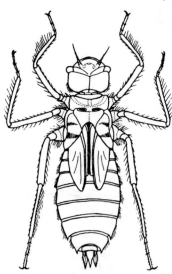

472. *Orthetrum cancellatum*

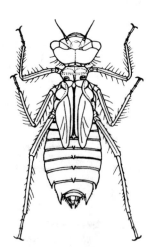

473. *Sympetrum nigrescens*

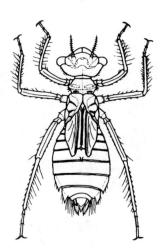

474. *Leucorrhinia dubia*

Figures 472–474 Larvae of Libellulidae

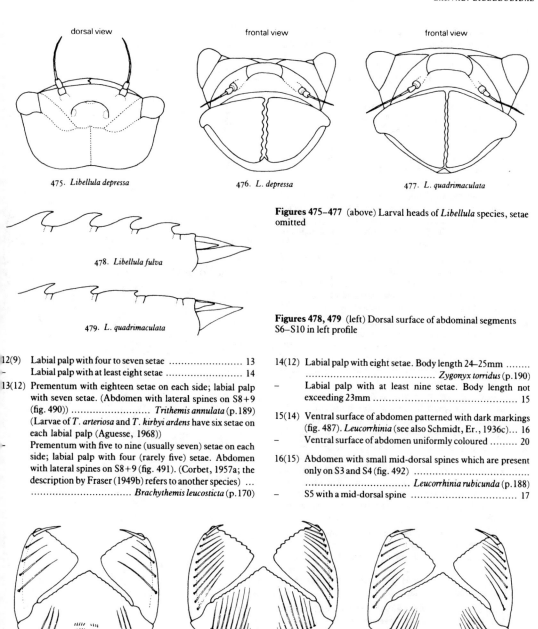

dorsal view

frontal view

frontal view

475. *Libellula depressa*

476. *L. depressa*

477. *L. quadrimaculata*

Figures 475–477 (above) Larval heads of *Libellula* species, setae omitted

478. *Libellula fulva*

479. *L. quadrimaculata*

Figures 478, 479 (left) Dorsal surface of abdominal segments S6–S10 in left profile

12(9) Labial palp with four to seven setae 13

– Labial palp with at least eight setae 14

13(12) Prementum with eighteen setae on each side; labial palp with seven setae. (Abdomen with lateral spines on S8+9 (fig. 490)) .... *Trithemis annulata* (p.189) (Larvae of *T. arteriosa* and *T. kirbyi ardens* have six setae on each labial palp (Aguesse, 1968))

– Prementum with five to nine (usually seven) setae on each side; labial palp with four (rarely five) setae. Abdomen with lateral spines on S8+9 (fig. 491). (Corbet, 1957a; the description by Fraser (1949b) refers to another species) ... . *Brachythemis leucosticta* (p.170)

14(12) Labial palp with eight setae. Body length 24–25mm .... *Zygonyx torridus* (p.190)

– Labial palp with at least nine setae. Body length not exceeding 23mm 15

15(14) Ventral surface of abdomen patterned with dark markings (fig. 487). *Leucorrhinia* (see also Schmidt, Er., 1936c)... 16

– Ventral surface of abdomen uniformly coloured 20

16(15) Abdomen with small mid-dorsal spines which are present only on S3 and S4 (fig. 492) .... *Leucorrhinia rubicunda* (p.188)

– S5 with a mid-dorsal spine 17

480. *Libellula fulva*

481. *L. depressa*

482. *L. quadrimaculata*

Figures 480–482 Larval labia of *Libellula* species

17(16) Abdomen with lateral spine on S9 at least as long as the median dorsal length of the segment and extending beyond the apex of S10 18
– Abdomen with lateral spine on S9 much shorter than the median dorsal length of the segment and not reaching the apex of S10 19

18(17) S3–7 with mid-dorsal spines (absent from S8 and S9); S8+9 with lateral spines (fig. 493) ..... *L. albifrons* (p.186)
– S3–9 with mid-dorsal spines; S5 or S6 to S9 with lateral spines (fig. 494) .... *L. caudalis* (p.186)

19(17) S3–8 with mid-dorsal spines (fig. 495) ... *L. pectoralis* (p.189)
– S4–7 with mid-dorsal spines (absent from S8). (Gardner (1953a) describes all larval instars) .... *L. dubia* (p.187)

20(15) Labial suture, with labium in resting position, at level of middle coxae. (Mid-dorsal spines on S3–8, lateral spines on S8+9) .... *Selysiothemis nigra* (p.193)
– Labial suture at level of middle and hind coxae. *Sympetrum* (excluding *fonscolombei*: couplet 11) 21

21(20) Mid-dorsal abdominal spines long and on S4–8, that on S7 at least half as long as the segment 22
– Mid-dorsal abdominal spines short, on S3–8 or on fewer segments, that on S7 at most one-third as long as the segment 23

22(21) Labial palp with nine setae. Lateral spines on S8 and S9 only about half as long as the median lengths of their segments .... *Sympetrum pedemontanum* (p.183)
– Labial palp with ten or eleven setae. Lateral spines on S8 and S9 as long as or longer than the median lengths of their segments (fig. 497) .... *S. depressiusculum* (p.182)

23(21) Lateral spine on S9 short, less than half as long as the mid-dorsal length of the segment (fig. 498) 24
– Lateral spine on S9 long, more than half the mid-dorsal length of the segment (figs 499, 500) 25

24(23) Labial palp with one or two short setae proximally and eight to ten long setae (fig. 488). Lateral spine on S9 only slightly longer than that on S8 (fig. 498). (Gardner (1951b) describes all larval instars) .... *S. danae* (p.182)
– Labial palp with ten or eleven long setae (no short setae) (fig. 489). Lateral spine on S9 almost twice as long as that on S8 .... *S. flaveolum* (p.180)

25(23) Abdomen with mid-dorsal spines on S3–8, that on S3 very small (fig. 500). (Prementum with fourteen or fifteen setae on each side, ten or eleven of which are long) ....
.... *S. vulgatum* (p.177)
– Abdomen with mid-dorsal spines on S4 or S5 to S8, that on S4 very small 26

26(25) Lateral spine on S9 reaching, or almost reaching, the apices of the paraprocts. Prementum with twelve to fourteen setae on each side, eight or nine of which are long
.... 27
– Lateral spine on S9 not nearly reaching to the apices of the paraprocts. Prementum with fourteen to sixteen setae on each side, ten or eleven of which are long 28

27(26) Labial palp with eleven setae. Mid-dorsal spines on S4 or S5 to S8, those on S6 and S7 the longest ....
.... *S. meridionale* (p.178)
– Labial palp with nine to eleven setae. Mid-dorsal spines on S4 or S5 to S8, that on S7 the longest (fig. 499). (Gardner (1950a) describes all larval instars) ....
.... *S. sanguineum* (p.181)

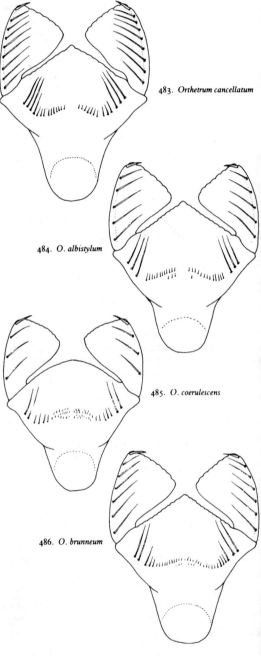

483. *Orthetrum cancellatum*

484. *O. albistylum*

485. *O. coerulescens*

486. *O. brunneum*

Figures 483–486 Larval labia of *Orthetrum* species

28(26) Mid-dorsal spines on S4–8, straight in profile (fig. 501) (Gardner, 1955a) .... *S. nigrescens* (p.177)
– Mid-dorsal spines on S5–8 (sometimes a rudimentary spine on S4), slightly curved in profile (fig. 502) (Gardner, 1950b; Corbet, 1951) .... *S. striolatum* (p.176)

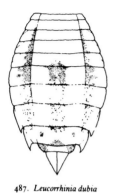

487. *Leucorrhinia dubia*

488. *Sympetrum danae*

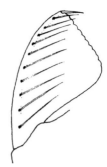

489. *S. flaveolum*

Figure 487 (above) *Leucorrhinia* species ventral surface of larval abdomen to show dark patterning

Figures 488, 489 Left labial palps of *Sympetrum* species

490. *Trithemis annulata*

491. *Brachythemis leucosticta*

492. *Leucorrhinia rubicunda*

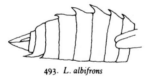

493. *L. albifrons*

494. *L. caudalis*

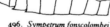

495. *L. pectoralis*

496. *Sympetrum fonscolombei*

497. *S. depressiusculum*

498. *S. danae*

499. *S. sanguineum*

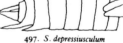

500. *S. vulgatum*

Figures 490–500 (above) Diagrams of abdomens of larval Libellulidae in right lateral view to show disposition and shapes of lateral and mid-dorsal spines on abdominal segments S3–S10

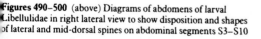

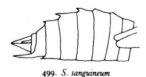

S8 S6 S4

501. *Sympetrum nigrescens*

Figures 501, 502 (right) Dorsal surface of abdominal segments n profile showing form of mid-dorsal spines

S8 S6 S4

502. *S. striolatum*

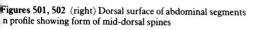

Supplement

All additions and corrections to the text of the first edition are cross-referenced, with the relevant page number or numbers, colour plate numbers and text figure references shown in parentheses.

Nomenclature

Calopteryx splendens (Harris, 1776) (p.57)
The year of publication was 1776, not 1782.

Calopteryx haemorrhoidalis (Vander Linden, 1825) (p.57)
C. h. asturica Ocharan, 1983 is a subspecies with an extensive clear apex to the male forewing, occurring from the extreme south-west of France, along the Spanish Atlantic coast and into northern Portugal.

Sympecma paedisca (Brauer, 1882) (p.66)
The European taxon *S. paedisca annulata* (Sélys, 1887), which was described from Turkey, is confirmed as a subspecies of *S. paedisca* after reexamination of type material (Jödicke, 1993a).

Coenagrion caerulescens (Fonscolombe, 1838) (p.81)
It has been claimed that the spelling could be interpreted as *coerulescens* (e.g. Davies & Tobin, 1984), but the original typography is unclear and the more widely accepted *caerulescens* is retained here.

Aeshna isosceles (Müller, 1767) (p.111)
A recent phylogenetic analysis of extant genera of Aeshnidae (von Ellenrieder, 2002) places *Aeshna*, *Anaciaeschna*, *Anax* and *Hemianax* together in one of the three clades comprising the large sister-group (Aeshnodea) of the Gomphaeschninae. The generic placement of *A. isosceles*, however, is still uncertain since it appears to lie outside both *Aeshna* and *Anaciaeschna*.

Cordulegaster Leach, 1815 (p.136 *et seq.*)
Lohmann (1992) described the genus *Sonjagaster*, now more often considered a subgenus of *Cordulegaster*. *C. (S.) insignis* belongs to this taxon as does *C. (S.) helladica* (Lohmann, 1993), the typical subspecies of which is a new European dragonfly from Greece. A new subspecies of *C. boltonii* (Donovan), *C. b. iberica* Boudot & Jacquemin, 1995, has been described from Albacete, Spain. It is close to *C. b. algirica*.

Somatochlora meridionalis Nielsen, 1935 (p.149)
Treated as a subspecies of *S. metallica* (Vander Linden) in the first edition, *S. meridionalis* is now more usually considered a distinct species. In Bulgaria it inhabits small streams in shady places whereas *S. metallica* frequents montane bog pools up to an altitude of 2100m (Marinov, 1999).

Orthetrum anceps (Schneider, 1845) (p.168)
O. ramburii (Sélys, 1848) is now considered to be a junior synonym of *O. anceps* and the latter name should be used. *O. anceps* is, however, very closely related to *O. coerulescens* and reliably distinguishable from it only by the form of the anterior lamina of the male secondary genitalia. The two taxa are allopatric with a zone of hybridization running through Sicily, the Balkans and Hungary, but 'pure' *coerulescens* and *anceps* persist alone, respectively, on Corsica and Sardinia (Mauersberger, 1994).

Sympetrum nigrescens Lucas, 1912 (p.177)
Doubts about the validity of this taxon, at least as a full species, are increased by the finding that the larva cannot be separated from that of *S. striolatum* on features of those parts of the body which differ between *S. striolatum*, *S. sanguineum* and *S. vulgatum* (Jödicke, 1993b).

Sympetrum sinaiticum Dumont, 1977 (p.179)
Plate 30 (217, 218)
Described from Tunisia as a subspecies of *Sympetrum decoloratum* (Sélys) but now regarded as a separate species. Jödicke (1994) synonymized *S. d. decoloratum* under *S. vulgatum* (L., 1758) and recognized a number of subspecies of *S. sinaiticum* of which *S. s. sinaiticum* is confined to North Africa and *S. s. tarraconensis* Jödicke, 1994 is the Spanish form described from Tarragona, Catalonia. In the key to species of *Sympetrum* (p.173), *S. sinaiticum* will run out as *S. decoloratum*. It is a pale dragonfly with much reduced black markings and can be distinguished from the similarly coloured *S. meridionale* by the sexual characters given in the key for *S. decoloratum*.

Orthography

To conform with the original authors' spelling, *Cercion lindenii*, *Ischnura graellsii*, *Cordulegaster boltonii*, *Oxygastra curtisii*, *Orthetrum ramburii* (a junior synonym of *O. anceps*, see above) and *Sympetrum fonscolombii* should appear thus, terminating in a second '*i*'.

Thermoregulation

Sternberg (1996) published an important paper on thermoregulation in *Aeshna* species in which the significance of abdominal colour pattern was studied. Internal heat gain from incident light was found to be higher in more darkly coloured abdominal segments because of light absorption. Light coloured spots reflect light and serve as overheating protectors. In male *A. caerulea* the reversible colour change ensures that under high light intensity, when the Tyndall-blue spots develop, internal heating is reduced.

American species observed in Europe

ISCHNURA (ANOMALAGRION) HASTATA
(Say, 1839) (Coenagrionidae)

A breeding population of this species has been present, though misidentified, on the Azores since at least 1938 (Belle & van Tol, 1990). It is a widespread damselfly ranging from Canada, United States, Central America and the West Indies to South America. The male is unmistakable, very small, at most 25mm long, with the triangular pterostigma separated from the forewing margin by a petiole. No male, however, has been found on the Azores, where *I. hastata* seems to be parthenogenetic (Cordero Rovera *et al.*, 2001). *I. pumilio* is the only other zygopteran known on the islands.

ANAX JUNIUS (Drury, 1770) (Aeshnidae)
Plate 30 (212)

The first seven confirmed European records of this North American species were made in September 1998 in Cornwall and the Isles of Scilly (Corbet, 2000); they are attributed to hurricane displacement of migrants from the eastern United States seaboard. The species may be distinguished from the similar *A. imperator* in the male by the absence of a central black line on the second abdominal segment (which is blue with a pair of submedian dark marks), and by a pointed tooth on the outer apical angle of the superior abdominal appendage. The female has a pair of small occipital tubercles, closer to each other than are

those of *A. parthenope*. In both sexes, the dorsal face of the frons is yellow with a roundish, black median spot bracketed in front and at the sides by a blue mark. (See also p.30)

PACHYDIPLAX LONGIPENNIS (Burmeister, 1839) (Libellulidae)

A single specimen was found dead on an oil-rig off Shetland in 1999, but there is no certainty that it reached this position unaided. The species is distributed from Canada across the United States to Mexico, Bermuda and the Bahamas.

Accidental introductions into Europe

In the 1970s *Crocothemis servilia* (Drury, 1770), a very close Asian relative of the European *C. erythraea*, was accidently introduced into Florida, probably amongst aquatic plant material. It became successfully established in the wild, spreading some 250km in seven years. Several odonate species have been similarly accidentally introduced into Europe with aquatic plants and have been found as larvae, adults or both in nursery tanks (Lieftinck, 1978; Agassiz, 1981; Brooks, 1988). None of these introductions, listed below, seems to have resulted in the establishment of wild populations.

	Origin	Imported into
Coenagrionidae		
Agriocnemis femina Brauer, 1868	Australasia	Netherlands
Argia fumipennis (Burmeister, 1839)	N. America	Great Britain
Enallagma signatum Hagen, 1861	N. America	Great Britain
Ischnura posita (Hagen, 1861)	N. America	Great Britain
I. senegalensis (Rambur, 1842)	Asia, Africa	Great Britain, Finland
Ceriagrion cerinorubellum (Brauer, 1865)	S.E. Asia	Great Britain
Pseudagrion microcephalum (Rambur, 1842)	Australasia	Finland
Pseudagrion sp.		Great Britain
Aeshnidae		
Anax gibbosulus Rambur, 1842	Australasia	Great Britain
A. guttatus (Burmeister, 1839)	S.E. Asia	Great Britain, Finland
A. imperator Leach, 1815	Palaearctic	Great Britain
Gomphidae		
Ictinogomphus decoratus melaenops (Sélys, 1857)	S.E. Asia	Netherlands
Libellulidae		
Orthetrum sabina (Drury, 1770)	Asia etc.	Great Britain
O. japonicum internum McLachlan, 1894	S.E. Asia	Finland
Crocothemis servilia (Drury, 1770)	Asia	Great Britain
Erythemis simplicicollis (Say, 1839)	N. & C. America	Great Britain
Neurothemis fluctuans (Fabricius, 1793)	S.E. Asia	Finland, Netherlands
Rhodothemis rufa (Rambur, 1842)	S.E. Asia	Great Britain
Tramea transmarina euryale Sélys, 1878	S.E. Asia	Great Britain
Urothemis bisignata Brauer, 1868	Australasia	Great Britain

Other species newly recognized in Europe

LESTES (CHALCOLESTES) PARVIDENS
(Artobelevski, 1929)

This is an eastern form, long considered a subspecies of *L. viridis* (see p. 61), which has been found in Bulgaria, Croatia (Krk Island), Greece and Italy. It lives syntopically with *L. viridis* in ponds near Rome in Italy with only limited hybridization (Dell'Anna *et al.*, 1999). The two species can be distinguished morphologically both as adults and larvae. The male cercus of *L. parvidens* (fig. 503) is yellow with a sharply defined black tip, that of *L. viridis* (fig. 504) is more diffusely darkened at apex and also at base. In addition, the inner subapical tooth is smaller in *L. parvidens*, but contrastingly, the teeth on the ventral tip of the ovipositor are more strongly raised in *L. parvidens* (fig. 505) than *L. viridis* (fig. 506). Larvae reportedly differ in that *L. parvidens* has a smaller tooth on the first labial palp segment than *L. viridis*. There is a difference in phenology between the two species, adult *L. parvidens* peaking in emergence about the third week of June and *L. viridis* peaking almost a month later. Diurnal flight activity of *L. parvidens* is at a maximum about 12.00h whilst that of *L. viridis* is greatest some two hours later. Electrophoretic analysis of multilocus enzymes further supports the separation of *L. parvidens* and *L. viridis* as closely related but distinct species.

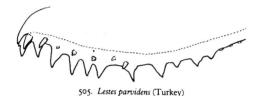

505. *Lestes parvidens* (Turkey)

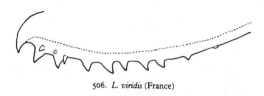

506. *L. viridis* (France)

Figures 505, 506 Apical toothed part of ovipositor in left lateral view of female Lestidae, under high magnification (cf. Figures 44–50, p. 60)

COENAGRION INTERMEDIUM Lohmann, 1990

C. intermedium was originally described as a subspecies of *Coenagrion ponticum* Bartenef, 1929. *C. ponticum* has been removed from synonymy under *C. syriacum* (Morton, 1924) (see p.87) and *C. p. intermedium*, which may be endemic to Crete, is considered to be a full species in the *C. puella*-complex (Battin, 1991, 1993). The male has a stalked black U-shaped mark on S2 (as in *C. pulchellum*), each superior abdominal appendage has a small internal tooth (as in *C. puella*), and the inferior appendages terminate in narrow, finger-like processes (cf. fig. 507, *C. ponticum*), as in *C. pulchellum* but directed horizontally and not upwards. Females of *C. intermedium* (not examined) are said to somewhat resemble those of *C. scitulum*, having a pterostigma twice as long as broad. Larvae of *C. intermedium* have been found in small rivers amongst tree roots, in the same places as *Boyeria cretensis* larvae.

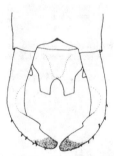

503. *Lestes parvidens* (Cyprus)

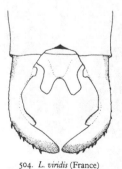

504. *L. viridis* (France)

Figures 503, 504 Male *Lestes* (*Chalcolestes*) species, terminal abdominal appendages in dorsal view (cf. Figures 38–43, p. 59)

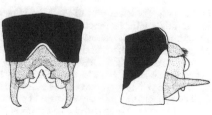

507. *C. ponticum*

Figure 507 Male *Coenagrion ponticum*, abdominal appendages and segment S10 from above (left) and in left lateral view (right). Dark areas of appendages are stippled. (cf. Figures 85–95, p. 78) Drawn from a specimen collected at Mersin, Turkey (K. M. Guichard)

ISCHNURA FOUNTAINEI Morton, 1905 (p.95)

This species, briefly discussed in the first edition, ranges from Central Asia through the Middle East to North Africa and, following discovery on the island of Pantelleria (Italy) in the Sicilian Channel, it may now be considered a European damselfly.

ANAX IMMACULIFRONS Rambur, 1842 (p.115)
Plate 30 (211)

This large aeshnid breeds in streams on the island of Karpathos in the Dodecanese (Greece) and was briefly considered in the first edition. It was inadvertently omitted from the list of European species of Odonata on p.50. It may at once be distinguished from *A. imperator*, *A. parthenope* and *Hemianax ephippiger* by its abdominal pattern of yellow and pinkish yellow spots. The frons is entirely without dark marking.

BOYERIA CRETENSIS Peters, 1991

The insect found on Crete that was previously referred to *B. irene* (see p.120) has since been described as a distinct species (Peters, 1991). Differences in abdominal colour pattern between *cretensis* and *irene* are illustrated by Boudot (1998); the anterior transverse pale mark on S3–8 is larger in *B. cretensis* than *B. irene*, occupying about one third of the lengths of these segments, and S2 has a lateral black spot. The larva has been found in small rivers amongst tree roots.

SOMATOCHLORA BORISI Marinov, 2001

The exciting discovery by Milen Marinov of a new European dragonfly was made in the eastern Rhodopes Mountains of southern Bulgaria. The holotype male and three male paratypes (plus more than twenty specimens released after capture) were collected from the Deimin dere River near Byal Gradetz in May and June 1999 and 2000 (Marinov, 2001). The species was subsequently found in adjacent north-eastern Greece on the River Diavolorema. It is apparently a riverine insect with a spring emergence. In morphology and coloration, *S. borisi* most closely resembles *S. metallica* and *S. meridionalis*. The three species have similar venation (one cubito-anal crossvein in forewing and divided hindwing discoidal cell) and only abdominal segments S2 and S3 with lateral yellow spots. In *S. borisi*, however, the lower part of the frons is metallic green in the middle with two separated lateral yellow spots (i.e. not with a complete, transverse yellow bar as in *S. metallica* and *S. meridionalis*), and the postclypeus is yellow laterally (not entirely metallic green as in the other two species) (see figures 508, 509). The thorax is densely hairy and has no lateral yellow spot, thus differing from *S. meridionalis*. The tips of the male superior anal appendages (fig. 510) are bluntly rounded and curved downwards and outwards, not pointed and curved upwards as in *S. metallica* and *S. meridionalis*, and the inferior appendage is truncated and weakly bilobed.

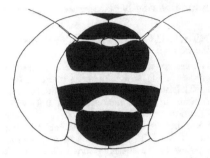

508. *Somatochlora metallica*

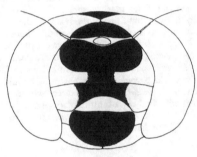

509. *S. borisi*

Figures 508, 509 Head in frontal view of *Somatochlora* species to show the difference in facial pattern. Metallic green areas are shown solid.
S. borisi is drawn from photographs of a paratype, kindly supplied by M. Marinov.

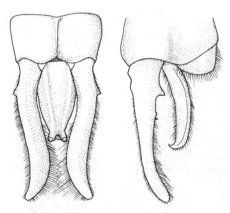

510. *Somatochlora borisi*

Figure 510 Male *Somatochlora borisi*, abdominal appendages and segment S10 in dorsal view (left) and right lateral view (right) (cf. Figures 262–266, pp. 146–147)
Partly after H. Wildermuth in Marinov (2001)

ORTHETRUM TAENIOLATUM (Schneider, 1845) (p.162)
Plate 30 (215, 216)

Known from Rhodes and described in the first edition, this species should now also be included in the European list. It is a small *Orthetrum* (body length about 39mm) with abdomen maximally about as long as hind wing and S2 and S3 about 2.5 times as deep as S4, pterostigma pale, membranule infuscate with a small, white proximal spot and the sides of the thorax, before pruinosity develops, are pale-striped. In the key to species of *Orthetrum*, these characters lead to *O. chrysostigma* from which *O. taeniolatum* may be separated by the presence on the side of the thorax of two pale stripes, one behind the humeral and one behind the metapleural sutures (in *O. chrysostigma* only the stripe behind the former suture is represented but this is much more conspicuous than in *O. taeniolatum*), and accessory male genitalia (fig. 511) with anterior lamina short (long in *O. chrysostigma*).

ORTHETRUM SABINA (Drury, 1770)
Plate 30 (213, 214)

There is a short account of this wide-ranging species in the first edition (see p. 162) since its distribution includes North Africa. Subsequently, its discovery on the Greek islands of Kos (Dell'Anna, 1994) and Samos (Grand, 2001) has given it European status. The species can be recognized by the following combination of characters: in both sexes the abdomen is slightly shorter than the hindwing and considerably swollen at the base (S2 and S3 maximally fully four times as deep as S4), the pterostigma is pale, the radial vein is black, the arculus is usually between the second and third antenodals, or under the second, the row of cells between veins IR_3 and Rspl is partly double, the membranule is infuscate with a small, pale spot proximally, and body length is 47–49mm. The female vulvar scale is similar to that of *O. chrysostigma*, but the male accessory genitalia (fig. 512) are distinctive with a dense tuft of golden brown setae on the anterior lamina. Other features are the absence of blue pruinosity, the conspicuous pale stripes on the thorax and the dark brown abdomen with white to pale yellow spots.

TRITHEMIS FESTIVA (Rambur, 1842) (p.190)
Plate 30 (219)

This species was briefly mentioned in the first edition as occurring on Cyprus and Rhodes. The mature male *T. festiva* differs from other western Palaearctic *Trithemis* in having a rich, deep blue body, an amber patch at the base of the hind wing and sometimes a dark tip to the front wing. Male accessory genitalia (fig. 513) are very similar to those of *T. annulata* but with a somewhat narrower genital lobe. Both sexes have mostly black wing veins, in contrast to those of *T. annulata* which are mostly light reddish to yellowish brown. Females of the two species differ in abdominal pattern with *T. festiva* having a mid-lateral dark brown stripe which expands posteriorly in S4–7, S8 is mainly dark brown with small yellow spots and S9 and S10 are entirely dark brown. In female *T. annulata*

there is no mid-lateral brown stripe with S4–6 entirely yellowish except for a fine, dark lateral carina, and S7–10 are darkened only mid-dorsally, S7 slightly and S10 most broadly.

511. *Orthetrum taeniolatum* (India)

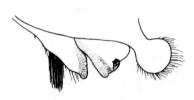

512. *O. sabina* (Oman)

Figures 511, 512 Male *Orthetrum* species, accessory genitalia in left profile (cf. Figures 288–295, p. 165). Apex of anterior lamina in frontal view is shown only for *O. taeniolatum*

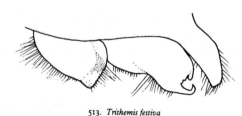

513. *Trithemis festiva*

Figure 513 Male *Trithemis festiva*, accessory genitalia in left lateral view (cf. Figure 364, p. 190)
Drawn from a specimen collected in Cyprus

Species new to British Isles

The list of species recorded from the British Isles has increased, although breeding is known for only two of the additions. As well as the American *Anax junius* (see above), new discoveries are *Lestes barbarus* (Norfolk 2002, three), *Anax parthenope* (Gloucestershire 1996, and every year since in England; breeding confirmed in Cornwall in 1999; recorded in Ireland in 2000 and 2001; Orkney 2000), *Aeshna affinis* (Kent 1952, one), possibly *Epitheca bimaculata* (presently unconfirmed report from Leicestershire in 2002), *Crocothemis erythraea* (five records – Cornwall 1995 and 1998, Isle of Wight 1999, Devon 2000, Hampshire 2002), *Sympetrum pedemontanum* (Gwent 1995, one), *Pantala flavescens* (Kent 1989, the fourth British record) and, most spectacularly, *Erythromma viridulum*.

Colonies of *E. viridulum* were first located in Essex in 1999 and on the Isle of Wight in 2000 and have since consolidated. Breeding has now been confirmed and it seems that the species is established in England. Following a large influx of migrants to the coasts of East Anglia and Kent in 2001 there was penetration inland as far as Bedfordshire and Buckinghamshire. In 2002 further sites in Bedfordshire were discovered.

Expansion of recorded ranges within Europe

Discrimination between actual range expansion, and an apparent increase resulting from more intensive recording, may not be easy, unless the area is historically well-documented as is Britain and some other areas in western Europe. The spread of *Erythromma viridulum* in England is associated with the situation in the Netherlands where the species was rare and of spasmodic occurrence until the late 1970s when records burgeoned. *E. viridulum* is now found throughout the country. Colonization of the Netherlands appears to have been chiefly from Belgium and from the Rhine valley.

A general spread northwards in Europe, often with a strong easterly or westerly component, has been reported for several other species. These include:

Lestes barbarus (Map 6, p.62): Belgium, first recorded breeding in 1994 and bred at same two localities (Merchtem and Wilrijk) for following three years at least; France, reported in 1996 from Pas de Calais (Nord); Germany, Heligoland (Korn, 1988), Konstanz and Mecklenburg (Vossfeld, 1995); Netherlands, found again after a very long absence in regions of Meinweg and Groote Peel in hot summer of 1995, reported in 2002 to be abundant on coastal dunes; Switzerland, Solothurn, a population persisted for 3 years (Schwaller, 1989); see also under 'Species new to British Isles' (p. 217).

Sympecma paedisca (Map 12, p.66, as *S. annulata*): Germany, Brandenburg, 50km north of Berlin; Poland, Bydgoszcz, Tucholskie; Sweden, Baltic island of Gotland (Dannelid & Ekestubbe, 2001).

Aeshna affinis (Map 44, p.108): Belgium, population at Willebroek reported in 1995; Germany, increase in records (1998) from Nordrhein-Westfalen (e.g. River Lippe), found in 1994 in Schleswig-Holstein (Nehms) and Niedersachsen, in 1995 near Leipzig, in 1996 in Niederlausitz and Rheinland-Pfalz (Deidesheim, breeding), breeding in Brandenburg (1994, several localities in valley of River Elbe), Saxony, Thüringia and Sachsen-Anhalt (1994) in eastern Germany; Netherlands, Hook of Holland, Limburg, Meuse Valley, North Brabant (near Tilburg), Walcheren Island and many other localities in hot summer of 1995 after apparent 44-year absence; Guerold *et al.* (2001) provide proof of breeding in Lorraine, France; see also under 'Species new to British Isles' (p.217).

Anax parthenope (Map 50, p.115): Seen ovipositing in France, Vienne (near Montmorillon) in 1995 (R. Rogers,

pers. comm.). Reported to be increasingly abundant in Thüringer Wald (Kipping, 1997) and Nordrhein-Westfalen (1998), Germany and in Netherlands (1999); see also under 'Species new to British Isles' (p.217).

Gomphus flavipes (Map 55, p.123): Returned to Netherlands in 1998 (last previously recorded 1902) with several sites found on River Waal near Nijmegen; reappeared in western Germany from Bavaria to Schleswig Holstein (1997 after absence of 87 years) and Niedersachsen, with breeding sites on the Rivers Elbe, Weser (Bremen) and Rhine (15 between Bingen and Köln (Cologne)) (Freyhof, 1998); extended range in eastern Germany from Saxony to Mecklenburg with breeding reported in 1994 in the middle reaches of the River Elbe in Sachsen-Anhalt; recorded (1999) from France in Allier (1992) and Loire-Atlantique (1999, the westernmost record).

Gomphus pulchellus (Map 59, p.127): In Germany, became established in Niedersachsen in the 1980s, breeding in Baden-Württemberg in 1994 (Göppingen), first record for Schleswig-Holstein in 1992 and reported in 1997 from Thüringia (Hildburghausen, Eisenach). First record for Austria (Vorarlberg) in 1988 (Gächte, 1988).

Orthetrum brunneum (Map 89, p.166): Spreading north, in Germany seen near Koblenz in 1989, breeding at Braunschweig, Niedersachsen, and the first breeding record for Nordrhein-Westfalen in 2001. In eastern Germany found breeding since 1992 in waters associated with open-cast mining south of Leipzig, often in company with *O. coerulescens*. Czech Republic, near Děčin in 1995. France, Somme, valley of Evoissons. Netherlands, returned in 1995 after 93 years to Zuid-Limburg (suspected breeding) and Winterswijke. Reported from Lublin, Poland in 2000; confirmed on Guernsey, Channel Islands in 2001.

Crocothemis erythraea (Map 94, p.172): Many observations in the last decade from Belgium, Netherlands and western Germany where the species has become increasingly abundant; first record for Saxony, eastern Germany in 1999; found on the Belgian coast (West Vlaanderen) in 1990, breeding in 1993 and said to be common throughout the country by 1995; France, Alsace (1994), Euure (1988), Pas de Calais (Nord) (1996) and Calvados (1992); Netherlands, 3 sites in hot summer of 1995, Walcheren Island in 1996; Poland, Bialowiéza (Kalkman & Dijkstra, 2000); Czech Republic from Moravia (reports 1990, 1997) and eastern Bohemia (1998); Slovakia, River Hron, Ipel Hills and Mochovce; Switzerland, breeding in canton Zurich; see also under 'Species new to British Isles' (p.217).

Attention must be drawn to the close similarity between *C. erythraea* and the Asiatic *C. servilia*; the two species can be distinguished with certainty only after microscopic examination of the male genitalia (Lohmann, 1981). Since sight records are not always reliable, voucher material should be provided from populations in newly-discovered localities.

Sympetrum depressiusculum (Map 103, p.182): Increasingly reported from Netherlands with large numbers in North Brabant in 1997 and the first documented breeding in 1998. Cited by Kalkman & Dijkstra (2000) from Bialowiéza, Poland. New records from southern Europe are Italy, Calabria and Umbria (Perugia, Lake Trasimeno); Slovenia, known only from western part of River Drava lowlands; Switzerland, Valais, Martigny. There was a very large immigration in 1994 to the Camargue, France.

Sympetrum pedemontanum (Map 105, p.183): Large numbers in Netherlands (North Brabant) in 1997 and sightings and breeding were increasingly reported northwest to Zeeland towards the end of the century; France, found in Vaucluse, Aveyron, and Haut-Rhin (Vosges and near Mulhouse); Germany, Niedersachsen, Nordrhein-Westfalen, Schleswig Holstein and Sachsen-Anhalt; Poland, increasingly recorded from lowlands (Bernard & Labedzki, 1993), noted in 1998 from Podlasie and by Kalkman & Dijkstra (2000) from Bialowiéza. Records from the Czech Republic (Děčin, 1999); Italy (Romagna); Romania (southern Dobrogea, 1989 and near Bucharest) indicate a *southerly* extension; see also under 'Species new to British Isles'(p.218).

Within the British Isles a corresponding expansion has been seen in *Aeshna mixta* (reached Cheshire in 1993 and found in Northumberland in 2002, first Irish record in 2000 and extensive immigration to southern Ireland in 2002), *Anax imperator* (bred Edinburgh, 1994 and Wicklow, 2002) and *Sympetrum sanguineum* (now widespread in Ireland and in England breeding as far north as Northumberland).

Global warming may well be the driving force for these northward range expansions, but the undoubtedly complex interaction between dragonfly physiology, physical and biological factors in the environment and rising temperature will affect species differently and in a way which we have hardly begun to understand. Climatic changes have been invoked to explain range extensions of some dragonflies in Japan and the United States as well as in Europe, and it has been suggested that dragonflies such as *Crocothemis erythraea* could be sensitive monitors of climatic change. Other factors may, however, be implicated in recorded range expansions; an increase in the number of competent observers (mentioned above) and an increase in some areas of man-made ponds and reservoirs must also be considered.

Migration of dragonflies can lead to their unexpected appearance at new sites, but it is exceptional for such migrants to found breeding colonies which persist for any length of time. In 1995 there was a remarkable invasion of Britain by migrant *Sympetrum*, particularly *S. flaveolum* but large numbers also of *S. fonscolombii* and some *S. vulgatum*. In 1996, 1998, 2000 and 2002 *S. fonscolombii* again appeared in numbers and the species is now regarded as a regular rather than a rare visitor to Britain. From none of these influxes was a breeding population established that persisted beyond a second year, but the frequency of recorded immigrant dragonflies is certainly increasing. *Hemianax ephippiger* had been noted only seven times in the British Isles before 1988 but since then it has been reported almost annually. In 1995 there were especially large invasions of northern Europe (*H. ephippiger* bred as far north as Denmark) and eastern Spain (Muñoz Pozo & Blasco-Zumeta, 1996), and in 1998 an influx to Madeira was reported. *Pantala flavescens*, another notable migrant, was found in Kent in 1989, the third British record.

Other new distribution records

Other new national or significant regional distribution records, not previously mentioned, are as follows:

Calopteryx virgo: Greece, known distribution (Stobbe, 1990).

C. splendens: France, the nominate subspecies and *C. s. xanthostoma* are allopatric in Provence and *C. s. caprai* is absent from south-eastern France (Papazian, 1995); Germany, Sachsen-Anhalt, increasing abundance since 1991 in lowlands of River Bode area; Greece, known distribution and subspp. (Stobbe, 1990); *C. s. xanthostoma* in Italy, Sicily (Galletti *et al.*, 1987); *C. s. caprai*: Switzerland, Tessin, 1 site (Maibach & Meier, 1987).

C. haemorrhoidalis: France, southern Charente.

Epallage fatime: Map 4 (p.58) in the first edition erroneously indicated a site in Austria, the spot in the present edition being repositioned in Hungary; Greece, Dráma, Mt Falakrón (Wakeham-Dawson *et al.*, 1999) and Kos (Dell'Anna, 1994).

Lestes viridis: British Isles, Channel Islands, Jersey, breeding 1990; Poland, Bydgoszcz (Bory Tucholskie).

L. parvidens: Italy, near Rome where it coexists with *L. viridis*, but in south-eastern Europe (east of the Adriatic) probably only *L. parvidens* occurs.

L. macrostigma: France, Charente-Maritime, Yves and Vendée, Île de Noirmoutier (Machet, 1990) and Olonne (Landemaine, 1991), and Charente-Maritime (Lebioda, 1987); Slovenia, Carniola.

L. dryas: Ireland, Counties Longford, Meath, Roscommon, Tipperary and Wicklow (extinct?); Italy, Calabria (Terzani, 1987).

Sympecma fusca: Netherlands, Drenthe; Sweden, Gotland, Södermanland and Uppland (Dannelid & Ekestubbe, 2001).

Platycnemis subdilitata: Canary Islands.

Pyrrhosoma nymphula: Turkey, Istranca Mts.

Erythromma najas: Italy, Lazio, Lake Ventina.

E. viridulum: Austria, near Linz, 1991; British Isles, see under 'Species new to British Isles' (see p.218); Germany, Nordrhein-Westfalen, significant increase in records reported in 1996; Netherlands (see under 'Expansion of recorded ranges' (see p.218); Poland, Bialowiéza (Kalkman & Dijkstra, 2000); Slovenia.

Coenagrion mercuriale: Germany, Brandenburg, Potsdam, 1989.

C. m. castellani: Morocco, co-existing with *C. m. hermeticum* and considered a full species (Ben Azzouz *et al.*, 1989).

C. scitulum: Recently spread to Belgium; France, Lot et Garonne (Duras) and Dordogne (Ste Alvère) (pers. obs.), Eure; Italy, Calabria; Luxembourg (Eischtal, 1996); Serbia, Mt Golija; Slovenia, River Drava and Bela Krajina (Carniola).

C. caerulescens: France, Ardèche (Monnerat, 1992), Aude and Tarn.

C. hastulatum: France, Île-de-France; Netherlands, 10 populations (Ketelaar, 1998); Serbia, Mt Golija; Spain, Gerona (Martin-Casacuberta, 1997).

C. lunulatum: Ireland (Nelson, 1999), reported in 2002 from Counties Donegal, Leitrim and Mayo; Netherlands,

Zeeland (Baaijens, 2001), Drenthe, reported in 2002 to be common on the coast including some islands of Waddensea; Switzerland, persists in one locality (Maibach & Meier, 1987).

C. armatum: Netherlands, 1999; Poland, 38 localities (Buczyńsky, 2000); Slovakia, Poprad-Štufy (David, 2000).

C. ornatum: France, départements Saône-et-Loire, Nièvre, Allier and Cher (westernmost record) (Lett, 2001); Germany, Bavaria (Eichstätt), Niedersachsen (Lever Wald) and Nordrhein-Westfalen (Rahden); Romania, River Mures.

C. hylas freyi: Austria, North Tyrol, a site in the same region as that found by Heidemann (1974) might be the only remaining locality, Heidemann's population having disappeared and a reported site in Switzerland (Berne) never having been confirmed (Kiauta & Kiauta, 1991). Recently, however, other sites in Austria have been intimated.

C. puella: Finland, Ulko-Tammio Is., 2000.

Cercion lindenii: France, Corsica (Papazian, 1987); Germany, Hessen (Giessen), Niedersachsen (Harz Mts and Holzminden), Nordrhein-Westfalen, significant increase in records reported in 1996; Poland, national distribution and habitats (Bernard, 1995), Wielkopolska, northernmost record; Romania, River Mures.

Enallagma cyathigerum: France, Corsica (Dommanget, 1987); Great Britain, Shetland Islands, the only breeding species of Odonata (1995); Greece, Crete (Brändle & Rödel, 1994); Italy, Sicily (Galletti *et al.*, 1987).

Ischnura pumilio: Denmark (H. Pedersen, pers. comm.); France, Île-de-France; Netherlands, Drenthe; Poland, Bialowiéza (Kalkman & Dijkstra, 2000).

I. graellsii: Italy, Sardinia, new (Burmeister, 1989).

I. senegalensis: Madeira (PT) (Stauder, 1991), where it occurs with *I. pumilio*. [*I. saharensis* is thought to be the only damselfly on the Canary Islands.]

Nehalennia speciosa: Austria, Upper Austria, Ibm and Steiermark; Denmark, near Elsinore, 1990 (Holmen, 1991); Germany, Niedersachsen; Poland, distribution reviewed (Bernard, 1988).

Ceriagrion tenellum: Germany, Niedersachsen, near Essen; Netherlands, Drenthe, Dwingeloo and Uffelte; Switzerland, Neuchâtel. Represented in Greece by *C. t. georgfreyi*.

Aeshna caerulea: Czech Republic, Krkonoše Mts.

A. juncea: Bulgaria, Prin Mts; Italy, Lombardy, Brescia; Slovenia, Bloke Plateau; Spain, Gerona.

A. subarctica: France, Lorraine, Hautes-Vosges (Boudot *et al.*, 1990), Jura Mts (Grand, 1993); Germany, Brandenburg, 50km north of Berlin; Netherlands, Drenthe, 1992, now common but only in Friesland (Dingemanse, 1998); Slovenia (Bedjanič, 2000); Switzerland, Jura Mts, Praz Rodet.

A. crenata: Finland, Helvetinjärvi; Lithuania, new national record (Bernard, 2002).

A. serrata: Norway, 1995.

A. mixta: Estonia, Kabli (1994) and Puhtu (1951); Finland, first recorded 2002.

A. affinis: Austria, Salzburg; France, Corsica (Papazian, 1987); Germany, Bavaria (Eichstätt); Italy, island of Ustica north of Sicily; Poland, 25km west of Poznań, 1992

at species northern limit (Bernard & Samolag, 1994); Slovenia, Styria (Petišovci, 1997); Spain, Balearic Islands, Minorca (Ocharan, 1987); Switzerland, St Gallen (Meier, 1988).

A. cyanea: Albania (Dumont *et al.*, 1993); Ireland, Co. Cork, one found dead, 1988 (Bond, 1989); Turkey, Istranca Mts.

A. viridis: Germany, Niedersachsen (Bremen), Rheinland-Pfalz (Nassau/Lahn) and Bavaria, Passau; Hungary, Bakony Mts (Toth, 1985) and River Drava region; Slovenia, River Drava.

A. grandis: Croatia, region unknown; Hungary, Bakony Mts, River Drava region and Kecskemét; Serbia, Montenegro, Durmitor Mts.

A. isosceles: Germany, Sachsen-Anhalt, Colbe and near Aschersleben; Italy, Abruzzi.

Anax imperator: Cyprus (Monnerat, 1999); Denmark, first record 1994 (Nielsen, 1994) and again in 1995 at same locality; Italy, island of Ustica north of Sicily.

A. parthenope: France, Loir-et-Cher and Maine-et-Loire; Germany, near Zeitz, Brandenburg, 50km north of Berlin, Borna south of Leipzig; Luxembourg, first record 1997; Hungary, Bakony Mts (Toth, 1985); Portugal, Albufeira, 1990 (E. M. Marsh, pers. comm.).

A. immaculifrons: Greece, island of Karpathos (Battin, 1990; Dell'Anna, 1994).

Hemianax ephippiger: There were large immigrations of this species into Europe in 1990 and 1995. Austria, Vienna and Neusiedler See, observed 1989, and exuviae found Rivers Salzach near Salzburg and Danube near Linz in 1990; Croatia, Vojvodina, Osijek; Czech Republic, eastern Bohemia (1998) and five localities in Moravia and Silesia (1995); France, Aude, Jura (bred 1997), Nièvre, Provence (very large immigration in 1990 (Papazian, 1992)), Saône-et-Loire; Germany, Bonn, Brandenburg and Kaiserslauten; Greece, River Nestos delta, several islands in the Aegean (Dell'Anna, 1994); Hungary, several records in 1989; Italy, island of Ustica (1995) and Lipari Is. (1996) north of Sicily; Netherlands, Noord Brabant, Budel-Dorplein in 1995 (new national record); Norway, 1995; Poland, many records in 1995 and breeding confirmed in south-west; Italy, 1990 immigration reported from Latium, Campania, Puglia, Sardinia; Romania, River Mures; Slovenia; Spain, migration of 'thousands' into Aragon, Los Monegros in 1995 (Muñoz Pozo & Blasco-Zumeta, 1996); Sweden, Öland, 1995; Switzerland, Aargau, bred 1989 (1 generation), St Gallen, Valais, emergence recorded at Verney, 1990; Turkey, Istanbul.

Brachytron pratense: France, Pyrénées-Orientales and Corsica (Papazian, 1990); Greece, Macedonia, Prespa Lake (Brändle & Rödel, 1994); Switzerland, St Gallen (Meier, 1988).

Boyeria irene: France, Côte d'Or, Finistere, Île-de-France and Yonne; Italy, Arezzo, Basilicata, Calabria, Campania and Tuscany.

Gomphus vulgatissimus: Germany, distribution reviewed (Mauersberger & Zessin, 1990), Schwäbisch Hall, Niedersachsen (breeding in tributaries of River Aller); Italy, Campania in 1989 and Basilicata reported in 1995.

G. simillimus: France, Saône-et-Loire.

G. graslini: France, Sarthe and Loir-et-Cher.

Ophiogomphus cecilia: France, départements Bouches-du-Rhône (Canal de Vergières, larvae), Loire-Atlantique, Sarthe and Vosges; Germany, Rheinland-Pfalz, breeding 1992; Niedersachsen (breeding populations increasing in number on Rivers Elbe and Weser), Rivers Aller (breeding) and Ise and Luneburg Heath, Saarland, Bavaria, River Danube at Deggendorf and River Aurach in Neustadt/Bad Winsheim, Franconian Jura; Greece, single male collected on River Nestos is first national record (Rödel, 1991); Italy, Lombardy (Brescia); Netherlands, Limburg, River Roer (Geraeds & Hermans, 2000); Spain, new (Picazo & Alba-Tercedor, 1990).

Onychogomphus forcipatus: Distribution of subspecies in Europe and Asia Minor (Boudot *et al.*, 1990b); Germany, Bavaria (Spessart), Brandenburg and Schwäbisch Hall; Italy, Campania (Salerno) and Basilicata; Netherlands, Meuse Valley.

O. uncatus: France, Orne and Nièvre; Italy, Calabria.

O. costae: Spain, Sevilla, 1994, Spanish distribution reviewed (Salamanca-Ocaña *et al.*, 2001).

Lindenia tetraphylla: Slovenia (one record).

Cordulegaster boltonii: Germany, south of Leipzig; Hungary, occurrence in Hungary questionable since previous citations supported by voucher specimens are all referable to *C. heros* (Ambrus *in litt.*, 1991).

C. heros: Austria, Carinthia, 1995, larval distribution in streams compared with that of *C. bidentata* (Lang *et al.*, 2001); Bulgaria, Prin Mts (Beutler, 1987); Croatia, Zagreb; Hungary, Sopron; Romania, River Mures; Serbia, Valjevo; Macedonia (Adamović *et al.*, 1982); Slovenia, Ljubljana district and Styria (Ptuj).

C. bidentata: Bulgaria, Prin Mts (Beutler, 1987); France, Ardennes; Germany, Sachsen-Anhalt, Hessen (Witzenhausen) and Harz Mts (seen in 1995 after 83-year absence), Nordrhein-Westfalen (Holzwickede) and Sachsen; Hungary, Bakony Mts, last larval instar distinguished from that of *C. heros* (Ambrus *et al.*, 1992b); Italy, Campania, Salerno; Luxembourg, reported 1994; Slovenia, River Drava.

Cordulia aenea: Albania (Dumont *et al.*, 1993); Greece, first record (Ottolenghi, 1991); Italy, Emilia-Romagna (Salsi, 1993) and Lazio, Lake Ventina; Slovenia, Styria.

Somatochlora metallica: Italy, Lombardy, Mantova; Spain, Lérida, new (Dantart & Martin, 1999).

S. meridionalis: Bulgaria (Marinov, 1999); France, Provence, westernmost record (Kotarac, 1996); Italy, Friuli (occurs with *S. metallica*), Campania, Marche, Piedmont (Cuneo); Romania, Banat, 1987 (Beutler, 1988); Slovakia, north and south-west (David, 2000); Slovenia, Kozjansko and Tolmin, upper valley of River Soča.

S. alpestris: France, Lorraine, Hautes-Vosges (Boudot *et al.*, 1990); Germany, Thüringer Wald and Erzgebirge, Chemnitz; Slovenia, Šijec and Pokljuka; Switzerland, distribution details (Wildermuth, 1999).

S. arctica: France, Hautes-Vosges (Boudot *et al.*, 1990), Lorraine, Savoie; Germany, Saxony (Erzgebirge, Chemnitz), Thüringia (Thüringer Wald, breeding 1996); Poland, Bialowiéza (Kalkman & Dijkstra, 2000), Janowskie Forest, Carpathian and Sudetan Mts; Slovenia, Bloke Plateau.

S. sahlbergi: Finland, Inari (Butler, 1992), is not a new regional record but the cited paper includes biological data on larva and adult; Norway, Finnmark, Varanger, one male near Bugøynes in 1990 (Pedersen, 1992); Sweden, Torne Lappmark, near Karesuando (Sahlén 1994).

S. flavomaculata: Germany, Saarland (near Altstadt) and Nordrhein-Westfalen; Netherlands, Drenthe and island of Vlieland, 1982 after 9-year absence (Beukeboom, 1988); Norway (Olsvik, 1990), Østfold; Slovakia, Záhorská, 1989, claimed to be first record.

Epitheca bimaculata: Austria, Burgenland; Croatia, Zagreb; France, Allier, Creuse, Hautes-Vosges (Boudot *et al.*, 1990), Île de France Indre (Brenne) (Male-Malherbe, 1998; Male-Malherbe & Deberge, 1993), Isère (Grand, 1988), Lorraine, Meuse (Argonne, breeding) and Nièvre; Germany, Saarland (reported in 1992 to be rapidly expanding) and adjacent parts of Luxembourg and France, Rheinland-Pfalz and Uckermark (Trockur & Mauersberger, 2000); Hungary, River Drava region; Norway, Vestfold (Hof) and Østfold; Slovenia, near Lenart; Switzerland, Jura, Porrentruy.

Macromia splendens: Spain, Galicia and Cadiz, upper catchment of River Hozgarganta near Jimena de la Frontera.

Oxygastra curtisii: France, Alsace, 50km south of Strasbourg; Italy, Umbria; Spain, Cadiz, River Hozgarganta near Jimena de la Frontera; Germany and Luxembourg, rediscovered 1999.

Libellula quadrimaculata: France, Corsica, Ponte Leccia, (Benstead & Jeffs, 1991); Greece, Macedonia, Prespa Lake (Brändle & Rödel, 1994).

L. fulva: Germany, Rheinland-Pfalz, increasing in 1993; Italy, Abruzzi; Turkey, Marmaris Peninsula, 1990 (C. Best, pers. comm.).

L. depressa: Finland, first recorded 1919, range expanded to 1940s then retracted from 1950s and almost extinct in 1980s, but expanded again through 1990s (Valtonen, 1995); Norway, Østfold, 1977 (last recorded in Norway in 1896).

Orthetrum cancellatum: Cyprus (Monnerat, 1999); Ireland, Co. Waterford, 2002.

O. albistylum: Czech Republic; France, Gironde (David, 1990); Italy, coastal Emilia-Romagna; Poland, Bialowiéza (Kalkman & Dijkstra, 2000).

O. coerulescens: Germany, Niedersachsen near Gifhorn (River Ise) and 40km north of Osnabruck, Brandenburg, Luckau and associated with open-cast mining at Lausitz and Borna south of Leipzig; Lithuania, Švenčionys; Poland, Lublin.

O. anceps: Romania, Oradea (Arnold, 1988).

Brachythemis leucosticta: Italy, Sicily (Galletti *et al.*, 1987).

Sympetrum striolatum: Greece, Crete (Battin, 1989).

S. nigrescens: Norway, near Trondheim; Sweden (Ivarsson, 1998).

S. vulgatum: Sweden, Västerbotten.

S. meridionale: Germany, Rheinland-Pfalz, 1992; Luxembourg (Gerend & Proess, 1994); Netherlands, second record (Tromp & Wasscher, 2000).

S. sinaiticum: Spain, Andalusia (1989), Cordoba, Malaga, Tarragona, Valencia and Zaragoza (Los Monegros, 1989 and 1990) (Muñoz Pozo & Blasco-Zumeta, 1996). Eighteen localities cited by Jödicke (1997).

S. fonscolombii: France, Calvados; Germany, Baden-Württemberg (bred near Göppingen, 1996), Rheinland-Pfalz (bred in hot summer of 1991), Nordrhein-Westfalen (reported as bivoltine in 1990 and 1991 near Göttingen), Niedersachsen, Bavaria, Thüringia, Mecklenburg, Brandenburg, Saxony; Netherlands, Zeeland; Sweden, Öland.

S. sanguineum: Italy, island of Ustica north of Sicily in 1995; Spain, Mallorca (Albufera) in 1997.

S. depressiusculum: France, Alsace, 30km south of Strasbourg.

S. danae: France, Maine-et-Loire.

Leucorrhinia caudalis: Croatia, Posavina; Denmark, a few localities (Pedersen & Holmen, 1994); France, Alsace, 30km south of Strasbourg, Île de France and Vienne, Pinail (Prévost & Durepaire, 1994); Germany, Brandenburg (all 5 European *Leucorrhinia* species); Hungary, River Drava region; Luxembourg, 1997; Norway, Vestfold, Hof and Østfold; Slovenia, River Mura region.

L. albifrons: Denmark, a few localities (Pedersen & Holmen, 1994); France, Gironde (David, 1990), Landes and near mouth of Gironde (Ruddele, 1998); Germany, Saxony, near Chemnitz and Niederlausitz; Norway, Østfold; Poland, Koszalin, Bytow; Sweden, Västerbotten.

L. dubia: Netherlands, Drenthe, near Assen and Zeeland; Serbia, Mt Golija and Durmitor Mts; Slovenia, Pohorje Mts.

L. rubicunda: Czech Republic, northern Moravia; France, Pas de Calais and Île-de-France; Germany, Eifel Mts, near Dürren and northern Bavaria; Luxembourg, 1989; Netherlands, Drenthe (near Assen) and Wadden See island of Terschelling; Slovakia, Orava; Switzerland, Baselland.

L. pectoralis: Belgium, Flanders in 2000, first recent record; Czech Republic (and Slovakia), 12 sites detailed; France, Rhône (Grand *et al.*, 2001); Germany, Bavaria (Spessart), Rheinland-Pfalz; Hungary, River Drava region; Netherlands, Zeeland and Wadden See island of Terschelling; Norway, Hordaland and Vestfold, Hof; Poland, Bialowiéza (Kalkman & Dijkstra, 2000), Poland, western Europe and Baltic States (Pedersen & Holmen, 1994); Slovenia; Spain, Lérida, new (Dantart & Martin, 1999); 'Yugoslavia', Macedonia, Prespa Lake (Karanan, 1987).

Trithemis annulata: France, Roussillon, 2001 (J. Tinning, pers. comm.), Pyrénées-orientales, Corsica (Roche, 1989); Greece, Crete (Battin, 1989); Italy, Calabria (Terzani, 1987), Marche, Toscana near Monterotondo Maritimo (Terzani, 1991); Spain, Aragon and Catalonia (Grand, 1990).

Zygonyx torridus: Turkey, Koysegiz, Caunos, probable sighting (P. W. Swire, pers. comm.).

Pantala flavescens: Morocco in 1995, a new national record.

Selysiothemis nigra: Bulgaria (Beshovski & Gashtarov, 1997); Italy, Marche and Umbria (Perugia, Lake Trasimeno); Spain, Aragon, Monegros in 1994 (Grand, 1995).

Contraction of recorded ranges

The many examples of range extensions might give the impression that dragonflies as a whole are doing well. Unfortunately this is far from the truth. Some species may be increasing their distributional range, but at the same time populations within that range may be lost. Recognition of population loss requires proving a negative and is therefore much less frequently reported than population gain.

In a comprehensive survey of Odonata in Switzerland, Maibach & Meier (1987) conclude that of 70 resident species one is now extinct (*Coenagrion ornatum*, last seen 1957), 18 are endangered or nearly extinct, 26 are vulnerable and threatened and the survival of only 25 is considered not to be in jeopardy. The list of more or less threatened species includes many which develop as larvae in flowing water or oligotrophic pools, whilst survival of few such species is judged to be secure. It seems that the most important factor contributing to dragonfly disappearance is biotope degradation as a result of water pollution. Eutrophication of pools and lakes is held responsible for the reduction in number of sites in Swizerland occupied by *Epitheca bimaculata* from 14 to four.

Similar data available for Luxembourg (Proess & Gerend, 1998) show 13 of 61 species to be extinct and only 17 not threatened.

Coenagrion hastulatum: Germany, Rheinland-Pfalz, Kaiserslautern, reported in 1993 as becoming much scarcer.

Ceriagrion tenellum: Slovenia, reported in 1992 to be restricted to just a few localities in Istria.

Ophiogomphus cecilia: Denmark, although several new localities have been recently reported, management and pollution of larger streams is considered responsible for the species' reduction in status to rare (Pedersen & Holmen, 1994).

Aeshna juncea: Germany, Rheinland-Pfalz, a reported (1993) decrease in number and strength of populations.

A. viridis: Denmark, considered to be decreasing in 1994.

A. grandis: Bulgaria, removed from fauna list (Beshovski, 1993).

Orthetrum anceps (= *ramburii*): Due to misidentification, erroneously indicated from Corsica where only *O. coerulescens* occurs (Mauersberger, 1994).

Leucorrhinia albifrons: Switzerland, extinct in canton Zurich.

L. caudalis: Switzerland, extinct in canton Zurich.

L. pectoralis: Denmark, only one breeding population known in 1994.

References

(references to Supplement follow on pages 235–238)

Aagaard, K. & Dolmen, D., 1977. Vann-nymfer i Norge. *Fauna Oslo* **30** : 61–74.

Abro, A., 1982. The effects of parasitic water mite larvae (*Arrenurus* spp.) on Zygopteran imagoes (Odonata). *J. Invertebr. Pathol.* **39** : 373–381.

Adamovic, Z.R., 1967. Odonata collected in Dubrovnik district, Yugoslavia. *Dt. ent. Z.* **14** : 285–302.

Adetunji, J.F. & Parr, M.J., 1974. Colour changes and maturation in *Brachythemis leucosticta* (Burmeister) (Anisoptera: Libellulidae). *Odonatologica* **3** : 13–20.

Aguesse, P., 1958. Une sous-espèce nouvelle d'*Ischnura* en Afrique du Nord (Odonata). *Revue fr. Ent.* **25** : 149–57.

———, 1968. Les Odonates de l'Europe Occidentale, du Nord de l'Afrique et des Iles Atlantiques. *Faune de l'Europe et du Bassin Méditerranéen* **4**. 258pp.

Aguiar, C. & Aguiar, S., 1983. *Brachthemis* (sic) *leucosticta* (Burm.) and *Trithemis annulata* (P. de Beauv.) in Portugal (Anisoptera: Libellulidae). *Notul. odonatol.* **2** : 8–9.

Aguiar, S. & Aguiar, C., 1985. *Paragomphus genei* (Sel.), *Hemianax ephippiger* (Burm.) and *Trithemis annulata* (P. de Beauv.) in Portugal (Anisoptera: Gomphidae, Aeshnidae, Libellulidae). *Notul. odonatol.* **2** : 82–3.

Aguilar, J. d', Dommanget, J.-L. & Préchac, R., 1985. *Guide des Libellules d'Europe et d'Afrique du Nord.* 341pp. Delachaux et Niestlé, Neuchatel & Paris. (English edition 1986, Collins, London.)

Alcock, J., 1979. Multiple mating in *Calopteryx maculata* (Odonata: Calopterygidae) and the advantage of non-contact guarding by males. *J. nat. Hist.* **13** : 439–46.

Ali Khan, R., 1973. The copulatory complex of *Hemianax ephippiger* (Burmeister) (Aeshnidae: Anisoptera). *Zool. J. Linn. Soc.* **52** : 1–7.

Allioni, C., 1766. Manipulus insectorum Taurinensium. *Mél. philos. math. Soc. r. Turin* **3** : 185–98.

Ander, K., 1944. *Aeschna subarctica* (E.M. Walker) subsp. *interlineata* subsp. nov. (Odon.). *Opusc. ent.* **9** : 164.

———, 1950. Zur Verbreitung und Phänologie der boreoalpinen Odonaten der Westpaläarktis. *Opusc. ent.* **15** : 53–71.

———, 1963. *Lestes barbarus* F. funnen i Sverige. *Opusc. ent.* **28** : 196–197.

Asahina, S., 1973. Discovery of *Aeschna subarctica* Walker in Japan. *Tombo* **15** : 9–10.

Askew, R.R., 1982. Roosting and resting site selection by coenagrionid damselflies. *Adv. Odonatol.* **1** : 1–8.

Baez, M., 1985. Las libellulas de las Islas Canarias. *Enciclopedia canaria* No. 28 : 48pp.

Balestrazzi, E., Bozzetti, R. & Bucciarelli, I., 1977. Odonati di Borgoratto Alessandrino (Piemonte). (Ricerche entomologiche in un territorio con caratteristiche xerotermiche. I). *Boll. Soc. ent. ital.* **109** : 11–30.

——— & Bucciarelli, I., 1971a. Ricerche faunistiche sulle Torbiere d'Iseo. I. Sulla presenza di una colonia di *Leucorrhinia pectoralis* (Charp.). (I contributo alla conoscenza degli Odonata). *Boll. Soc. ent. ital.* **103** : 159–66.

——— & ———, 1971b. Ricerche faunistiche sulle Torbiere d'Iseo. II. *Nehalennia speciosa* (Charp.), genere nuovo per la fauna Italiana. (III contributo alla conoscenza degli Odonata). *Boll. Soc. ent. ital.* **103** : 195–8.

———, ——— & Galletti, P.A., 1982. Sulla variabilita di *Cordulegaster pictus* (?) *trinacriae* Waterston, 1976, con descrizione della femina e dell'exuvia ninfale (Odonata Cordulegasteridae). *G. ital. Entomologia* **1** : 63–71.

———, Galletti, P.A. & Pavesi, M., 1983. Sulla presenza in Italia di *Cordulegaster boltoni immaculifrons* Selys, 1850 e considerazioni sulle specie italiane congeneri (Odonata Cordulegasteridae). *G. ital. Entomologia* **1** : 153–68.

Bartenef, A.N., 1915. *Faune de la Russie et des Pays limitrophes. Insectes Pseudoneuroptères. Libellulidae 1.* : 1–352. Petrograd.

———, 1919. *Faune de la Russie et des Pays limitrophes. Insectes Pseudoneuroptères. Libellulidae 2.* : 353–576. Petrograd.

———, 1929. Neue Arten und Varietäten der Odonata des West-Kaukasus. *Zool. Anz.* **85** : 54–68.

Belle, J., 1982. Odonata collected in the Canary Islands. *Ent. Ber. Amst.* **42** : 75–6.

———, 1983. Some interesting Odonata Anisoptera from the Tarn, France. *Ent. Ber. Amst.* **43** : 93–5.

———, 1984. *Orthetrum trinacria* (Selys) new to the fauna of Spain with records of three other Afrotropical Odonata Anisoptera. *Ent. Ber. Amst.* **44** : 79–80.

———, 1985. *Gomphus vulgatissimus* (Linnaeus) new to the fauna of Spain, with records of other interesting Odonata Anisoptera. *Ent. Ber. Amst.* **45** : 14–15.

Belyshev, B.F., 1973. *Strekozy Sibiri.* 1(1) : 1–336. Nauk. Sibir. otd. Novosibirsk.

——— & Haritonov, A.Y., 1977. (On the history of the boreal dragonfly species *Nehalennia speciosa* Charp. 1840 and the centre of origin of the genus *Nehalennia* Selys 1850.) *Ent. Obozr.* **56** : 776–80.

——— & ———, 1980. On the reasons for a sharp curve in the western boundary of the ranges of some eastern dragonfly species in the north of western Siberia. *Odonatologica* **9** : 317–19.

Benitez Morera, A., 1950. *Los Odonatos de España.* 101pp. Inst. Español Entom., Madrid.

Beshovski, V., 1964. Odonata from the Bulgarian Black Sea coast. Odonata from south Bulgaria. *Bull. Inst. Mus. Zool.* **15** : 115–29, **17** : 109–24.

Beutler, H., 1980. Ein weiterer Beleg der Libelle *Aeshna affinis* v.d. Lind. aus der Mark Brandenburg von 1952. *Ent. Nachr. Dresden* **24** : 60–61.

———, 1985. Biometrische und variationsstatistische Untersuchungen an der Kleinlibelle *Cercion lindeni* (Selys 1840), mit Beschreibung einer neuen Unterart (Odonata, Coenagrionidae). *Ent. Abh.* **49** : 69–83.

Bick, G.H. & Bick, J.C., 1981. Heterospecific pairing among Odonata. *Odonatologica* **10** : 259–70.

Bilek, A., 1954. Eine neue Agrionide aus Bayern (Odonata). *NachrBl. bayer. Ent.* **3** : 97–9.

———, 1966. Ergebnisse der Albanien-Expedition 1961 des Deutschen Entomologischen Instituts, 46. Beitrag : Odonata. *Beitr. Ent.* **16** : 327–46.

———, 1967. Beitrag zur Odonatenfauna Griechenlands. Ergebnisse meiner Reise 1965. *Dt. ent. Z.* **14**(n.f.) : 303–12.

Bischof, A., 1971. Die Odonaten des Kantons Graubuenden. *Mitt. ent. Ges. Basel* **21** : 1–7.

———, 1973. Die Odonaten des Kantons Graubuenden. *Mitt. ent. Ges. Basel* **23** : 24–6.

———, 1976. Die Odonaten des Kantons Graubuenden. *Mitt. ent. Ges. Basel* **26** : 1–5.

Boudot, J.-P., Goutet, P. & Jacquemin, G., 1987. *Somatochlora arctica* (Zett.) new for the French Pyrenees and for the southern French Alps, with further records of some rare Odonata in France. *Notul. odonatol.* **2** : 150–2.

Boulard, M., 1981. Les bases morphologiques de l'attelage en tandem chez *Coenagrion scitulum* R. (Odonata, Zygoptera). *Annls Soc. ent. Fr.* **17**(n.s.) : 429–40.

Brauer, F., 1865. Bericht über die von Herrn Baron Ransonnet am rothen Meere und auf Ceylon gesammelten Neuropteren (L.). *Verh. zool.-bot. Ges. Wien* **15** : 1009–18.

——, 1867. Neue exotische Odonaten. *Verh. zool.-bot. Ges. Wien* **17** : 811–16.

——, 1868. Verzeichniss der bis jetzt bekannten Neuropteren im Sinne Linné's. *Verh. zool.-bot. Ges. Wien* **18** : 359–416, 711–42.

——, 1882. *Sympycna paedisca* m. Zur Richtigstellung dieser neuen Art. *Verh. zool.-bot. Ges. Wien* **32** : 75–6.

Brittinger, C.C., 1845. Beschreibung einer neuen Libellula. *Ent. Ztg. Stettin.* **6** : 205–7.

——, 1850. Die Libelluliden des Kaiserreichs Österreich. *Sber. Akad. Wiss. Wien* **4** : 328–36.

Brongniart, C., 1893. *Recherches pour servir à l'histoire des Insectes fossiles des temps primaires précédées d'une Etude sur la nervation des ailes des Insectes.* **1** : 493pp. St. Etienne.

Brullé, A., 1832. *Expédition scientifique de Morée.* **3** 395pp. (Odonata 99–106). Levrault, Paris.

Bucciarelli, I., 1971. Presenza in Sicilia di una colonia di *Selysiothemis nigra* (V. D. Lind.) e note su altre specie raccolte nell'isola e nell Italia meridionale. *Boll Soc. ent. ital.* **103** : 175–85.

——, 1972. Interessanti reperti in Basilicata (IV Contributo alla conoscenza degli Odonati). *Boll. Soc. ent. ital.* **104** : 86–7.

——, 1976. Allevamento, studio e distribuzione degli Odonati Italiani. *Boll. Soc. ent. ital. suppl.* **17** : 1–4.

——, 1977. Dati preliminari sul popolamento odonatoligico di Calabria, Sicilia e Sardegna (VIII Contributo alla conoscenza degli Odonati). *Annali Mus. civ. Stor. nat. Giacomo Doria* **81** : 374–86.

Buchecker, H., 1876. *Systema Entomologiae sistens Insectorum classes, genera, species.* **1** Odonata (Fabr.) europ. iv+16pp. München.

Buchholtz, C., 1951. Untersuchungen an der Libellen Gattung *Calopteryx* Leach unter besonderer Berücksichtigung ethologischer Fragen. *Z. Tierpsychol.* **8** : 273–93.

——, 1955. Eine vergleichende Ethologie der Orientalischen Calopterygiden (Odonata) als Beitrag zu ihrer systematischen Deutung. *Z. Tierpsychol.* **12** : 363–86.

——, 1956. Eine Analyse des Paarungsverhaltens und der dabei wirkenden Auslöser bei den Libellen *Platycnemis pennipes* Pall. und *Pl. dealbata* Klug. *Z. Tierpsychol.* **13** : 13–25.

Buchholz, K.F., 1954. Zur Kenntnis der Odonaten Griechenlands. *Bonn. zool. Beitr.* **5** : 51–71.

——, 1955. Morphologische Differenzierungen bei der Rassenbildung von *Anax parthenope* Selys (Odonata: Aeschnidae). *Bonn. zool. Beitr.* **6** : 118–31.

Bulimar, F., 1984. *Agrion lindeni* Selys 1840 (Insecta-Odonata), specie nouă pentru fauna Românei. *Studii Cerc. Biol.* (Biol. Anim.) **36** : 75–80.

Burmeister, H.C.C., 1839. *Handbuch der Entomologie.* **2** : 805–62, 1016–17. Eslin, Berlin.

Byers, C.F., 1941. Notes on the emergence and life history of the dragonfly *Pantala flavescens*. *Proc. Fla Acad. Sci.* **6** : 14–25.

Calvert, P., 1907. Odonata. In *Biologia Centrali Americana 2 Insecta Neuroptera*. xxx+420pp. London.

——, Longfield, C., Cowley, J. & Schmidt, Er., 1949. *Agrion* versus *Calopteryx*. *Ent. News* **60** : 145–51.

Cammaerts, R., 1979. Les odonates de Belgique et des régions limitrophes. In *Atlas provisoire des insectes de Belgique*, cartes 1333–1400, eds. Leclerq, J. & Verstraeten, C., Gembloux.

Cannings, S.G. & Cannings, R.A., 1985. The larva of *Somatochlora sahlbergi* Trybôm, with notes on the species in the

Yukon territory, Canada (Anisoptera: Corduliidae). *Odonatologica* **14** : 319–30.

Capra, F., 1934. Su alcuni Odonati e Mirmeleonidi di Sicilia. *Annali Mus. civ. Stor. nat. Giacomo Doria* **57** : 92–7.

——, 1937. Raccolte entomogische nell'Isola di Capraia fatte da C. Mancini e F. Capra (1927–1931) V. Odonati e Neurotteri con note sulla memoria di B. Angelini: Ascalafi Italiani. *Boll. Soc. ent. ital.* **69** : 50–58.

——, 1945. Odonati di Liguria. Res. Ligusticae LXXI. *Annali Mus. civ. Stor. nat. Giacoma Doria* **62** : 253–75.

——, 1976. Quanto si sa sugli Odonati e Neurotteri dell' Arcipelago Toscano. (Studi sulla Riserva Naturale dell'Isola di Montecristo. XIII). *Soc. ital. biogeogr.* **5** : 541–47.

Carchini, G., 1983. A key to the Italian odonate larvae. *Soc. int. odonatol. rapid Comm. (suppl.)* **1** : 101pp.

—— & Nicolai, P., 1984. Food and time resource partitioning in two co-existing *Lestes* species (Zygoptera: Lestidae). *Odonatologica* **13** : 461–6.

Carfi, S., Del Centina, P. & Terzani, F., 1980. Odonati raccolti in Sicilia, Calabria e Basilicata negli anni 1963–1977. *Redia* **63** : 37–47.

Carle, F.L., 1982. The wing vein homologies and phylogeny of the Odonata: a continuing debate. *Soc. int. odonatol. rapid Comm.* **4** : 66pp.

Castellani, O., 1951. Quinto contributo alla conoscenza della Fauna odonatologica d'Italia – Odonati di Sardegna, ecc. ed importanti catture nel Lazio meridionale. *Boll. Ass. romana Ent.* **6** : 9–12.

Charpentier, T. de, 1825. *Horae entomologicae.* xvi+255pp. (Odonata xii, 1–50). Gosohorsky, Bratislavia.

——, 1840. *Libellulinae europaeae descriptae ac depictae.* 180pp. Leopold Voss, Leipzig.

Cirdei, F. & Bulimar, F., 1965. *Fauna Republicii Populare Române. Insecta Odonata.* **7**(5) : 274pp. Bucureşti.

Clausen, W., 1984. The exuviae of *Aeshna juncea* (L.) and *Aeshna subarctica* (Wlk.). *J. Br. Dragonfly Soc.* **1** : 59–67.

Compte Sart, A., 1960. Biografia de la "Selysiothemis nigra" V.D.L. (Odonatos). *Graellsia* **18** : 73–115.

——, 1962. Un Odonato nuevo para España, *Brachythemis leucosticta* (Burm.). *Vie Milieu* **13** : 604–7.

——, 1963. Revisión de los Odonatos de Baleares. *Inst. Biol. aplicada* **35** : 33–81.

——, 1965. Distribución, ecologia y biocenosis de los Odonatos ibéricos. *Inst. Biol. aplicada* **39** : 33–64.

——, 1979. Nueva cita de *Diplacodes lefebvrei* (Ramb.) en Espana (Odonata, Libellulidae). *Graellsia* **33** : 227–36.

Comstock, J.H. & Needham, J.G., 1898. The wings of insects. *Am. Nat.* **32** : 43–8, 81–9, 231–57, 335–40, 413–24, 561–5, 769–77, 903–11.

Conci, C., 1949. L'*Agrion mercuriale Castellani* Roberts in Italia (Odonata: Agrionidae). *Boll. Soc. ent. ital.* **79** : 62–4.

—— & Nielsen, C., 1956. Odonata. *Fauna d'Italia.* **1** : xi+298pp.

Conesa Garcia, M.A., 1985a. A new record of *Orthetrum trinacria* (Sel.) in the Iberian Peninsula (Anisoptera: Libellulidae). *Notul. odonatol.* **2** : 83–4.

——, 1985b. Aportaciones a la biologia de *Diplacodes lefebvrei* (Rambur, 1842) (Odon., Anisop., Libellulidae) en la Peninsula Ibérica. *Bol. Asoc. esp. Ent.* **9** : 321–30.

—— & Garcia Raso, J.E., 1982. Aportaciones a la biologia de *Brachythemis leucosticta* (Burmeister, 1839) (Anisoptera, Libellulidae) en el sud de España. *Mon. Trab. zool. Malaga 3/4* : 21–4.

Consiglio, C., 1950. Cattura del *Gomphus flavipes* Charp. a Roma. *Boll. Soc. ent. ital.* **80** : 16.

——, 1952. Odonati dell'Italia meridionale e degli Abruzzi. *Memorie Soc. ent. ital.* **31** : 96–108.

——, 1976. Some observations on the spacing pattern of *Anax imperator* Leach (Anisoptera: Aeshnidae). *Odonatologica* **5** : 11–14.

Corbet, P.S., 1951. The development of the labium of *Sympetrum striolatum* (Charp.) (Odon., Libellulidae). *Entomologist's mon. Mag.* **87** : 289–96.

——, 1952. An adult population study of *Pyrrhosoma nymphula* (Sulzer): (Odonata: Coenagrionidae). *J. Anim. Ecol.* **21** : 206–22.

——, 1953. A terminology for the labium of larval Odonata. *Entomologist* **86** : 191–6.

——, 1955a. The immature stages of the Emperor Dragonfly, *Anax imperator* Leach (Odonata, Aeshnidae). *Entomologist's Gazette* **6** : 189–204.

——, 1955b. The larval stages of *Coenagrion mercuriale* (Charp.) (Odonata, Coenagriidae). *Proc. R. ent. Soc. Lond. (A)* **30** : 115–26.

——, 1956. The life histories of *Lestes sponsa* (Hansemann) and *Sympetrum striolatum* (Charpentier) (Odonata). *Tijdschr. Ent.* **99** : 217–29.

——, 1957a. Larvae of East African Odonata. 6–8. 6. *Brachythemis lacustris* Kirby 7. *Brachythemis leucosticta* (Burmeister) 8. *Zyxomma flavicans* (Martin). *Entomologist* **90** : 28–34.

——, 1957b. The life-history of the Emperor Dragonfly *Anax imperator* Leach (Odonata: Aeshnidae). *J. Anim. Ecol.* **26** : 1–69.

——, 1962. *A Biology of Dragonflies.* xvi+247pp. Witherby Ltd., London.

——, 1983. Odonata in phytotelmata. In *Phytotelmata: terrestrial plants as hosts for aquatic insect communities* : 29–54. Frank, J.H. & Lounibos, L.P. eds. Plexus, Marlton, New Jersey.

——, Longfield, C. & Moore, N.W., 1960. *Dragonflies.* xii+260pp. Collins, London.

Costa, A., 1884. Diagnosi di nuova artropodi trovati in Sardegna. *Boll. Soc. ent. ital.* **15**(1883) : 332–41.

Cotton, D.C.F., 1982. *Coenagrion lunulatum* (Charpentier) (Odonata: Coenagrionidae) new to the British Isles. *Entomologist's Gazette* **33** : 213–14.

Cowley, J., 1933. The larvae of the European species of *Gomphus* Leach (Odonata). *Entomologist's mon. Mag.* **19**(3rd ser.) : 251–2.

——, 1934. Changes in the generic names of the Odonata. *Entomologist* **67** : 200–5.

——, 1935. Remarks on the names of some British Odonata. *Entomologist* **68** : 154–6.

——, 1937. Some Odonata from Livornia (Latvia). *Entomologist* **70** : 61–3.

——, 1940. A list of the Odonata of the islands of the western Mediterranean area. *Proc. R. ent. Soc. Lond. B* **9** : 172–8.

Crozier, W.J., Wolf, E. & Zerrahn-Wolf, G., 1937. Critical illumination and critical frequency for response to filtered light, in dragonfly larvae. *J. gen. Physiol.* **20** : 363–410.

Crucitti, P., Galletti, P.A. & Pavesi, M., 1981. Un interessante reperto sardo: *Brachythemis leucosticta* (Burmeister), genere nuovo per la fauna Italiana (Anisoptera: Libellulidae). *Notul. odonatol.* **1** : 115–17.

Curtis, J., 1838. *British Entomology.* **15** : no. 512. London.

——, 1839. *British Entomology.* **16** : no. 732. London.

Dale, J.C., 1834. *Cordulia Curtisii* Dale, a new species hitherto undescribed, characterised by Mr. Dale. *Mag. nat. Hist.* **7** : 60–1.

Dapling, J.G. & Rocker, C., 1969. A population study on the Scarce Ischnura (Odon., Zygoptera). *Bull. amat. Ent. Soc.* **28** : 15–20.

Davies, D.A.L., 1981. A synopsis of the extant genera of the Odonata. *Soc. int. odonatol. rapid Comm.* **3** : 59pp.

—— & Tobin, P., 1984. The dragonflies of the world: a systematic list of the extant species of Odonata. Vol. 1. Zygoptera, Anisozygoptera. *Soc. int. odonatol. rapid Comm. (suppl.)* **3** : 127pp.

—— & ——, 1985. The dragonflies of the world: a systematic list of the extant species of Odonata. Vol. 2. Anisoptera. *Soc. int. odonatol. rapid Comm. (suppl.)* **5** : 151pp.

Degrange, C. & Seassau, M.D., 1970. Odonates de quelques hautes toubières et étangs à sphaignes du Dauphiné. *Trav. Lab. hydrobiol. Grenoble* **61** : 89–106.

De Haan, W., 1826. P.L. Van der Linden, Monographiae Libellulinarum Europaearum Specimen. Bruxellis apud J. Frank, 1825. *Bijd. Natuurk. Wetensch.* **1**(2) : 45–9.

De Marmels, J., 1975. Die Larve von *Hemianax ephippiger* (Burmeister, 1839) (Anisoptera: Aeshnidae). *Odonatologica* **4** : 259–63.

——, 1979a. Liste der in der Schweiz bisher Nachgewiesenen Odonaten. *Notul. odonatol.* **1** : 37–40.

——, 1979b. Libellen (Odonata) aus der Zentral- und Ostschweiz. *Mitt. schweiz. ent. Ges.* **52** : 395–408.

—— & Schiess, H., 1977. Zum Vorkommen der Zwerglibelle *Nehalennia speciosa* (Charp. 1840) in der Schweiz (Odonata: Coenagrionidae). *Vierteljahrsschr. naturforsch. Ges. Zurich* **122** : 339–48.

—— & ——, 1978. Le Libellule del canton Ticino e delle zone limitrofe. *Boll. Soc. tic. Sci. nat.* **1977/78** : 29–83.

Djakonov, A.M., 1922. (On a new species of the genus *Aeschna* from northern Russia (*Aeschna elisabethae*, sp. n.)). *Izv. sev. oblast. Sta.Zashch. Rast. Vredit.* **3** : 123–30.

Dommanget, J.-L., 1981. Captures interessantes d'odonates en France. *Notul. odonatol.* **1** : 120–21.

——, 1984. *Somatochlora arctica* (Zett.) et *Leucorrhinia albifrons* (Burm.) en France Centrale (Anisoptera: Corduliidae, Libellulidae). *Notul. odonatol.* **2** : 46–8.

—— & Martinez, M., 1984. Les odonates de Corse: considérations générales et synthèse des données actuelles. *Entomologiste* **40** : 27–36.

Donath, H., 1981. Verbreitung und Ökologie von *Lestes barbarus* (F.) in der nordwestlichen Niederlausitz (Odonata, Lestidae). *Novius, Berlin* **3** : 33–6.

Donovan, E., 1807. *The natural history of British Insects.* **12** : 102pp. Rivington, London.

——, 1811. *The natural history of British Insects.* **15** : 83pp. Rivington, London.

Dreyer, W., 1978. Etho-ökologische untersuchungen an *Lestes viridis* (v.d. Linden) (Zygoptera: Lestidae). *Odonatologica* **7** : 309–22.

Drury, D., 1770. *Illustrations of Natural History; wherein are exhibited two hundred and forty forms of exotic insects.* **1** : 130pp. White, London.

Dufour, C., 1982. Odonates menacés en Suisse Romande. *Adv. Odonatol.* **1** : 43–54.

Dumont, H.J., 1972. The taxonomic status of *Calopteryx xanthostoma* (Charpentier, 1825) (Zygoptera: Calopterygidae). *Odonatologica* **1** : 21–9.

——, 1973. Contribution a la connaissance des Odonates du Maroc. *Bull. Soc. Sci. nat. phys. Maroc* **52** : 149–79.

——, 1974. *Ischnura intermedia* spec. nov. from Turkey, and its relations to *I. forcipata* Morton, 1907 and *I. pumilio* (Charpentier, 1825) (Zygoptera: Coenagrionidae). *Odonatologica* **3** : 153–65.

——, 1975. Endemic dragonflies of Late Pleistocene age of the Hula Lake area (Northern Israel), with notes on the Calopterygidae of the rivers Jordan (Israel, Jordan) and Litani (The Lebanon) and description of *Urothemis edwardsi hulae* subspec. nov. (Libellulidae). *Odonatologica* **4** : 1–9.

——, 1976. *Aeschna charpentieri* Kolenati, 1846, a synonym of *Cordulegaster insignis* Schneider, 1845, and on the correct status of *Cordulegaster charpentieri* auctorum (Anisoptera: Cordulegasteridae). *Odonatologica* **5** : 313–21.

——, 1977a. Redécouverte d'*Oxygastra curtisi* (Dale, 1834) en Belgique (Odonata). *Bull. Annls Soc. r. ent. Belg.* **113** : 26.

——, 1977b. An analysis of the Odonata of Tunisia. *Bull. Annls Soc. r. ent. Belg.* **113** : 63–94.

——, 1977c. A review of the dragonfly fauna of Turkey and adjacent Mediterranean islands (Insecta Odonata). *Bull. Annls Soc. r. ent. Belg.* **113** : 119–71.

Dumont, H.J., 1977d. Sur une collection d'Odonates de Yougoslavie, avec notes sur la faune des territoires adjacents de Roumanie
et de Bulgarie. *Bull. Annls Soc. r. ent. Belg.* **113** : 187–209.

——, 1978. Odonates d'Algerie, principalement du Hoggar et
d'oasis du sud. *Bull. Annls Soc. r. ent. Belg.* **114** : 99–106.

—— & Hinnekint, B.O.N., 1973. Mass migration in dragonflies, especially in *Libellula quadrimaculata* L.: a review, a new
ecological approach and a new hypothesis. *Odonatologica* **2** :
1–20.

—— & Schneider, W., 1984. On the presence of *Cordulegaster
mzymtae* Bartenef, 1929 in Turkey, with a discussion of its
geographic distribution and taxonomic position (Anisoptera:
Cordulegastridae). *Odonatologica* **13** : 467–76.

Dunkle, S.W., 1979. Ocular mating marks in female nearctic
Aeshnidae (Anisoptera). *Odonatologica* **8** : 123–7.

Eichwald, E. von, 1837. *Reise auf den Capischen Meere und in den
Caucasus, 1825–1826.* **1**(2) : (272).

Evans, M.W.F., 1845. *British Libellulinae or Dragon Flies.* Private
publication, London.

Eversmann, E., 1836. Libellulinae, Wolgam fluvium inter et
montes Uralenses observatae, *and* Libellulinarum species
novae, quas inter Wolgam fluvium et montes Uralenses observ-
avit Dr. Eduard Eversmann. *Bull. Soc. imp. Moscou* **9** : 233,
235–48.

——, 1841. Quaedam insectorum species novae in Rossia orien-
tali observatae. *Bull. Soc. imp. Moscou* **14** : 351–60.

——, 1854. Beiträge zur Lepidopterologie Russlands, und
Beschreibung einiger anderen Insecten aus den südlichen
Kirgisensteppen, den nördlichen Ufern des Aral Sees und des
Sir Darjas. *Bull. Soc. imp. Moscou* **27** : 174–205.

Fabricius, J.C., 1775. *Systema Entomologiae.* 832pp. (Odonata
420–26). Kortius, Flensburg & Lipsia.

——, 1781. *Species Insectorum.* **1** : viii+552pp. Hamburgi &
Kilonii.

——, 1798. *Supplementum Entomologiae systematicae.* 572pp.
(Odonata 283–7). Proft & Storch, Hafniae (Copenhagen).

Falchetti, E. & Utzeri, C., 1974. Preliminary observations on the
territorial behavior of *Crocothemis erythraea* (Brullé) (Odonata,
Libellulidae). *Fragm. ent.* **10** : 295–300.

Ferreras Romero, M., 1981a. Un odonato nuevo para la fauna
Ibérica *Trithemis annulata* (Palisot de Beauvois, 1805)
(Anisoptera, Libellulidae). *Boln Asoc. esp. Entomologia* **4** :
191–3.

——, 1981b. La larve d'*Ischnura graellsi* Rambur, 1842 (Zygoptera: Coenagrionidae). *Odonatologica* **10** : 223–6.

——, 1983. Notas sobre la fauna odonatologica de la Laguna de
Zoñar, Andalucia, España. *Notul. odonatol.* **2** : 11–12.

—— & Puchol Caballero, V., 1984. *Los insectos Odonatos en
Andalucia, Bases para su estudio faunistico.* 152pp.

Fonscolombe, M. Boyer de, 1837. Monographie des Libellulines
des environs d'Aix. *Annls Soc. ent. Fr.* **6** : 129–50.

——, 1838. Monographie des Libellulines des environs d'Aix.
Annls Soc. ent. Fr. **7** : 75–106, 547–75.

Förster, F. 1906. Forschungsreise durch Südschoa, Galla und die
Somaliländer von Carlo Freiherr von Erlanger. Libellen. *Jb.
nassau Ver. Naturk.* **59** : 301–44.

Fourcroy, A.F. de, 1785. *Entomologia Parisiensis, ...* **2** : 233–
544. Paris.

Francez, A.-J., 1985. Les odonates d'Auvergne: répartition de
quelques espèces rares ou peu connues essai de zoogéographie
régionale. *Entomologiste* **41** : 101–11.

—— & Brunhes, J., 1983. Odonates des toubières d'Auvergne
(Massif Central Français) et répartition en France des odonates
d'altitude. *Notul. odonatol.* **2** : 1–8.

Franke, U., 1979. Bildbestimmungsschlüssel mitteleuropäischer
Libellen-Larven (Insecta: Odonata). *Stuttg. Beitr. Naturk.*
(A) **333** : 1–17.

Frantsevich, L.I. & Mokrushov, P.A., 1984. Visual stimuli releasing attack of a territorial male in *Sympetrum* (Anisoptera:
Libellulidae). *Odonatologica* **13** : 335–50.

Fraser, F.C., 1929. A revision of the Fissilabioidea (Cordulegasteridae, Petaliidae and Petaluridae) (Order Odonata). Part 1.
Cordulegasteridae. *Mem. Indian Mus.* **9** : 69–167.

——, 1936. *The Fauna of British India, including Ceylon and
Burma. Odonata* **3** : xi+461pp. Taylor & Francis, London.

——, 1943. The function and comparative anatomy of the
oreillets in the Odonata. *Proc. R. ent. Soc. Lond.* A **18** :
50–56.

——, 1947. Is *Somatochlora alpestris* Selys (Odon., Corduliidae)
a British species? *Entomologist's mon. Mag.* **83** : 86–7.

——, 1949a. The nymph of *Ischnura pumilio* Charpentier
(Order Odonata). *Proc. R. ent. Soc. Lond.* A **24** : 46–50.

——, 1949b. *Exploration du Parc National Albert Mission G. F.
de Witte (1933–1935). Odonata.* **61** : 21pp. Bruxelles.

——, 1949c. Odonata. *Handbk Ident. Br. Insects* **1**(10) : 49pp.
Royal Entomological Society of London.

——, 1950. The nymphal stages of *Coenagrion scitulum* (Rambur), a recent addition to the British fauna (Odonata). *Entomologist's mon. Mag.* **86** : 97–102.

——, 1952. Methods of exophytic oviposition in Odonata.
Entomologist's mon. Mag. **88** : 261–2.

——, 1954. Nomenclature of the European species of Odonata. *Entomologist's mon. Mag.* **90** : 211.

——, 1955. A note on the determination of the nymph of *Aeshna
caerulea* (Ström). *Entomologist's Gazette* **6** : 85.

——, 1956. Name proposed for *Cordulia aenea* (Linnaeus,
1746, No. 769 Nec 768) (Odon., Corduliidae). *Entomologist's
mon. Mag.* **92** : 20–21.

Fry, C.H., 1984. *The Bee-eaters.* 304pp. T. & A.D. Poyser,
Calton, England.

Fudakowski, J., 1930a. Odonaten aus Central-
Albanien. *Fragm. faun.* **1** : 187–92.

——, 1930b. Über die Formen von *Calopteryx splendens* Harr.
aus Dalmatien und Herzegowna (Odonata). *Annls Mus. zool.
pol.* **9** : 57–63.

Gadeau de Kerville, H., 1905. Les insectes odonates de la Nor-
mandie. Première liste. *Bull. Soc. amis Sci. nat. Rouen* **1904** :
165–74.

Galletti, P.A., 1978. Nuovo reperti di *Lindenia tetraphylla*
(V.d.L.) in Italia (Odonata Gomphidae). *Bol. Soc. ent. ital.*
110 : 223–4.

—— & Pavesi, M., 1983. Su alcuni Odonati di Grecia. *G. ital.
Entomologia* **1** : 247–60.

—— & ——, 1985. Ulteriori considerazioni sui *Cordulegaster*
italiani (Odonata Cordulegasteridae). *G. ital. Entomologia* **2** :
307–26.

Gardner, A.E., 1950a. The life-history of *Sympetrum sanguineum*
Müller (Odonata). *Entomologist's Gazette* **1** : 21–6.

——, 1950b. The life-history of *Sympetrum striolatum striolatum*
(Charpentier) (Odonata). *Entomologist's Gazette* **1** : 53–60.

——, 1950c. The life-history of *Aeshna mixta* Latreille (Odonata). *Entomologist's Gazette* **1** : 128–38.

——, 1951a. The life-history of *Sympetrum fonscolombii* Selys.
Odonata-Libellulidae. *Entomologist's Gazette* **2** : 56–66.

——, 1951b. The life-history of *Sympetrum danae* (Sulzer) = *S.
scoticum* (Donovan) (Odonata). *Entomologist's Gazette* **2** : 109–
27.

——, 1951c. The early stages of Odonata. *Proc. S. Lond. ent.
nat. Hist. Soc.* 1950–51 : 83–8.

——, 1952. The life-history of *Lestes dryas* Kirby. Odonata. *Entomologist's Gazette* **3** : 4–26.

——, 1953a. The life-history of *Leucorrhinia dubia* (Van der
Lind.) (Odonata). *Entomologist's Gazette* **4** : 45–65.

——, 1953b. The life-history of *Libellula depressa* Linn. (Odonata). *Entomologist's Gazette* **4** : 175–201.

——, 1954a. The life-history of *Coenagrion hastulatum* (Charp.)
(Odonata; Coenagriidae). *Entomologist's Gazette* **5** : 17–40.

——, 1954b. A key to the larvae of the British Odonata. Part I.
Zygoptera. Part II. Anisoptera. *Entomologist's Gazette* **5** :
157–71, 193–213.

————, 1955a. A study of the genitalia of the two species *Sympetrum nigrescens* Lucas and *S. nigrifemur* (Selys) with notes on their distribution. (Odonata: Libellulidae). *Entomologist's Gazette* **6** : 86–108.

————, 1955b. An apparent migration of *Sympetrum danae* (Sulz.) (Odonata) in Ireland. *Entomologist's Gazette* **6** : 149.

————, 1963. Report on the insects collected by E. W. Classey and A. E. Gardner Expedition to Madeira in December 1957 (With special reference to the life-history of *Sympetrum nigrifemur* (Selys)). Part III. Odonata. *Proc. S. Lond. ent. nat. Hist. Soc.* **1962** : 62–85.

———— & MacNeill, N., 1950. The life-history of *Pyrrhosoma nymphula* (Sulzer) (Odonata). *Entomologist's Gazette* **1** : 163–82.

Geest, W., 1905. Beiträge zur Kenntnis der bayerischen Libellen-fauna. *Z. wiss. InsektBiol.* **1** : 254–6.

Geijskes, D.C. & Tol, J. van, 1983. *De libellen van Nederland (Odonata).* 368pp. Koninklijke Nederlandse Natuurhistorische Vereniging, Hoogwoud (N.H.).

Gerstäcker, C.E.A., 1891. Die von Herrn Dr. F. Stuhlmann in Ostafrika gesammelten Termiten, Odonaten und Neuropteren. *Jb. hamb. wiss. Anst.* **9** : 183–91.

Ghiliani, V., 1874. Sopra alcune invasioni di Libelluline nell'Italia superiore. *Boll. Soc. ent. ital.* **6** : 227–8.

Gibbons, R.B., 1986. *Dragonflies and Damselflies of Britain and Northern Europe.* 144p. Country Life Books, Hamlyn, Twickenham.

Gibo, D.L., 1981. Some observations on slope soaring in *Pantala flavescens* (Odonata: Libellulidae). *J. N.Y. ent. Soc.* **89** : 184–7.

Gmelin, J.F., 1788. *In* Linnaeus, C., *Systema Naturae* **1**(5) 13th edition : (2225–3020). Beer, Lipsiae.

Goyvaerts, P., 1979. *Cordulia aenea* (Odonata: Corduliidae) a new county record. *Ir. Nat. J.* **19** : 329.

Greenewalt, C.H., 1962. Dimensional relationships for flying animals. *Smithson. misc. Collns* **144** : 46pp.

Guérin-Méneville, F.E., 1838. Description de la *Petalura Selysii. Guérin Magas. Zool.* : no. 201.

Hagen, H.A., 1839. Verzeichniss der Libellen Ostpreussens. *Preuss. Provinzialbl.* **21** : 54–8.

————, 1840. *Synonymia Libellularum Europaearum.* Diss. inaug. Regimontii Prussorum. 81pp. Dalkowski.

————, 1856. Die Odonaten-Fauna des russischen Reichs. *Ent. Ztg. Stettin.* **17** : 363–81.

————, 1861. Synopsis of the Neuroptera of North America. With a list of the South American species. *Smithson. misc. Collns* : 347pp. (Odonata 55–187).

————, 1867. Notizen beim Studium von Brauer's Novara-Neuropteren. *Verh. zool.-bot. Ges. Wien* **17** : 31–62.

Hämäläinen, M., 1981. *Nehalennia speciosa* (Charp.) (Coenagrionidae) Tammisaaresta. *Notul. ent.* **61** : 221.

————, 1983. *Aeshna viridis* (Evers.) (Aeshnidae) Pojois-Karjalalle uusi sudenkorentolaji. *Notul. ent.* **63** : 211.

————, 1984a. *Orthetrum coerulescens* – sudenkorennon (Odonata: Libellulidae) levinneisyydestä ja esiintymisestä Suomessa. *Notul. ent.* **64** : 74–5.

————, 1984b. *Coenagrion puella* – tytönkorennon (Odonata: Coenagrionidae) esiintymisestä Suomessa. *Notul. ent.* **64** : 75–6.

————, 1984c. Odonata of Inari Lapland. *Kevo Notes* **7** : 31–8.

————, 1985a. Note on "nigrescens-like" specimens of *Sympetrum striolatum* (Odonata, Libellulidae) in the Aaland Islands. *Notul. ent.* **65** : 68.

————, 1985b. *Ischnura pumilio* (Odonata, Coenagrionidae) Suomelle uusi sudenkorentolaji. *Notul. ent.* **65** : 155.

Hammond, C.O., 1977. *The Dragonflies of Great Britain and Ireland.* 115pp. Curwen Press, London. (Revised edn 1983, Harley Books.)

Hansemann, J.W.A., 1823. Anfang einer Auseinandersetzung der deutschen Arten der Gattung *Agrion. Weidmann's zool. Mag.* **2** : 148–61.

Harris, M., 1782. *An Exposition of English Insects.* viii+170pp. White & Robson, London.

Harz, K., 1978. *Coenagrion freyi* Bilek ist eine gute Art (Odonata, Zygoptera, Coenagrionidae). *Articulata* **1** : 61–4.

Heidemann, H., 1974. Ein neuer Europäischer fund von *Coenagrion hylas* (Tryböm) (Zygoptera: Coenagrionidae). *Odonatologica* **3** : 181–5.

————, 1982. Geschlechtliche Verirrungen einheimischer Libellen. *Libellula* **1982** : 49–51.

Heinrich, B. & Casey, T.M., 1978. Heat transfer in dragonflies: 'fliers' and 'perchers'. *J. exp. Biol.* **74** : 17–36.

Heymer, A., 1964. Ein Beitrag zur Kenntnis der Libelle *Oxygastra curtisii* (Dale 1834). *Beitr. Ent.* **14** : 31–44.

————, 1966. Étude comparées du comportement inné de *Platycnemis acutipennis* Selys 1841 et de *P. latipes* Rambur 1842 (Odon., Zygoptera). *Annls Soc. ent. Fr.* **2**(n.s.) : 39–73.

————, 1967. *Hemianax ephippiger* en Europe (Odon., Anisoptera). *Annls Soc. ent. Fr.* **3**(n.s.) : 787–95.

————, 1968. Contribution à la connaissance de la morphologie et de la répartition du genre *Platycnemis* Burmeister, 1839, en Europe et en Asie Mineure. *Beitr. Ent.* **18** : 605–23.

————, 1969. Fortpflanzungsverhalten und Territorialität bei *Orthetrum coerulescens* (Fabr., 1798) und *O. brunneum* (Fonsc., 1837) (Odonata; Anisoptera). *Revue Comport. Animal* **3** : 1–24.

————, 1972. Comportements social et territorial des Calopterygidae (Odon. Zygoptera). *Annls Soc. ent. Fr.* **8**(n.s.) : 3–53.

————, 1973. Ethologische Freiland-beobachtungen an der Kleinlibelle *Agrion lindeni* Selys, 1840. *Revue Comport. Animal* **7** : 183–9.

Hofslund, P.B., 1977. Dragonfly attacks and kills a Ruby-throated Hummingbird. *Loon* **49** : 238.

Huijs, L.G.J. & Peters, H.P.J., 1984. *Sympetrum pedemontanum* (Allioni, 1766) in Nederland waargenomen (Odonata: Libellulidae). *Ent. Ber. Amst.* **44** : 21–4.

Jacobs, M.E., 1955. Studies on territorialism and sexual selection in dragonflies. *Ecology* **36** : 566–86.

Jacobson, G. & Bianchi, V., 1905. *Die Orthopteren und Pseudoneuropteren des Russischen Reiches und der angrenzenden Gebiete.* (pp.635–846). St. Petersburg.

Johannsson, O.E., 1978. Co-existence of larval Zygoptera (Odonata) common to the Norfolk Broads (U.K.). I. Temporal and spatial separation. *Oecologia* **32** : 303–21.

Johansson, C.H., 1859. *Odonata Sueciae. Sveriges Trollsländor.* 122pp. Bergh, Västeras.

Johnson, C., 1964. The evolution of territoriality in the Odonata. *Evolution* **18** : 89–92.

————, 1975. Polymorphism and natural selection in ischnuran damselflies. *Evol. Theory* **1** : 81–90.

Jurzitza, G., 1964. A propos de quelques espèces rares d'Odonates en Camargue. *Bull. Annls Soc. Hort. Montpellier* **4**.

————, 1965. Gedanken zu einigen Problemen des "Rassenkreises *Cordulegaster boltonii* (Donovan)" (Odonata, Anisoptera). *NachrBl. bayer. Ent.* **14** : 4–8.

————, 1966. Über interspezifische Paarungsversuche bei Odonaten. *Beitr. naturk. Forsch. SüdwDtl.* **20** : 71–2.

————, 1969. Eiablage von *Chalcolestes viridis* (Van der Linden) in postcopula und ohne Begleitung durch das Mannchen sowie Gedanken zur Evolution des Fortpflanzungsverhaltens bei den Odonaten. *Tombo, Tokyo* **12** : 25–7.

————, 1970. Beobachtungen zur Oekologie und Ethologie von *Ischnura pumilio* (Charp.). *Beitr. naturk. Forsch. SüdwDtl.* **24** : 151–3.

————, 1978. *Unsere Libellen.* 72pp+120 colour photographs. Keller & Co., Stuttgart.

Kaiser, H., 1968. 'Zeitliches Territorialverhalten' bei der Libelle *Aeschna cyanea. Naturwissenschaften* **55** : 657–8.

————, 1969. Regulation der Individuendichte am Paarungsplatz bei der Libelle *Aeschna cyanea* durch 'zeitliches Territorialverhalten'. *Zool. Anz., suppl.* **33** : 79–85.

Kaiser, H., 1974a. Verhaltensgefüge und Temporiaverhalten der Libelle *Aeschna cyanea* (Odonata). *Z. Tierpsychol.* **34** : 398–429.

———, 1974b. Die tägliche Dauer der Paarungsbereitschaft in Abhängigkeit von der Populationsdichte bei den Männchen der Libelle *Aeschna cyanea* (Odonata). *Oecologia* **14** : 375–87.

———, 1974c. Intraspezifische Aggression und räumliche Verteilung bei der Libelle *Onychogomphus forcipatus* (Odonata). *Oecologia* **15** : 223–34.

———, 1982. Do *Cordulegaster* males defend territories? A preliminary investigation of mating strategies in *Cordulegaster boltoni* (Donovan) (Anisoptera: Cordulegasteridae). *Odonatologica* **11** : 139–52.

Karsch, F., 1890. Ueber Libellulinen der Sammlung des Herrn Dr. Heinrich Dohrn. *Ent. Ztg. Stettin.* **51** : 295–8.

Kemp, R.G. & Kemp, G.S., 1985. Some records of Odonata from the departments of Halkidiki and Thessaloniki, northern Greece. *Notul. odonatol.* **2** : 75–7.

Kennedy, C.H., 1915. Notes on the life history and ecology of the dragonflies (Odonata) of Washington and Oregon. *Proc. U.S. natn Mus.* **49** : 259–345.

———, 1917. A new species of *Somatochlora* (Odonata) with notes on the *cingulata* group. *Can. Ent.* **49** : 229–36.

———, 1920. The phylogeny of the zygopterous dragonflies as based on the evidence of the penes. *Ohio J. Sci.* **21** : 19–32.

Kiauta, B., 1964. Notes on some field observations on the behaviour of *Leucorrhinia pectoralis* Charp. (Odonata: Libellulidae). *Ent. Ber. Amst.* **24** : 82–6.

———, 1968. Variation in size of the m-chromosome of the dragonfly, *Calopteryx virgo* (L.), and its significance for the chorogeography and taxonomy of the *Calopteryx virgo* superspecies. *Tijdschr. ned. gen. Vereen.* **12** : 11–16.

——— & Kiauta-Brink, M.A.J.E., 1975. The chromosomes of the dragonfly, *Sympecma annulata braueri* (Yakobsen & Bianki, 1905) from the Netherlands, with a note on the classification of the family Lestidae (Odonata, Zygoptera). *Genen Phaenen* **18** : 39–48.

Killington, F.J., 1924. *Ischnura elegans*, Lind.: Its teneral colour phases, and its mature varieties and aberrations. *Entomologist* **57** : 1–6.

———, 1925. Notes on the prey of dragonflies. *Entomologist* **58** : 181–3.

Kirby, W.F., 1889. A revision of the subfamily Libellulinae with descriptions of new genera and species. *Trans. zool. Soc. Lond.* **12** : 249–348.

———, 1890. *A Synonymic catalogue of Neuroptera Odonata, or Dragonflies: with an appendix of fossil species.* ix + 202pp. Gurney & Jackson, London.

———, 1898. On some dragonflies obtained by Mr. and Mrs. Lort Phillips in Somaliland. *Proc. zool. Soc. Lond.* **1896** : 521–3.

Klausnitzer, B., Friese, G., Heinicke, W., Joost, W. & Müller, G., 1978. Bedrohte Insektenarten in der Deutschen Demokratischen Republik. I. *Beitr. Ent. Berlin* **27** : 81–7.

Koch, S., 1979. Libellenfunde in Algerien und Tunesien (Odonata). *Ent. Z. Frankf. a. M.* **89** : 77–80.

Kolbe, H.J., 1885. Beitrag zur Kenntnis der Pseudoneuroptera Algeriens und der Ostpyrenäen. *Berl. ent. Z.* **29** : 151–7.

Kolenati, F.A., 1846. *Meletemata Entomologica.* **5** : 114–15.

———, 1856. Meletemata Entomologica. *Bull. Soc. imp. Moscou* **29** : 419–512.

Komnick, H., 1978. Osmoregulatory role and transport ATPases of the rectum of dragonfly larvae. *Odonatologica* **7** : 247–62.

Königstedt, D., 1980. Zur Verbreitung der Keilflecklibelle (*Anaciaeschna isoceles*) in Mecklenburg (Odonata). *Faun. abh. st. Mus. tierk. Dresden* **7** : 175–8.

——— & Schmidt, D., 1980. Zur Kenntnis der Libellenfauna der Umgebung von Greifswald (Insecta, Odonata). *Faun. abh. st. Mus. tierk. Dresden* **8** : 51–61.

Kormondy, E.J., 1961. Territoriality and dispersal in dragonflies (Odonata). *J. N.Y. ent. Soc.* **69** : 42–52.

Krieger, F. & Krieger-Loibl, E., 1958. Beiträge zum Verhalten von *Ischnura elegans* und *Ischnura pumilio* (Odonata). *Z. Tierpsychol.* **15** : 82–93.

Krüner, U., 1977. Revier- und Fortpflanzungsverhalten von *Orthetrum cancellatum* (Linnaeus) (Anisoptera: Libellulidae). *Odonatologica* **6** : 263–70.

Kumar, A., 1971. The larval stages of *Orthetrum brunneum brunneum* (Fonscolombe) with a description of the last instar larva of *O. taeniolatum* (Schneider) (Odonata: Libellulidae). *J. nat. Hist.* **5** : 121–32.

Landmann, A., 1981. Beitrag zur Odonatenfauna Nordtirols (Insecta: Odonata, Libellulidae). *Ber. naturwiss.-med. Innsbruck* **68** : 107–9.

Latreille, P.A., 1805. *Histoire naturelle, générale et particulière, des Crustacés et des Insectes.* **13** : 432pp. (Odonata 5–16). Dufart, Paris.

———, 1810. *Considérations générales sur l'ordre naturel des composant les classes des Crustacés, des Arachnides et des Insectes.* 444pp. Paris.

Lawton, J.H., 1970. A population study on larvae of the damselfly *Pyrrhosoma nymphula* (Sulzer) (Odonata: Zygoptera). *Hydrobiologia* **36** : 33–52.

Leach, W.E., 1815. Entomology. *Brewster's Edinburgh Encycl.* **9** : 57–172.

Le Quesne, W.J., 1946. The dragonflies of Jersey. *Entomologist's mon. Mag.* **82** : 112–14.

Lieftinck, M.A., 1926. Odonata Neerlandica. De libellen of waternimfen van Nederland en het aangrenzend gebied. Tweede gedeelte: Anisoptera. *Tijdschr. Ent.* **69** : 85–226.

———, 1929. *Aeschna subarctica* in Europa, eene nieuwe aanwinst voor de Nederlandsche en Belgische Odonatenfauna. *Tijdschr. Ent.* **72** : 169–86.

———, 1964. Aantekeningen over *Coenagrion hylas* (Trybom) in Midden-Europa (Odonata, Coenagrionidae). *Tijdschr. Ent.* **107** : 3.

———, 1966. A survey of the dragonfly fauna of Morocco (Odonata). *Bull. Inst. r. Sci. nat. Belg.* **42** : 1–63.

Linnaeus, C., 1758. *Systema Naturae.* **1** 10th edition. 824 + iiipp. (Odonata 543–6). Stockholm.

———, 1761. *Fauna Suecica.* Editio altera. 578pp. Stockholm.

Lödl, M., 1976. Die Libellenfauna Österreichs. *Linzer Biol. beitr.* **8** : 135–59, 383–7.

———, 1978. Zur Verbreitung und Ökologie von *Orthetrum coerulescens* (Fabricius, 1798) (Odonata: Libellulidae). *Linzer Biol. beitr.* **10** : 111–29.

Lohinai, G., 1982. A *Coenagrion vernale* (Hagen, 1839) előfordulásáról hazánkban (Odonata). *Folia ent. hung.* **43** : 245–7.

Lohmann, H., 1980. Faunenliste der Libellen (Odonata) der Bundesrepublik Deutschland und Westberlins. *Soc. int. odonatol. rapid Comm.* **1** : 34pp.

———, 1981. Zur Taxonomie einiger *Crocothemis*-Arten, nebst Beschreibung einer neuen Art von Madagaskar (Anisoptera: Libellulidae). *Odonatologica* **10** : 109–16.

Loibl, E., 1958. Zur Ethologie und Biologie der deutschen Lestiden (Odonata). *Z. Tierpsychol.* **15** : 54–81.

Longfield, C., 1937. *The Dragonflies of the British Isles.* 220pp. Warne & Co., London & New York.

Lucas, W.J., 1900. *British Dragonflies (Odonata).* xiv + 356pp. Upcott Gill, London.

———, 1912. British Odonata in 1911. *Entomologist* **45** : 171–3.

———, 1930. *The aquatic (naiad) stage of the British dragonflies (Paraneuroptera).* xii + 132pp. Ray Society, London.

McGeeney, A., 1986. *A Complete Guide to British Dragonflies.* x + 133 pp. Jonathan Cape, London.

McLachlan, R., 1880. Notes on the entomology of Portugal. II. Pseudo-Neuroptera (in part) & Neuroptera-Planipennia. *Entomologist's mon. Mag.* **17** : 103–8.

————, 1889a. *Aeschna borealis*, Zett. *Entomologist's mon. Mag.* 25 : 273–4.

————, 1889b. Neuroptera collected by Mr. J.J. Walker on both sides of the straits of Gibraltar. *Entomologist's mon. Mag.* 25 : 344–9.

————, 1896. On some Odonata of the subfamily Aeschnina. *Ann. mag. nat. hist.* 17 : 409–25.

————, 1903. *Hemianax ephippiger*, Burm., taken at Devonport in February. *Entomologist's mon. Mag.* 14 : 96.

MacNeill, N., 1951a. Separation characters for nymphs of *Lestes dryas* Kirby and *L. sponsa* (Hansemann) (Odonata-Lestidae). *Entomologist* 84 : 40–42.

————, 1951b. Further notes on the nymphs of *Lestes dryas* and *L. sponsa* (Hansemann) (Odonata: Lestidae). *Entomologist* 84 : 156–7.

Macan, T.T., 1974. Twenty generations of *Pyrrhosoma nymphula* (Sulzer) and *Enallagma cyathigerum* (Charpentier) (Zygoptera: Coenagrionidae). *Odonatologica* 3 : 107–19.

Machado, A.B.M. & Martinez, A., 1982. Oviposition by egg-throwing in a zygopteran, *Mecistogaster jocaste* Hagen, 1869 (Pseudostigmatidae). *Odonatologica* 11 : 15–22.

Maibach, A., 1985. Révision systématique du genre *Calopteryx* Leach (Odonata, Zygoptera) pour l'Europe occidentale. I. Analyses biochimiques. *Mitt. schweiz. ent. Ges.* 58 : 477–92.

————, 1986. Révision systématique du genre *Calopteryx* Leach (Odonata, Zygoptera) pour l'Europe occidentale. II. Analyses morphologiques et synthèse. *Mitt. schweiz. ent. Ges.* 59 : 389–406.

————, 1987. Révision systématique du genre *Calopteryx* Leach pour l'Europe occidentale (Zygoptera: Calopterygidae). 3. Révision systématique, étude bibliographique, désignation des types et clé de determination. *Odonatologica* 16 : 145–74.

Martens, K., 1982a. *Onychogomphus uncatus* (Charpentier, 1840) (Odonata, Gomphidae), a new species for the Belgian fauna. *Bull. Annls Soc. r. ent. Belg.* 118 : 193–4.

————, 1982b. New localities for *Epitheca bimaculata* (Charp.), with a review of its status in western Europe (Anisoptera: Corduliidae). *Notul. odonatol.* 1 : 157–9.

————, 1984. Courtship display in *Trithemis arteriosa* (Burm.) (Anisoptera: Libellulidae). *Notul. odonatol.* 2 : 67–8.

Martin, R., 1900. Odonates nouveaux ou peu connus. *Bull. Mus. Hist. nat., Paris* 6 : 103–8.

————, 1908. Aeschnines. *Collections Zoologiques du Baron Edm. de Sélys Longchamps.* 18 : 84pp. Brussels.

————, 1909. Notes sur trois Odonates de Syrie. *Bull. Soc. ent. Fr.* 12 : 212–14.

————, 1910. Contribution à l'étude des Neuroptères de l'Afrique. II. Les Odonates du département de Constantine. *Annls Soc. ent. Fr.* 79 : 95–104.

Martynov, A.B., 1932. (New permian Palaeoptera with a discussion of some problems of their evolution.) *Trav. Inst. paléontol. Acad. Sci. URSS* 1 : 1–44.

May, E., 1933. Libellen oder Wasserjungfern (Odonata). *Die Tierwelt Deutschlands* 27 : 124pp. Fischer ed., Jena.

May, M.L., 1976. Thermoregulation and adaptation to temperature in dragonflies (Odonata: Anisoptera). *Ecol. Monogr.* 46 : 1–32.

————, 1978. Thermal adaptations of dragonflies. *Odonatologica* 7 : 27–47.

————, 1982. Heat exchange and endothermy in Protodonata. *Evolution* 36 : 1051–8.

Mayer, G., 1961. Studien an der Heidelibelle *Sympetrum vulgatum* (L.). *Naturkund. Jb. Stadt Linz* 7 : 201–17.

Midttun, B., 1977. Observations of *Somatochlora arctica* (Zett.) (Odonata) in western Norway. *Norw. J. Ent.* 24 : 117–19.

Mielewczyk, S., 1972. Über das vorkommen von *Lestes barbarus* (Fabricius) in Polen (Zygoptera: Lestidae). *Odonatologica* 1 : 37–40.

————, 1974. Bemerkungen über die Synonymie von *Coenagrion lunulatum* (Charpentier, 1840) – *C. vernale* (Hagen, 1839,

nomen nudum) (Zygoptera: Coenagrionidae). *Odonatologica* 3 : 267–8.

————, 1978. Libellen (Odonata) der Pieninen. *Fragm. faun.* 12 : 265–94.

Mierzejewski, W., 1913. *Aeschna osiliensis* nov. sp. *Bull. Acad. Sci. Cracovie B* 1913 : 79–87.

Mihajlović, L., 1974. Contribution to the study of diffusion of the species *Hemianax ephippiger* (Burmeister) in Europe. *Beitr. Ent.* 24 : 105–6.

Mikkola, K., 1968. *Hemianax ephippiger* (Burm.) (Odonata) carried to Iceland from the Eastern Mediterranean by an air-current? *Opusc. ent.* 33 : 111–13.

Miller, A.K. & Miller, P.L., 1985a. Simultaneous occurrence of crepuscular feeding and sexual activity in *Boyeria irene* (Fonsc.) in southern France (Odonata, Aeshnidae). *Entomologist's mon. Mag.* 121 : 123–4.

———— & ————, 1985b. Flight style, sexual identity and male interactions in a non-territorial dragonfly, *Onychogomphus forcipatus unguiculatus* (Van der Linden) (Odonata: Gomphidae). *Entomologist's mon. Mag.* 121 : 127–32.

————, ———— & Siva-Jothy, M.T., 1984. Pre-copulatory guarding and other aspects of reproductive behaviour in *Sympetrum depressiusculum* (Sélys) at rice fields in southern France (Anisoptera: Libellulidae). *Odonatologica* 13 : 407–14.

Miller, P.L., 1983. The duration of copulation correlates with other aspects of mating behaviour in *Orthetrum chrysostigma* (Burmeister) (Anisoptera: Libellulidae). *Odonatologica* 12 : 227–38.

————, 1987. An examination of the prolonged copulations of *Ischnura elegans* (Vander Linden) (Zygoptera: Coenagrionidae). *Odonatologica* 16 : 37–56.

————, & Miller, C.A., 1981. Field observations on copulatory behaviour in Zygoptera, with an examination of the structure and activity of the male genitalia. *Odonatologica* 10 : 201–18.

Miyakawa, K., 1977. On growth processes in the dragonfly antenna. *Odonatologica* 6 : 173–80.

————, 1983. Status of *Calopteryx japonica* Sélys of *C. virgo* – group (Odonata, Zygoptera). *Kontyû* 51 : 192–202.

Mokrushov, P.A., 1982. (Territorial behaviour of the four-spotted dragonfly *Libellula quadrimaculata* (Odonata, Anisoptera)). *Vest. Zool.* 2 : 58–62.

Montgomery, B.E., 1954. Nomenclatural confusion in the Odonata: the *Agrion-Calopteryx* problems. *Ann. ent. Soc. Am.* 47 : 471–83.

Moore, N.W., 1952a. On the so-called "territories" of dragonflies (Odonata – Anisoptera). *Behaviour* 4 : 85–100.

————, 1952b. Notes on the oviposition behaviour of the dragonfly *Sympetrum striolatum* Charpentier. *Behaviour* 4 : 101–3.

————, 1953. Population density in adult dragonflies (Odonata – Anisoptera). *J. Anim. Ecol.* 22 : 344–59.

————, 1964. Intra- and interspecific competition among dragonflies (Odonata). An account of observations and field experiments on population density control in Dorset, 1954–60. *J. Anim. Ecol.* 33 : 49–71.

Morton, K.J., 1895. Neuroptera observed in Glen Lochay. *Entomologist's mon. Mag.* 6 : 260–63.

————, 1905. Odonata collected by Miss Margaret E. Fountaine in Algeria, with description of a new species of *Ischnura*. *Entomologist's mon. Mag.* 16 : 146–9.

————, 1907. Odonata collected by Lt.-Colonel Nurse, chiefly in North-Western India. *Trans. ent. Soc. Lond.* 1907 : 303–8.

————, 1915. Notes on Odonata from the environs of Constantinople. *Entomologist* 48 : 129–34.

————, 1916. Some palaearctic species of *Cordulegaster*. *Trans. ent. Soc. Lond.* 1915 : 273–90.

————, 1924. The dragonflies of Palestine, based primarily on collections made by Dr. P. A. Buxton, with notes on the species of adjacent regions. *Trans. ent. Soc. Lond.* 1924 : 25–44.

————, 1925. *Macromia splendens* at last: an account of dragon-fly hunting in France. *Entomologist's mon. Mag.* 61 : 1–5.

Morton, K.J., 1927. *Aeschna subarctica* Walker in Europe. *Entomologist's mon. Mag.* **13** : 86–9.

——, 1928. Odonata collected in Austrian Tirol, the Trentino and Tuscany. *Entomologist's mon. Mag.* **14** : 254–60.

Müller, J., 1980. Libellenfunde (Insecta, Odonata) in Naturschutzgebieten des Bezirkes Magdeburg, DDR. *Arch. Nat-Schutz LandschForsch.* **20** : 145–53.

——, 1984. DDR-Erstnachweis der Späten Adonislibelle *Ceriagrion tenellum* (De Villers) im Naturschutzgebiet Mahlpfuhler Fenn, Kreis Tangerhütte (Bez. Magdeburg) (Insecta, Odonata, Coenagrionidae). *Faun. Abh. st. Mus. Tierk. Dresden* **12** : 3943.

Müller, O.F., 1764. *Fauna Insectorum fridrichsdalina.* xxiv+96pp. (Odonata 59–63). Gleditsch, Hafniae & Lipsiae.

——, 1767. Enumeratio ac descriptio Libellularum agri Friedrichsdalensis. *Nova Acta Acad. Caesar. Leop. Carol.* **3** : 122–31.

Münchberg, P., 1930. Zur Biologie der Odonaten-genera *Brachytron* Ev. und *Aeschna* Fbr. *Z. Morph. Ökol. Tiere* **20** : 172–232.

Navas, L., 1907. Neurópteros de Espana y Portugal. *Broteria (zool.)* **6** : 185–242.

Needham, J.G., 1897. *Libellula deplanata* of Rambur. *Can. Ent.* **29** : 144–6.

Newman, E., 1833. Entomological notes. *Ent. Mag.* **1** : 505–14.

Nielsen, C., 1935a. Note odonatologiche. *Boll. Soc. ent. ital.* **67** : 59–62.

——, 1935b. Odonati del Fezzan raccolti dal Prof. G. Scortecci (Missione della R. Società Geografica) e Catalogo delle specie finora catturate. *Atti Soc. ital. Sci. nat.* **74** : 373–82.

——, 1941. Odonati di Sardegna. *Memorie Soc. ent. ital.* **19** : 235–58.

——, 1955. Esuvia ninfale del genere 'Orthetrum Newmann' (sic) riferibile, probabilmente, all' 'O. nitidinerve Selys'. *Boll. Ist. Ent. Univ. Bologna* **21** : 197–200.

——, 1956. Odonati del Sahara nord occidentale. *Revue fr. Ent.* **23** : 191–5.

—— & Conci, C., 1951. Note su Odonati italiani. *Boll. Soc. ent. ital.* **81** : 76–9.

Nielsen, P., 1979. *Lestes barbarus* (Fabricius) fundet i Danmark (Odonata: Lestidae). *Ent. Meddr.* **47** : 96.

Norling, U., 1967. *Hemianax ephippiger* (Burm.) found in Iceland (Odonata). *Opusc. ent.* **32** : 99–100.

——, 1971. The life-history and seasonal regulation in *Aeshna viridis* Eversm. in southern Sweden (Odonata, Aeshnidae). *Entomologica scand.* **2** : 170–90.

——, 1976. Seasonal regulation in *Leucorrhinia dubia* (Vander Linden) (Anisoptera: Libellulidae). *Odonatologica* **5** : 245–63.

——, 1984a. The life cycle and larval photoperiodic responses of *Coenagrion hastulatum* (Charpentier) in two climatically different areas (Zygoptera: Coenagrionidae). *Odonatologica* **13** : 429–49.

——, 1984b. Photoperiodic control of larval development in *Leucorrhinia dubia* (Vander Linden): a comparison between populations from northern and southern Sweden (Anisoptera: Libellulidae). *Odonatologica* **13** 529–50.

Ocharan, F.J., 1983. *Brachythemis leucosticta* (Burm.) (Odonata; Libellulidae) en el norte de España. *Bol. Cien. Naturaleza I.D.E.A.* **32** : 3–9.

——, 1984. Captura de *Gomphus vulgatissimus* (L.) en el norte de España (Odonata: Gomphidae). *Bol. Cien. Naturaleza I.D.E.A.* **34** : 3–6.

Overbeek, H., 1970. A record of *Gomphus graslini* Rambur, 1842 (Odonata) from Spain. *Ent. Ber. Amst.* **30** : 16–17.

Pajunen, V.I., 1962. Studies on the population ecology of *Leucorrhinia dubia* v.d. Lind. (Odon., Libellulidae). *Annls zool. Soc. Vanamo* **24** : 1–79.

——, 1964a. Mechanism of sex recognition in *Leucorrhinia dubia* v.d. Lind., with notes on the reproductive isolation between *L. dubia* and *L. rubicunda* L. (Odon., Libellulidae). *Annls zool. Fenn.* **1** : 55–71.

——, 1964b. Aggressive behaviour in *Leucorrhinia caudalis* Charp. (Odon., Libellulidae). *Annls zool. Fenn.* **1** : 357–69.

——, 1966a. The influence of population density on the territorial behaviour of *Leucorrhinia rubicunda* L. (Odon., Libellulidae). *Annls zool. Fenn.* **3** : 40–52.

——, 1966b. Aggressive behaviour and territoriality in a population of *Calopteryx virgo* L. (Odon., Calopterygidae). *Annls zool. Fenn.* **3** : 201–14.

Palisot de Beauvois, A., 1805–1821. *Insectes recuellis en Afrique et en Amérique.* xvi+276pp. Levrault, Paris.

Pallas, P.S., 1771. *Reise durch verschiedene Provinzen des Russischen Reiches in den Jahren 1768–1774.* **1**. Petersburg.

Parr, M.J., 1973. Ecological studies of *Ischnura elegans* (Vander Linden) (Zygoptera: Coenagrionidae). *Odonatologica* **2** : 139–74.

——, 1976. Some aspects of the population ecology of the damselfly *Enallagma cyathigerum* (Charpentier) (Zygoptera: Coenagrionidae). *Odonatologica* **5** : 45–57.

——, 1983a. An analysis of territoriality in libellulid dragonflies (Anisoptera: Libellulidae). *Odonatologica* **12** : 39–57.

——, 1983b. Some aspects of territoriality in *Orthetrum coerulescens* (Fabricius) (Anisoptera: Libellulidae). *Odonatologica* **12** : 239–57.

—— & Parr, M., 1979. Some observations on *Ceriagrion tenellum* (De Villers) in southern England (Zygoptera: Coenagrionidae). *Odonatologica* **8** : 171–94.

Paulson, D.R., 1974. Reproductive isolation in damselflies. *Syst. Zool.* **23** : 40–49.

Pecile, L., 1981. Una nuova stazione italiana di *Nehalennia speciosa* (Charp.). *Gortania* **2** : 173–80.

Pickup, J., Thompson, D.J. & Lawton, J.H., 1984. The life history of *Lestes sponsa* (Hansemann): larval growth (Zygoptera: Lestidae). *Odonatologica* **13** : 451–9.

Pictet, F.J., 1843. (Description of *Cordulia splendens*). *Magasin Zool. Paris* 2nd ser., 5th year, Insectes, Pl. 117.

Pinhey, E.C.G., 1961. *A Survey of the Dragonflies (Order Odonata) of Eastern Africa.* 214pp. British Museum, London.

——, 1970. Monographic study of *Trithemis* (Libellulidae). *Mem. ent. Soc. sth. Africa* : 159pp.

Pritchard, G., 1965. Prey capture by dragonfly larvae (Odonata: Anisoptera). *Can. J. Zool.* **43** : 271–89.

Puschnig, R., 1935. Über das Vorkommen der Kleinlibelle *Nehalennia speciosa* Charp. *Mitt. Ver. naturk. Landesmus. Kärnten* **125** : 96–100.

Rambur, J.P., 1842. *Histoire naturelle des insectes. Néuroptères.* xvii+534pp. Roret, Paris.

Retzius, A.J., 1783. *Caroli De Geer genera et species Insectorum .....* vi+220pp. Lipsiae.

Riek, E.F. & Kukalova-Peck, J., 1984. A new interpretation of dragonfly wing venation based upon Early Upper Carboniferous fossils from Argentina (Insecta: Odonatoidea) and basic character states in pterygote wings. *Can. J. Zool.* **62** 1150–66.

Ris, F., 1897. Notes sur quelques Odonates de l'Asie centrale. *Annls Soc. ent. Belg.* **41** : 42–50.

——, 1909. Odonata. In *Die Süsswasserfauna Deutschlands* **9** : v+67pp. Jena.

——, 1911a. Übersicht der mitteleuropäischen Cordulinen-Larven. *Mitt. schweiz. ent. Ges.* **12** : 25–41.

——, 1911b. Libellulinen monographisch bearbeitet. *Collections Zoologiques du Baron Edm. de Selys Longchamps. Catalogue systématique et descriptif* **13** : 529–700.

——, 1916. *Aeschna coerulea* in der Schweiz. *Mitt. schweiz. ent. Ges.* **12** : 348–53.

——, 1927. *Aeschna subarctica* Walker, eine für Deutschland und Europa neue Libelle. *Ent. Mitt.* **16** : 99–103.

Risso, A., 1826. Histoire naturelle des principales productions de l'Europe meridionale, ... 5. Paris.

Robert, P.-A., 1958. *Les Libellules (Odonates).* Delachaux et Niestlé, Neuchâtel & Paris. 364pp.

Roberts, J.E.H., 1948. *Coenagrion castellani*, a new species of dragonfly in Europe (Order Odonata, Zygoptera). *Proc. R. ent. Soc. Lond. (B)* 17 : 63–6.

Robertson, H.M. & Paterson, H.E.H., 1982. Mate recognition and mechanical isolation in *Enallagma* damselflies (Odonata: Coenagrionidae) *Evolution* 36 : 243–50.

—— & Tennessen, K.J., 1984. Precopulatory genital contact in some Zygoptera. *Odonatologica* 13 : 591–5.

Rousseau, E., 1909. Etude monographique des larves des Odonates d'Europe. *Annls Biol. lacustre* 3 : 300–64.

Rudolph, R., 1976. Some aspects of wing kinetics in *Calopteryx splendens* (Harris) (Zygoptera: Calopterygidae). *Odonatologica* 5 : 119–27.

——, 1979. Bemerkungen zur Ökologie von *Ischnura pumilio* (Charpentier) (Zygoptera: Coenagrionidae). *Odonatologica* 8 : 55–61.

——, 1980. Die Ausbreitung der Libelle *Gomphus pulchellus* Selys 1840 in Westeuropa. *Drosera* 2 : 63–6.

Sahlén, G., 1985. *Sveriges Trollsländor (Odonata)*. Fältbiologerna, Sollentuna. 152pp.

——, 1987. A new site for *Somatochlora sahlbergi* Trybom in Inari Lapland (Odonata, Corduliidae). *Notul. entomol.* 67 : 3–4.

St Quentin, D., 1938. Die europäischen Odonaten mit boreoalpiner Verbreitung. *Zoogeographica, Jena* 3 : 485–93.

——, 1952. Der Rassenkreiss *Cordulegaster boltoni* (Donovan) (Odonata). *Ent. NachrBl. Wien* 4 :73–5.

——, 1957. Zwei bemerkenswerte *Cordulegaster*-Formen (Odonata) aus der Sammlung des Naturhistorischen Museums in Wien. *Annln naturh. Mus. Wien* 61 : 295–6.

——, 1959. *Catalogus faunae Austriae. XII, Odonata.*

——, 1963. Die infraspezifischen Formen von *Sympecma paedisca* Brauer (Odonata). *Annln naturh. Mus. Wien* 66 : 381–3.

——, 1971. Zum vorkommen von *Cordulegaster insignis* Schneider in Rumänien. *Studii communicari Sibiu* 16 : 205–208.

Sándor, U., 1957. Szitakötök Odonata. *Fauna Hung.* 5 Insecta 1(6) : 44pp.

Say, T., 1839. Descriptions of new North American neuropterous insects, and observations on some already described. *J. Acad. nat. Sci. Philad.* 8 : 9–46.

Scheffler, W., 1973. Libellen aus Bulgarien. *Dt. ent. Z.* 20 : 357–62.

Schiemenz, H., 1953. *Die Libellen unserer Heimat.* 154pp. Urania, Jena.

Schiess, H., 1973. Beitrag zur Kenntnis der Biologie von *Nehalennia speciosa* (Charpentier, 1840) (Zygoptera: Coenagrionidae). *Odonatologica* 2 : 33–7.

Schmidt, Eb., 1964a. Biologisch-ökologische Untersuchungen an Hochmoorlibellen (Odonata). *Z. wiss. Zool.* 169 : 313–86.

——, 1964b. Zur Verbreitung und Biotopbindung von *Aeshna subarctica* Walker in Schleswig-Holstein (Odonata). *Faun. Mitt. NordDtl.* 2 : 197–201.

——, 1965. Zum Paarungs- und Eiablageverhalten der Libellen. *Faun. Mitt. NordDtl.* 2 : 313–9.

——, 1974. Faunistisch-ökologische Analyse der Odonatenfauna der Nordfriesischen Inseln Amrum, Sylt und Föhr. *Faun.-ökol. Mitt.* 4 : 401–18.

——, 1975. *Aeshna viridis* Eversmann in Schleswig-Holstein, Bundesrepublik Deutschland (Anisoptera: Aeshnidae). *Odonatologica* 4 : 81–8.

——, 1977. Ausgestorbene und bedrohte Libellenarten in der Bundesrepublik Deutschland. *Odonatologica* 6 : 97–103.

——, 1978a. Odonata. In *Limnofauna Europaea*. Ed. J. Illies, Gustav Fischer, Stuttgart & New York, Swets & Zeitlinger, Amsterdam.

——, 1978b. Die Verbreitung der Kleinlibelle *Coenagrion armatum* Charpentier, 1840, in Nordwestdeutschland (Odonata: Coenagrionidae). *Drosera* 2 : 39–42.

——, 1980a. *Orthetrum albistylum* und andere südliche Libellenarten (Odonata) an einem Badeteich in den Nordalpen bei Mittenwald (Tennsee bei Krün). *Ent. Z. Frankf. a.M.* 90 : 145–7.

——, 1980b. Das Artenspektrum der Libellen der Insel Helgoland unter dem Aspekt der Fund- und Einwanderungswahrscheinlichkeit (Odonata). *Entomologia gen.* 6 : 247–50.

——, 1981. Quantifizierung und Analyse des Rückganges von gefährdeten Libellenarten in der Bundesrepublik Deutschland (Ins. Odonata). *Mitt. dt. Ges. allg. angew. Ent.* 3 : 167–70.

——, 1982. Odonaten-Zönosen kritisch betrachtet. *Drosera* 6 : 85–90.

——, 1987. Generic reclassification of some westpalaearctic Odonata taxa in view of their nearctic affinities (Anisoptera: Gomphidae, Libellulidae). *Adv. Odonatol.* 3 : 135–45.

Schmidt, Er., 1929. 7. Ordnung: Libellen, Odonata. *Die Tierwelt Mitteleuropas* 4 : 1–66.

——, 1936a. Die mitteleuropäischen *Aeshna*-Larven nach ihren letzten Häuten. *Dt. ent. Z.* : 53–73.

——, 1936b. Die westpaläarktischen Gomphiden-Larven nach ihren letzten Häuten (Ins. Odon.). *Senckenbergiana biol.* 18 : 270–82.

——, 1936c. Die europäischen *Leucorrhinia*-Larven, analytisch betrachtet (Ordnung Odonata). *Arch. Naturgesch.* 5 : 287–95.

——, 1938. Odonaten aus Syrien und Palästina. *Sber. Akad. Wiss. Wien* 147 : 135–50.

——, 1948a. *Pyrrhosoma elisabethae* n.sp., eine neue Odonatenart aus Griechenland. *Opusc. ent.* 13 : 69–74.

——, 1948b. *Calopteryx* versus *Agrion*; again? (Odonata). *Ent. News* 59 : 197–201.

——, 1950. Über die Ausbildung von Steppenformen bei der Waldlibelle *Platycnemis pennipes* (Pall.) (Odonata, Zygoptera). *Ber. naturf. Ges. Augsburg.* 2 : 55–106.

——, 1951. Was ist *Libellula isoceles* O. F. Müller 1767? *Ent. Z. Frankf. a.M.* 60(1950) : 1–7, 13–14.

——, 1953. Zwei neue Libellen aus dem Nahen Osten. *Mitt. münch. ent. Ges.* 43 : 1–9.

——, 1957. Ist unser Vierfleck (*Libellula quadrimaculata* Linné) eine homogene Art? (Odonata). *Ent. Z. Frankf. a.M.* 67 : 73–91.

——, 1959. Versuch einer Analyse der Libelle *Agrion caerulescens* Fonsc. 1838. *Bull. Inst. r. Sci. nat. Belg.* 35 : 1–17.

——, 1960. *Agrion puella Kocheri* nov. subsp. (Odonata), eine Richtigstellung. *C. r. Séanc. mens. Soc. Sci. nat. phys. Maroc* 7 : 123–6.

——, 1964. *Zur Genealogie der Libelle* Agrion pulchellum (*v.d. Lind. 1825*) *Selys-Hagen 1850* (Odonata, Zygoptera). 4pp. Bonn.

——, 1965. Über den Wanderweg der *Boyeria* aus Kreta (Odonata, Aeschnidae). *NachrBl. bayer. Ent.* 14 : 43–6.

——, 1968. Versuch einer Analyse der *Ischnura elegans*-Gruppe (Odonata, Zygoptera). *Ent. Tidskr.* 88 : 188–216.

Schneider, E., 1972. *Somatochlora alpestris* Selys (Odonata) aus dem Retezat Gebirge – der zweite Nachweis aus den Rumänischen Karpathen. *Stud. Muz. Brukenthal Sti. natur.* 17 : 273–5.

Schneider, W., 1981. Eine Massenwanderung von *Selysiothemis nigra* (van der Linden, 1825) (Odonata: Macrodiplactidae) und *Lindenia tetraphylla* (van der Linden, 1825) (Odonata: Gomphidae) in Südjordanien. *Ent. Z. Frankf. a.M.* 91 : 97–102.

——, 1984. Zum Nachweis von *Gomphus pulchellus* Selys 1840 in Jugoslawien (Odonata: Anisoptera: Gomphidae). *Ent. Z. Frankf. a.M.* 94 : 109–11.

——, 1985. Dragonfly records from SE-Turkey (Insecta: Odonata). *Senckenbergiana biol.* 66 : 67–78.

——, 1986a. Designation des Lectotypus von *Somatochlora meridionalis* Nielsen 1935 (Odonata: Anisoptera: Corduliidae). *Ent. Z. Frankf. a.M.* 96 : 73–8.

——, 1986b. Erstnachweis von *Cordulia aenea* (Linnaeus 1758) für die Türkei (Odonata: Anisoptera: Corduliidae). *Ent. Z. Frankf. a.M.* **96** : 92–3.

Schneider, W.G., 1845. Verzeichniss der von Hrn. prof. Dr. Loew im Sommer 1842 in der Türkei und Kleinasien gesammelten Neuropte.ra, nebst kurzer Beschreibung der neuen Arten. *Ent. Ztg. Stettin.* **6** : 110–116, 153–5.

Scopoli, J.A., 1772. *Annus Historico Naturalis.* **5** : 128pp. Milscher, Lipsiae.

Sélys Longchamps, E. de, 1837. *Catalogue des Lépidoptères ou papillons de la Belgique, précédé du tableau des Libellulines de ce pays.* 31pp. Liége.

——, 1839. Description de deux nouvelles espèces d'Aeshna du sous-genre *Anax. Bull. Acad. r. Belg.* **6** : 386–93.

——, 1840. *Monographie des Libellulidées d'Europe.* 220pp. Paris & Bruxelles.

——, 1841. Nouvelles Libellulidées d'Europe. *Revue zool.,* Paris **4** : 243–6.

——, 1843. Note sur quelques Libellules d'Europe. *Annls Soc. ent. Fr.* **1** : 107–9.

——, 1848. Liste des Libellulidées d'Europe et diagnose de quatre espèces nouvelles. *Revue zool.,* Paris **11** : 15–19.

——, 1849. *In* Lucas: Exploration scientifique de l'Algérie. *Zool. (Hist. Nat. An. Art.)* **3** Neuroptères : 110–40.

——, 1853. Synopsis des Caloptérygines. *Bull. Acad. r. Belg.* **20** : 1–73.

——, 1854. Synopsis des Gomphines. *Bull. Acad. r. Belg.* **21** : 23–116.

——, 1862. Synopsis des Agrionines. 2e Légion : *Lestes. Bull. Acad. r. Belg.* **13**(2nd ser.) : 288–338.

——, 1863. Synopsis des Agrionines. 4e Légion : *Platycnemis. Bull. Acad. r. Belg.* **16**(2nd ser.) : 147–76.

——, 1869. Synopsis des Caloptérygines. Secondes additions. *Bull. Acad. r. Belg.* **27**(2nd ser.) : 1–36.

——, 1871a. Synopsis des Cordulines. *Bull. Acad. r. Belg.* **31**(2nd ser.) : 228–355.

——, 1871b. Nouvelle révision des Odonates de l'Algérie. *Annls Soc. ent. Belg.* **14** : 9–20.

——, 1872. Matériaux pour une faune Néuroptérologique de l'Asie Septentrionale. I. Odonates. *Annls Soc. ent. Belg.* **15** : 25–45.

——, 1873. Troisièmes additions au synopsis des Caloptérygines. *Bull. Acad. r. Belg.* **35**(2nd ser.) : 469–519.

——, 1876. Synopsis des Agrionines, 5e légion: *Agrion* (suite). *Bull. Acad. r. Belg.* **41**(2nd ser.) : 1–282.

——, 1878. Note sur deux Libellulines du genre *Urothemis. Annls Soc. ent. Belg.* **21** : c.r. séanc. lxiv-lxvi.

——, 1883. Synopsis des Aeschnines. 1e partie. Classification. *Bull. Acad. r. Belg.* **5**(3rd ser.) : 712–48.

——, 1884. Révision des *Diplax* paléarctique. *Annls Soc. ent. Belg.* **28** : 29–45.

——, 1885. Rectification concernant l'*Onychogomphus genei* Selys et signalement de deux Gomphines nouvelles. *Annls Soc. ent. Belg.* **29** : c.r. séanc. cxlvi-cxlvii.

——, 1887. Odonates de l'Asie mineure et révision de ceux des autres parties de la faune paléarctique (dite européenne). *Annls Soc. ent. Belg.* **31** : 1–49.

——, 1891. Odonates. Viaggio di Leonardo Fea in Birmania e regioni vicine. *Annali Mus. civ. Stor. nat. Genova* **30** : 433–518.

—— & Hagen, H.A., 1850. Revue des Odonates ou Libellules d'Europe. *Mém Soc. r. Sci. Liége* **6** : xxii+ 408pp.

—— & ——, 1854. Monographie des Caloptérygines. *Mém. Soc. r. Sci. Liége* **9**(sep.) : xi+291pp.

—— & ——, 1857. Monographie des Gomphines. *Mém. Soc. r. Sci. Liége* **11**(sep.) : viii+460pp.

Shaw, G., 1806. *General Zoology, or systematic natural history.* **6.** London.

Sherk, T.E., 1978a. Development of the compound eyes of dragonflies (Odonata). II. Development of the larval compound eyes. *J. exp. Zool.* **203** : 47–60.

——, 1978b. Development of the compound eyes of dragonflies (Odonata). III. Adult compound eyes. *J. exp. Zool.* **203** : 61–80.

Smith, E.M. & Smith, R.W.J., 1984. *Brachytron pratense* (Müller) and other Odonata of the Black Lochs, Argyll. *J. Br. Dragonfly Soc.* **1** : 51–4.

Sømme, S., 1937. Zoographische Studien über norwegische Odonaten. *Avh. Norske VidenskAkad.* **12** : 23pp.

Sonehara, I., 1965. Observations on *Sympetrum danae* on Mt. Yatsugatake, Central Japan. *Tombo* **8** : 2–9.

——, 1967. On the life history of the dragonflies of the genus *Epitheca* with special reference to *E. bimaculata sibirica* Selys in Mt. Yatsugatake. *Tombo* **10** : 2–24.

Stark, W., 1979. Mischformen von *Pyrrhosoma n. nymphula* (Sulzer, 1776) und *P. n. elisabethae* Schmidt, 1948 aus der Steiermark, Oesterreich (Zygoptera: Coenagrionidae). *Notul. odonatol.* **1** : 61–2.

Stein, J.P.E., 1863. Beitrag zur Neuropteren-Fauna Griechenlands (mit Berücksichtigung dalmatinischer Arten). *Berl. ent. Z.* **7** : 411–22.

Steinmann, H., 1984. Szitakötök Odonata. *Fauna Hung.* **160** : 112pp.

——, 1986. The odonate fauna of the Kiskunság National Park. *Fauna Kiskunság natn. Park* **1** : 85–91.

Stephens, J.F., 1835. *Illustrations of British Entomology. Mandibulata.* **6** : 240pp. Baldwin & Cradock, London.

Sternberg, K., 1987. On reversible, temperature-dependent colour change in males of the dragonfly *Aeshna caerulea* (Ström, 1783) (Anisoptera: Aeshnidae). *Odonatologica* **16** : 57–66.

Ström, H., 1783. Kort underretning om Eger Sognekald. *Nye Saml. k. Danske Vidensk. Selsk. skr.* **2** : 569–80.

Sulzer, J.H., 1761. *Die Kenntzeichen der Insecten nach Anleitung des Königl. Schwed. Ritters u. Leiburztes Carl Linnaeus .....* Heidegger, Zürich.

——, 1776. *Abgekürzte Geschichte der Insekten nach dem Linneischen System.* xxvii+345pp. Steiner, Winterthur.

Tanaka, Y. & Hisada, M., 1980. The hydraulic mechanism of the predatory strike in dragonfly larvae. *J. exp. Biol.* **88** : 1–19.

Testard, P., 1975. Note sur l'émergence, le sex-ratio et l'activité des adultes de *Mesogomphus genei* Selys, dans le sud de l'Espagne. *Odonatologica* **4** : 11–26.

Teyrovský, V., 1977. Odonata. Check list, enumeratio insectorum Bohemoslovakiae. *Acta ent. Mus. natn. Pragae* **15**, suppl.4 : 31–3.

Theischinger, G., 1979. *Cordulegaster heros* sp. nov. und *Cordulegaster heros pelionensis* ssp. nov., zwei neue Taxa des *Cordulegaster boltoni* (Donovan)-Komplexes aus Europa (Anisoptera: Cordulegasteridae). *Odonatologica* **8** : 23–38.

Thompson, D.J., 1978. Prey size selection by larvae of the damselfly *Ischnura elegans. J. Anim. Ecol.* **47** : 769–85.

Tiberghien, G., 1985. *Macromia splendens* (Pictet, 1843): Additions faunistiques, biologiques, et récapitulation des principales données connues (Odon. Anisoptera Corduliidae). *Bull. Soc. ent. Fr.* **90** : 8–13.

Tillyard, R.J., 1917. *The Biology of Dragonflies (Odonata or Paraneuroptera).* xii+396pp. University Press, Cambridge.

—— & Fraser, F.C., 1940. A reclassification of the order Odonata, based on some new interpretations of the venation of the dragonfly wing. Part III. *Aust. Zool.* **9** : 359–96.

Trybôm, F., 1889. Trollsländor (Odonater) insamlade under svenska expeditionen till Jenisei 1876. *Bih. K. svenska VetenskAkad. Handl.* **15** : 1–21.

Tuxen, S.L., 1976. 39a. Odonata : 7pp. In *The Zoology of Iceland.* Ejnar Munksgaard, Copenhagen & Reykjavik.

Ubukata, H., 1975. Life history and behavior of a corduliid dragonfly, *Cordulia aenea amurensis* Selys. II. Reproductive period with special reference to territoriality. *J. Fac. Sci. Hokkaido Univ. ser. VI, Zoology* **19** : 812–33.

Uhler, P.R., 1858. Descriptions of new species of Neuropterous Insects collected by the North Pacific Expedition under Capt. John Rodgers. *Proc. Acad. nat. Sci. Philad.* **1858** : 29–31.

Utzeri, C., 1985. Field observations on sperm translocation behaviour in the males of *Crocothemis erythraea* (Brullé) and *Orthetrum cancellatum* (L.) (Libellulidae), with a review of the same in the Anisoptera. *Odonatologica* **14** : 227–37.

—— & Belfiore, C., 1976. *Selysiothemis nigra* (Van der Linden) in Italia (Odonata Macrodiplactidae). *Fragm. ent.* **12** : 169–72.

——, Carchini, G., Falchetti, E. & Belfiore, C., 1984. Philopatry, homing and dispersal in *Lestes barbarus* (Fabricius) (Zygoptera: Lestidae). *Odonatologica* **13** : 573–84.

——, Falchetti, E. & Carchini, G., 1983. The reproductive behaviour in *Coenagrion lindeni* (Selys) in Central Italy (Zygoptera: Coenagrionidae). *Odonatologica* **12** : 259–78.

—— & Raffi, R., 1983. Observations on the behaviour of *Aeshna affinis* (Vander Linden) at a dried-up pond (Anisoptera: Aeshnidae). *Odonatologica* **12** : 141–51.

Valetta, A., 1949. A preliminary list of the Odonata of the Maltese Islands. *Entomologist* **82** : 85–7.

Valle, K.J., 1927. Zur Kenntnis der Odonatenfauna Finnlands III. Ergänzungen und Zusätze. *Acta Soc. Fauna Flora fenn.* **56** : 1–36.

——, 1931. Die Odonaten der Kanarischen Inseln. *Commentat. biol.* **6** : 1–7.

——, 1952. Die Verbreitungsverhältnisse der ostfennoskandischen Odonaten. (Zur Kenntnis der Odonatenfauna Finnlands VI). *Acta ent. fenn.* **10** : 1–87.

Valtonen, P., 1980. Die Verbreitung der finnischen Libellen (Odonata). *Notul. ent.* **60** : 199–215.

Vander Linden, P.L., 1820. *Agriones bononiensis descriptae.* 8pp. Bononiae; *Aeshnae bononiensis descriptae, adjecta annotatione ad Agriones bononienses descriptas.* 11pp. Bononiae.

——, 1823. Aeshnae bononienses descriptae, adjecta annotatione ad Agriones bononienses descriptas. *Opusc. scient.* **4** : 158–65.

——, 1825. *Monographiae Libellulinarum Europaearum Specimen.* 42pp. Frank, Brussels.

Verhoeven, J.T.A., 1980. The ecology of *Ruppia*-dominated communities in Western Europe. II. Synecological classification. Structure and dynamics of the macroflora and macrofauna communities. *Aquatic Bot.* **8** : 1–85.

Villers, C.J. de, 1789. *Caroli Linnaei Entomologia.* **3** : 657pp. (Odonata 1–15). Piestre & Delamollière, Lugduni.

Vogt, F.D. & Heinrich, B., 1983. Thoracic temperature variations in the onset of flight in dragonflies (Odonata: Anisoptera). *Physiol. Zoöl.* **56** : 236–41.

Waage, J.K., 1979a. Dual function of the damselfly penis: sperm removal and transfer. *Science (Wash.)* **203** : 916–8.

——, 1979b. Adaptive significance of postcopulatory guarding of mates and nonmates by male *Calopteryx maculata* (Odonata). *Behavl Ecol. Sociobiol.* **6** : 147–54.

Walker, E.M., 1908. A key to the North American species of *Aeshna* found north of Mexico. *Can. Ent.* **40** : 377–91.

——, 1958. *The Odonata of Canada and Alaska. Vol. 2.* University of Toronto Press. xi+318pp.

Wallengren, H.D.J., 1894. Öfversikt af Skandinaviens Pseudoneuroptera. *Ent. Tidskr.* **15** : 235–70.

Waringer, J., 1983. A study on embryonic development and larval growth of *Sympetrum danae* (Sulzer) at two artificial ponds in Lower Austria (Anisoptera: Libellulidae). *Odonatologica* **12** : 331–43.

Warren, R.G., 1964. Territorial behaviour of *Libellula quadrimaculata* L. and *Leucorrhinia dubia* Van Der L. (Odonata, Libellulidae). *Entomologist* **97** : 147.

Waterston, A.R., 1976. On the genus *Cordulegaster* Leach, 1815 (Odonata) with special reference to the Sicilian species. *Trans. R. Soc. Edinb.* **69** : 457–66.

——, 1985. Insects of Southern Arabia. Odonata from the Yemens and Saudi Arabia. In *Fauna of Saudi Arabia* **6**(1984) : 451–72.

Watson, J.A.L., 1982. A truly terrestrial dragonfly larva from Australia (Odonata: Corduliidae). *J. Aust. ent. Soc.* **21** : 309–11.

Weber, T. & Caillere, L., 1978. Thermistor telemetry of ventilation during prey capture by dragonfly larvae (*Cordulegaster boltoni*, Odonata). *J. comp. Physiol.* **128** : 341–5.

Wenger, O.P., 1959. Die beiden ♀ Formen von *Boyeria irene* (Odonata, Aeschnidae). *Mitt. schweiz. ent. Ges.* **32** : 304–11.

Whalley, P.E.S., 1979. New species of Protorthoptera and Protodonata (Insecta) from the Upper Carboniferous of Britain, with a comment on the origin of wings. *Bull. Br. Mus. nat. Hist. (Geol.)* **32** : 85–90.

——, 1980. *Tupus diluculum* sp. nov. (Protodonata), a giant dragonfly from the Upper Carboniferous of Britain. *Bull. Br. Mus. nat. Hist. (Geol.).* **34** : 285–7.

——, 1986. Bristly Oxtongue, *Picris echioides* L., a hazard for insects? *Bot. Soc. Br. Isles* **43** : 13–14.

Wildermuth, H., 1981. *Libellen, Kleinodien unserer Gewässer.* 26pp. Schweizer Naturschutz, Basel.

Yazicioğlu, T., 1982. Dragonflies from the Ergene river basin, Thrace, Turkey. *Notul. odonatol.* **1** : 148–50.

Zahner, R., 1959. Über die Bindung der Mitteleuropäischen *Calopteryx* Arten (Odonata, Zygoptera) an den Lebensraum des strömenden Wassers. I. Der Anteil der Larven an der Biotopbindung. *Int. Revue ges. Hydrobiol. Hydrogr.* **44** : 51–130.

——, 1960. Über die Bindung der Mitteleuropäischen *Calopteryx* Arten (Odonata, Zygoptera) an den Lebensraum des strömenden Wassers. II. Der Anteil der Imagines an der Biotopbindung. *Int. Revue ges. Hydrobiol. Hydrogr.* **45** : 101–23.

Zetterstedt, J.W., 1840. *Insecta Lapponica descripta.* 1140pp. (Odonata 1037–44). Voss, Lipsiae.

Zimmermann, W., 1973. Zur Kenntnis der Kleinen Pechlibelle, *Ischnura pumilio* (Charp.) (Odonata). *Ent. Ber., Berl.* **1972** : 108–12.

Recent Literature

The number of recent dragonfly books published in Europe is testimony to the growing interest in Odonata on the Continent as well as in Britain. Quite outstanding among these new books is Philip Corbet's *Dragonflies: Behaviour and Ecology of Odonata*, published in 1999 by Harley Books, Colchester. Other important titles of books, papers and CD-ROMs, published since the first edition was prepared and detailing distribution or identification of adults, exuviae or larvae of Odonata, are listed below on a regional basis. The advent in 1996 of a new Hungarian journal which is devoted to larvae, *Odonata-Studium Larvale*, and Jill Silsby's *Dragonflies of the World*, published in 2001 by the National History Museum, London, must also be mentioned.

Europe:

d'Aguilar, J. & Dommanget, J.-L., 1998. *Guide des libellules d'Europe et d'Afrique du Nord: l'identification et la biologie de toutes les espèces.* Delachaux & Niestlé, Lausanne-Paris. Revised edition of the book first published in 1985.

Bellman, H., 1987. *Libellen beobachten-bestimmen.* Neumann-Neudamm, Melsungen-Berlin-Basel-Wien. Revised edition 1993.

Boudot, J.-P., 2001. Les *Cordulegaster* du Paléarctique occidental: identification et répartition (Odonata, Anisoptera, Cordulegastridae). *Martinia* 17: 1–34.

Burmeister, E.-G., 1989. Spätsommeraspekt der Libellenfauna Sardiniens (Italien) (Insecta, Odonata). *NachrBl. bayer. Ent.* 38: 80–3.

Butler, S.G., 1993. Key to the larvae of European *Orthetrum* Newman (Anisoptera: Libellulidae). *Odonatologica* 22: 191–6.

Gerken, B. & Sternberg, K., 1999. *Die Exuvien Europäischer Libellen.* Huxaria Verlag, Höxter.

Heidemann, H. & Seidenbusch, R., 1993. *Die Libellenlarven Deutschlands und Frankreichs. Handbuch für Exuviensammler.* Bauer, Keltern. [covers all central Europe]

Jödicke, R., 1997. *Die Binsenjungfern und Winterlibellen Europas* (Lestidae). Westarp Wissenschaften, Magdeburg & Spektrum Akademischer Verlag, Heidelburg, Berlin, Oxford. [Neue Brehm-Bücherei, *Libellen Europas*].

Jurzitza, G., 1993. *Libellules d'Europe, Europe centrale et meridionale.* Delachaux & Niestlé, Lausanne & Paris.

——, 2000. *Der Kosmos Libellenführer. Die Arten Mittel-und Südeuropas.* Kosmos Verlag, Stuttgart.

Kählert, J., 1999/2001. *Die Libellen Europas.* CD-ROM. Burg, Germany.

Martens, A., 1996. *Die Federlibellen Europas* (Platycnemididae). Westarp Wissenschaften, Magdeburg & Spektrum Akademischer Verlag, Heidelburg, Berlin, Oxford. [Neue Brehm-Bücherei 626, *Libellen Europas*].

Müller, O., 1990. Mitteleuropäische Anisopterenlarven (Exuvien) – einige Probleme ihrer Determination (Odonata, Anisoptera). *Dt. ent. Z.* 37: 145–87.

Peters, G., 1987. *Die Edellibellen Europas. Aeschnidae.* Ziemsen Verlag, Wittenberg Lutherstadt.

Sandhall, Å., 1987. *Trollsländor i Europa.* Interpublishing, Stockholm.

Suhling, F. & Müller, O., 1996. *Die Flussjungfern Europas* (Gomphidae). Westarp Wissenschaften, Magdeburg & Spektrum Akademischer Verlag, Heidelburg, Berlin, Oxford. [Neue Brehm-Bücherei 628, *Libellen Europas*].

Albania:

Dumont, H. J., Mertens, J. & Miho, A., 1993. A contribution to the knowledge of the Odonata of Albania. *Opusc. zool. flumin.* 113: 1–10.

British Isles:

Brooks, S. (ed.) with illustrations by Lewington, R., 1997. *Field Guide to the Dragonflies and Damselflies of Great Britain and Ireland.* British Wildlife Publishing, Rotherwick.

Merritt, R., Moore, N. W. & Eversham, B. C., 1996. *Atlas of the Dragonflies of Britain and Ireland.* HMSO, London.

Powell, D. & Twist, C., 1999. *A Guide to the Dragonflies of Great Britain.* Arlequin Press, Chelmsford.

Separate accounts of the odonate fauna of several English counties have also been published.

Bulgaria:

Beschovski, V.L., 1994. Insecta: Odonata. *Fauna bulgarica* 23: 1–372.

Marinov, M., 2000. Pocket field guide to the dragonflies of Bulgaria. Eventus Publishing House, Sofia. [Cyrillic text but illustrated key usable without knowledge of Bulgarian].

Denmark:

Fogh Nielsen, O., 1998. *De Danske Guldsmede.* Apollo Books, Stenstrup. Also covers northern Germany and southern Norway and Sweden.

France (Corsica):

Michiels, N., 1988. Observations of dragonflies (Odonata) on Corsica. *Bull. Annls Soc. r. ent. Belg.* 124: 115–23.

Grand, D. & Papazian, M., 2000. Étude faunistique des odonates de Corse. *Martinia* 16: 31–50.

Germany:

Brock, V., Hoffmann, J., Kühnast, O., Piper, W. & Voss, K., 1997. Atlas der Libellen Schleswig-Holsteins. Landesamt für Natur und Umwelt des Landes Schleswig-Holstein, Flintbek.

Heidemann, H. & Seidenbusch, R., 1993, edn 2, 2002. *Die Libellenlarven Deutschlands.* Verlag Baur, Keltern. (Edn 2 published as Vol. 72 in *Die Tierwelt Deutschlands.*)

Kuhn, K. & Burbach, K., 1998. *Libellen in Bayern.* Verlag Eugen Ulmer, Stuttgart.

Sternberg, K. & Buchwald, R., 1999/2000. *Die Libellen Baden-Württembergs,* Vol. 1 (1999, Zygoptera), Vol. 2 (2000, Anisoptera). Verlag Eugen Ulmer, Stuttgart.

Greece (including islands, Cyprus):

Battin, T. J., 1989. Überblick über die Libellenfauna der Insel Kreta (Insecta: Odonata). *Z. ArbGem. öst. Ent.* 41: 52–64.

Dell'Anna, L., 1994. Contributo alla conoscenza degli odonati dell'Egeo (Odonata). *Fragm. entomol.* 25: 257–65.

Grand, D., 2001. Quelques observations de libellules en Grèce et à Chypre (Odonata). *Opusc. zool. flumin.* 196: 1–10.

Lopau, W., 1995a. Die Libellenfauna der Insel Lesbos, Griechenland. *Naturk. Reiseber.* 3: 1–81. Revised edition of work first published privately in 1991.

——, 1995b. Beitrag zur Kenntnis der Odonatenfauna der griechischen Inseln Rhodos, Kos, Samos und Chios.

Libellenbeobachtungen aus den Sommern 1992, 1993 und 1994 und andere naturkundliche Notizen. *Naturk. Reiseber.* **4**: 1-60.

—— & Wendler, A., 1995. Arbeitsatlas zur Verbreitung der Libellen in Griechenland und umliegenden Gebieten. Rasterkarten nach den in der Literatur vorhandenen Nachweisen sowie unveröffentlichten Beobachtungen. *Naturk. Reiseber.* **5**: 1-108. [Cyprus not covered]

Hungary:

Ambrus, A., Bánkuti, K. & Kovács, T., 1992a. *A Kisalföld és a Nyugat-Magyarországi peremvidék Odonata faunája.* Agrárker Rt., Györ.

Italy (Sicily):

Galletti, P. A., Pavesi, M. & Romano, F. P., 1987. *Brachythemis leucosticta* (Burm.) e considerazioni su altri odonati nuovi per la Sicilia (Insecta, Odonata). *Naturalista sicil.* (IV) **11**: 27-46.

Bedjanič, M. & Salamun, A., 1999. Contribution to the knowledge of the odonate fauna of Sicily, with some additional data from Basilicata, southern Italy. *Opusc. zool. flumin.* **169**: 1-4.

Luxembourg:

Proess, R. & Gerend, R., 1998. Rote Liste der Libellen Luxemburgs. *Bull. Soc. Nat. luxemb.* **99**: 137-48.

Netherlands:

Nederlandse Vereniging voor Libellenstudie, 2002. *De Nederlandse libellen (Odonata).* Nederlandse Fauna 4. Nationaal Natuurhistorisch Museum Naturalis, KNNV Uitgeverij & European Invertebrate Survey, Leiden.

Visser, H., 1997. *Libellenlarven van Nederland.* CD-ROM. Biodiversity Centre of ETI.

North Africa (Azores):

Belle, J., 1992. The Odonata of the Azores. *Ent. Ber., Amst.* **52**: 63-5.

North Africa (Morocco):

Jacquemin, G., 1994. Odonata of the Rif, northern Morocco. *Odonatologica* **23**: 217-37.

—— & Boudot, J.-P., 1999. *Les Libellules (Odonates) du Maroc.* Soc. Fr. Odonatol., Bois d'Arcy.

North Africa (Tunisia):

Jödicke, R., Arit, J., Kunz, B. Lopau, W. & Seidenbusch, R., 2000. The Odonata of Tunisia. *Pantala* **3**: 41-71.

Russia, Ukraine, Baltic States:

Spuris, Z. D., 1988. The species composition of the dragonfly fauna of the USSR. *Latv. Ent.* **31**: 5-24.

——, 1993. *Latvijas spāru (Odonata) noteicējs.* Zinātne, Riga. [in Latvian, Identification guide to Latvian dragonflies]

Stanionyte, A., 1993. The check-list of dragonflies (Odonata) of Lithuania. In: *New and rare Lithuanian Insect Species*, Institute of Ecology, Vilnius.

Slovakia:

Straka, V., 1990. Vážky (Odonata) slowenska. *Zbor. slov. nár. Muz.* **36**: 121-47.

Slovenia:

Kotarac, M., 1997. *Atlas kačjih pastirjev (Odonata) Slovenije, z Rdečim seznamen.* Center za kartografijo favne in flore, Miklavžna-Dravskem-polju.

Spain & Portugal:

Jödicke, R. [ed.], 1996. *Studies on Iberian dragonflies.* Ursus, Bilthoven.

Ocharan, F. J., 1988. Composicion de la odonatofauna iberica. *Revta Biol. Univ. Oviedo* **6**: 82-93.

Switzerland:

Maibach, A. & Meier, C., 1987. *Atlas de distribution des Libellules de Suisse (Odonata) (avec liste rouge).* Documenta Faunistica Helvetiae 3. CSCF, Neuchâtel.

Supplementary References (in addition to those given above)

Adamović, Ž., Andjus, L. & Mladenović, A., 1992. *Cordulegaster heros* Theischinger, 1979 in Serbia and Macedonia (Odonata: Cordulegastridae). *Opusc. zool. flumin.* **101**: 1-11.

Agassiz, D., 1981. Further introduced china mark moths (Lepidoptera: Pyralidae) new to Britain. *Entomologist's Gaz.* **32**: 21-6.

Ambrus, A., Bánkuti, K. & Kovacs, T., 1992b. Data on the anatomy of Hungarian *Cordulegaster* species. *Fol. Hist.-nat. Mus. Matr.* **17**: 177-80. [in Hungarian with English summary]

Arnold, A., 1988. Zur Libellen fauna (Odonata) um zwei Thermalbädern bei Oradea, Rumänien. *Ent. Nachr. Ber.* **32**: 91-2.

Baaijens, A., 2001. Een nieuwe waterjuffer voor Zeeland. *Zeeuwse Prikkebeen* **9**: 12.

Battin, T. J., 1990. *Anax immaculifrons* Rambur, 1842 from the island of Karpathos, Greece: an oriental representative in the European dragonfly fauna (Odonata: Aeshnidae). *Opusc. zool. flumin.* **47**: 1-10.

——, 1991. Description of the larva of *Coenagrion intermedium* Lohmann, 1990 from Crete, Greece (Zygoptera: Coenagrionidae). *Odonatologica* **20**: 333-6.

——, 1993. Revision of the *puella* group of the genus *Coenagrion* Kirby, 1890 (Odonata, Zygoptera), with emphasis on morphologies contributing to reproductive isolation. *Hydrobiologia* **262**: 13-29.

Bedjanič, M., 2000. *Aeshna subarctica elisabethae* Djakonov, 1922, new for the odonate fauna of Slovenia (Anisoptera: Aeshnidae). *Exuviae* **6**: 7-10.

Belle, J. & Tol, J. van, 1990. *Anomalagrion hastatum* (Say), an American damselfly indigenous to the Azores (Odonata, Coenagrionidae). *Tijdschr. Ent.* **133**: 143-7.

Ben Azzouz, B., Guemmouh, R. & Aguesse, P., 1989. A propos des *Coenagrion* du groupe *mercuriale* (Charpentier, 1840) et de la présence de *C. castellani* Roberts, 1948 au Maroc (Zygoptera: Coenagrionidae). *Odonatologica* **18**: 279-83.

Benstead, P. J. & Jeffs, C. J., 1991. Observation de *Libellula quadrimaculata* (L., 1758) près de Ponte-Leccia (Corse). *Martinia* **7**: 78.

Bernard, R., 1995. Wstepne dane o rozmieszczeniu i ekologii *Cercion lindenii* (Selys, 1840) (Odonata, Coenagrionidae) w Polsce. *Wiad. entomol.* **14**: 11-9.

——, 1998. Stan wiedzy o rozmieszczeniu i ekologii *Nehalennia speciosa* (Charpentier, 1840) (Odonata: Coenagrionidae) w Polsce. *Rocz. nauk. pol.Tow. Ochr. Przyr. "Salamandra"* 2: 67-93.

——, 2002. First records of *Aeshna crenata* Hagen, 1856 in Lithuania, with selected aspects of its biology (Odonata: Aeshnidae). *Opusc. zool. flumin.* 202: 1-21.

—— & Labedzki, A., 1993. The occurrence of *Sympetrum pedemontanum* (Allioni, 1766) (Odonata, Libellulidae) in Polish lowlands. *Wiad. ent.* 12: 163-71. [Polish with English summary]

—— & Samolag, J., 1994. *Aeshna affinis* (Vander Linden, 1820) in Poland (Odonata: Aeshnidae). *Opusc. zool. flumin.* 117: 1-7.

Beshovski, V. L., 1993. Critical notes on some Odonata species (Insecta, Odonata), reported by Bulgarian authors for the territories of Bulgaria, Greece and Macedonia. *Acta zool. bulg.* 46: 39-43.

—— & Gashtarov, V., 1997. *Selysiothemis nigra* new to Bulgaria. *Ent. Z.* 107: 309.

Beukeboom, I. W., 1988. Two new records of *Somatochlora flavomaculata* from the Netherlands (Odonata: Corduliidae). *Ent. Ber., Amst.* 48: 82-5.

Beutler, H., 1987. Ein Fund von *Cordulegaster heros* Theischinger, 1979 im Pirin-Gebirge in Bulgarisch-Mazedonien (Insecta, Odonata, Cordulegasteridae). *Faun, Abh. Mus. Tierkd. Dresden* 15: 11-4.

——, 1988. Libellen aus der Region Banat, Rumänien (Odonata). *Opusc. zool. flumin.* 30: 1-15.

Bond, K. G. M., 1989. *Aeshna cyanea* (Müller), Odonata Aeshnidae, a dragonfly new to Ireland. *Ir. Nat. J.* 23: 73-4.

Boudot, J.-P., 1998. Differences in male colour patterns between *Boyeria cretensis* Peters, 1991 and *B. irene* (Fonscolombe, 1838) (Odonata: Aeshnidae). *Opusc. zool. flumin.* 161: 1-3.

—— & Jacquemin, G., 1995. Revision of *Cordulegaster boltonii* (Donovan, 1807) in southwestern Europe and northern Africa, with description of *C. b. iberica* ssp.nov. from Spain (Anisoptera: Cordulegastridae). *Odonatologica* 24: 149-73.

——, —— & Goutet, P., 1990. Odonates des lacs toubières à sphaignes des Hautes-Vosges, France. *Opusc. zool. flumin.* 52: 1-11.

——, —— & Dumont, H. J., 1990b. Revision of the subspecies of *Onychogomphus forcipatus* (Linnaeus, 1758) in Europe and Asia Minor, and the true distribution of *Onychogomphus forcipatus unguiculatus* (Vander Linden, 1823) (Odonata, Gomphidae). *Bull. Annls Soc. r. belge Ent.* 126: 95-111.

Brändle, M. & Rödel, M.-O., 1994. Libellenfunde von Nordgriechenland und Kreta (Odonata). *Ent. Z., Essen* 104: 85-91.

Brooks, S. J., 1988. Exotic dragonflies in north London. *J. Br. Dragonfly Soc.* 4: 9-12.

Buczyński, P., 2000. On the occurrence of *Coenagrion armatum* (Charpentier, 1840) in Poland (Odonata: Coenagrionidae). *Opusc. zool. flumin.* 179: 1-10.

Butler, S.G., 1992. Notes on the collection and rearing out of the larva of *Somatochlora sahlbergi* Trybom from Finland (Odonata: Corduliidae). *Opusc. zool. flumin.* 84: 1-5.

Corbet, P.S., 2000. The first recorded arrival of *Anax junius* Drury (Anisoptera: Aeshnidae) in Europe: a scientist's perspective. *Int. J. Odonatology* 3: 153-162.

Cordero Rovera, A., Lorenzo Carballa, M.O. & Utzeri, C., 2001. Evidence for parthenogenetic reproduction in populations of *Ischnura hastata* from the Azores. *Abstracts from the 2nd WDA International Symposium of Odonatology*, Gaellivare, Sweden: 11.

Dannelid, E. & Ekestubbe, K., 2001. Vinterflickslända (*Sympecma fusca*) på spridning norrut? *Ent. Tidskr.* 122: 173-6.

Dantart, J. & Martin, R., 1999. *Somatochlora metallica* (Van der Linden, 1825) (Odonata: Corduliidae) and *Leucorrhinia pectoralis* (Charpentier, 1825) (Odonata: Libellulidae), dos nuevas especies de libélulas para la península ibérica. *Boln Asoc. esp. Ent.* 23: 147.

David, S, 1990. Deux nouveautés pour le département de la Gironde: *Leucorrhinia albifrons* (Burmeister, 1839) et *Orthetrum albistylum* (Selys, 1848) (Odonata, Anisoptera: Libellulidae). *Martinia* 6: 65-6.

——, 2000. New records of dragonflies (Insecta; Odonata) from Slovakia. *Biologia, Bratislava* 55: 444.

Dell'Anna, L., Utzeri, C., De Matthaeis, E. & Cobolli, M., 1999. Biological differentiation and reproductive isolation of syntopic central Italian populations of *Chalcolestes viridis* (Vander L.) and *C. parvidens* (Artobol.) (Zygoptera: Lestidae). *Anax, Wien* 2: 41.

Dingemanse, N. J., 1998. Vliegtijd von *Aeshna subarctica* in Nederland. *Amoeba, Amst.* 72: 108-10.

Dommanget, J.-L., 1987. *Enallagma cyathigerum* (Charpentier, 1840), nouvelle espèce pour la Corse (Odonata Zygoptera: Coenagrionidae). *Martinia* No. 6: 28-9.

Ellenrieder, N. von, 2002. A phylogenetic analysis of the extant Aeshnidae (Odonata: Anisoptera). *Syst. Ent.* 27: 437-67.

Freyhof, J., 1998. Exuvien- und Larvenfunde der Asiatischen Keiljungfer *Gomphus flavipes* (Charpentier) im Rhein von Nordrhein-Westfalen. *LöBF-Mitt.* 1998: 4.

Gächte, E., 1988. *Gomphus pulchellus* Selys, 1840 – Neu für Österreich (Anisoptera: Gomphidae). *Notul. odonatol.* 3: 6.

Geraeds, R. P. G. & Hermans, J.T., 2000. De gaffellibel (*Ophiogomphus cecilia* Fourcroy, 1785) langs de Roer. *Natuurh. Maandbl.* 89: 254-9.

Gerend, R. & Proess, R., 1994. Nachweis neuer und interessanter Libellen aus Luxemburg nebst einer provisorischen Fassung der Roten Liste der einheimischen Odonaten (Insecta, Odonata). *Bull. Soc. Nat. luxemb.* 95: 299-314.

Grand, D., 1988. Confirmation de la présence d'*Epitheca bimaculata* en Isère. *Sympetrum. Rev. Odonatol.* No. 2: 51-3.

——, 1990. *Trithemis annulata* in NE Spain (Anisoptera: Libellulidae). *Notul. odonatol.* 3: 75.

——, 1993. A propos de *Leucorrhinia albifrons* (Burmeister, 1839) et d'*Aeshna subarctica elisabethae* (Walker, 1908) dans les monts du Jura (Départements du Doubs et du Jura). *Martinia* 9: 19-20.

——, 1995. Sur la présence de *Selysiothemis nigra* (Vander L.) en Aragon, Espagne (Anisoptera: Libellulidae). *Notul. odonatol.* 4: 91.

——, Greff, N. & Dolcourt, G., 2001. *Leucorrhinia pectoralis* (Charpentier, 1825) nouveau pour le département du Rhône. *Martinia* 17: 107-9.

Holmen, M., 1991. The damselfly *Nehalennia speciosa* (Charpentier, 1840) new to Denmark. *Ent. Meddr.* 59: 1-3.

Ivarsson, T., 1998. "*Sympetrum nigrescens*" found in Sweden. *Nordisk Odonatologisk Forum Nyhetsbrev* 4: 6.

Jödicke, R., 1993a. Die Typen von *Sympecma paedisca annulata* (Sélys, 1887) (Odonata: Zygoptera: Lestidae). *Ent. Z., Essen* 103: 189-97.

——, 1993b. Die Bestimmung der Exuvien von *Sympetrum sanguineum* (Müll.), *S. striolatum* (Charp.) und *S. vulgatum* (L.) (Odonata: Libellulidae). *Opusc. zool. flumin.* 115: 1-8.

——, 1994. Subspecific division of *Sympetrum sinaiticum* Dumont, 1977, and the identity of *S. vulgatum decoloratum* (Sélys, 1884) (Anisoptera: Libellulidae). *Odonatologica* 23: 239-53.

——, 1997. Die Verbreitung von *Sympetrum sinaiticum tarraconense* Jödicke (Odonata, Anisoptera: Libellulidae). *Opusc. zool. flumin.* 155: 1-7.

Kalkman, V. J. & Dijkstra, K.-D. B., 2000. The dragonflies of the Bialowiéza area, Poland and Belarus (Odonata). *Opusc. zool. flumin.* 185: 1-19.

Karanan, B. S., 1987. Les odonates du Lac de Prespa (Macédoine, Yougoslavie). *God. Zh. prir. mat. Fak. Skopje* (Biol.) 37/38: 97-110.

Ketelaar, R., 1998. Speerwaterjuffe in het nauw. *Vlinders* 13: 19-21.

Kiauta, B. & Kiauta, M., 1991. Biogeographic considerations on

Coenagrion hylas freyi (Bilek, 1954), based mainly on the karyotype features of a population from North Tyrol, Austria (Zygoptera: Coenagrionidae). *Odonatologica* 20: 417–31.

Kipping, J., 1997. Zur Situation der Kleinen Königslibelle, *Anax parthenope* (Insecta, Odonata) in Thüringen. *Mauritania* 16: 462–4.

Korn, M., 1988. Erstnachweis der Südlichen Binsenjungfer (*Lestes barbarus*) auf Helgoland. *Seevögel* 9: 25.

Kotarac, M., 1996. *Somatochlora meridionalis* Nielsen, 1935, a new species for the odonate fauna of France. *Exuviae* 2(1995): 15–6.

Landemaine, D., 1991. *Lestes macrostigma* (Eversmann) dans le marais d'Olonne (Vendée). *Martinia* 7: 58.

Lang, C., Müller, H. & Waringer, J. A., 2001. Larval habitats and longitudinal distribution patterns of *Cordulegaster heros* Theischinger and *C. bidentata* Sélys in an Austrian forest stream (Anisoptera: Cordulegastridae). *Odonatologica* 30: 395–409.

Lebioda, B., 1987. Un mediterranéen exile en Charente-Maritime *Lestes macrostigma* (Eversmann, 1836). *Martinia* 6: 27–8.

Lett, J.-M., 2001. Première donnée de *Coenagrion ornatum* (Selys, 1850) dans la Région Centre, département du Cher (Odonata, Zygoptera, Coenagrionidae). *Martinia* 17: 94.

Lieftinck, M. A., 1978. Over een onopzettelijke kweek van een tropisch-Aziatische libel uit een verwamd aquarium in Nederland (Odonata, Gomphidae). *Ent. Ber., Amst.* 38: 145–50.

Lohmann, H., 1992. Revision der Cordulegastridae. 1. Entwurf einer neuen Klassifizierung der Familie (Odonata: Anisoptera). *Opusc. zool. flumin.* 96: 1–18.

Machet, P., 1990. Présence de *Lestes macrostigma* (Eversmann, 1836) dans l'Île de Noirmoutier, Vendée (Odonata, Zygoptera, Lestidae). *Martinia* 6: 17–8.

Male-Malherbe, E., 1998. Confirmation de la présence d'une population d'*Epitheca bimaculata* (Charpentier, 1825) dans le département de l'Indre (Odonata, Anisoptera, Corduliidae). *Martinia* 14: 30.

—— & Deberge, J., 1993. *Epitheca bimaculata* (Charpentier, 1825) nouveau pour le département de l'Indre. *Martinia* 9: 86.

Marinov, M., 1999. *Chalcolestes parvidens* (Artobolevski) and *Somatochlora meridionalis* Nielsen in Bulgaria (Zygoptera: Lestidae; Anisoptera: Corduliidae). *Notul. odonatol.* 5: 31–3.

——, 2001. *Somatochlora borisi* spec. nov., a new European dragonfly species from Bulgaria (Anisoptera: Corduliidae). *IDF-Report* 3: 9–16.

Martin-Casacuberta, R. M., 1997. Presencia de *Coenagrion hastulatum* (Charpentier, 1825) en la Peninsula Ibérica (Odonata: Coenagrionidae). *Boln Asoc. esp. Ent.* 21: 101.

Mauersberger, R., 1994. Zur wirklichen Verbreitung von *Orthetrum coerulescens* (Fabricius) und *O. ramburi* (Selys) = *O. anceps* (Schneider) in Europa und die Konsequenzen für deren taxonomischen Rang (Odonata, Libellulidae). *Dt. ent. Z.* 41: 235–56.

—— & Zessin, W., 1990. Zum Vorkommen und zur Ökologie von *Gomphus vulgatissimus* Linnaeus (Odonata, Gomphidae) in der ehemaligen DDR. *Ent. Nachr. Ber.* 34: 203–11.

Meier, C., 1988. Naturschutzkonzept Kahbrunner Riet. Die Libellen. *Anthos* (Spezial) 1: 57–9.

Monnerat, C., 1992. *Coenagrion caerulescens* (Fonscolombe, 1838) dans le département de l'Ardèche (Odonata, Zygoptera, Coenagrionidae). *Martinia* 8: 39–40.

——, 1999. Premieres observations de *Anax imperator* Leach et *Orthetrum cancellatum* (L.) pour Chypre (Anisoptera: Aeshnidae, Libellulidae). *Notul. odonatol.* 5: 25–40.

Muñoz Pozo, B. & Blasco-Zumeta, J., 1996. Contribución al conocimiento de los odonatos (Insecta: Odonata) de las aguas estacionales de Los Monegros (Zaragoza). *Zapateri. Revta aragon. ent.* 6: 141–5.

Nelson, B., 1999. The status and habitat of the Irish damselfly

Coenagrion lunulatum (Charpentier) (Odonata) in Northern Ireland. *Entomologist's mon. Mag.* 135: 59–68.

Nielsen, O. F., 1994. *Anax imperator* Leach, 1815 – ny dansk guldsmed (Odonata, Aeshnidae). *Ent. Meddr* 62: 97–9.

Ocharan, F. J., 1987. Nuevos datos sobre los odonatos de Menorca (Espana). *Bol. real Soc. esp. Hist. mat.* (Biol.) 83: 155–61.

Olsvik, H., 1990. *Somatochlora flavomaculata* (Van der Linden, 1825) (Odonata, Corduliidae) a new species to Norway. *Fauna norvegica* B 37: 111–2.

Ottolenghi, C., 1991. Contributo alla conoscenza degli odonati di Grecia. *Boll. Mus. civ. Stor. nat. Verona* 15(1988): 231–42.

Papazian, M., 1987. Trois nouvelles espèces pour la Corse. *Martinia* No. 5: 13–7.

——, 1990. *Brachytron pratense* (Müller, 1764): Nouvelle espèce pour la Corse (Odonata, Anisoptera: Aeshnidae). *Martinia* 6: 35.

——, 1992. Contribution à l'étude des migrations massives en Europe de *Hemianax ephippiger* (Burmeister, 1839) (Odon., Anisoptera, Aeshnidae). *Entomol. gall.* 3: 15–21.

——, 1995. Etude systématique et biogéographique de *Calopteryx splendens* (Harris, 1782) en Provence (Odonata: Zygoptera). *Bull. Soc. ent. Fr.* 100: 361–76.

Pedersen, H., 1992. *Somatochlora sahlbergi* Trybom, 1889 (Odonata: Corduliidae) – a new species to Norway. *Fauna norv.* (B) 39: 22.

—— & Holmen, M., 1994. Fredede insekter i Danmark. Del 4: Guldsmede. *Ent. Meddr.* 62: 33–58.

Peters, G., 1991. Die Schattenlibelle auf Kreta (*Boyeria cretensis* spec. nov.) und die Monophylie der "Gattung" *Boyeria* McLachlan, 1896 (Odonata Anisoptera, Aeshnidae). *Dt. ent. Z.* 38: 161–96.

Picazo, J. & Alba-Tercedor, J., 1990. First record of *Ophiogomphus cecilia* (de Fourcroy) in Spain (Anisoptera: Gomphidae). *Notul. odonatol.* 3: 171–2.

Prévost, O. & Durepaire, P., 1994. Etat de la population de *Leucorrhinia caudalis* (Charpentier, 1840) dans la réserve naturelle du Pinail (département de la Vienne). *Martinia* 10: 23–7.

Roche, B., 1989. *Trithemis annulata* (Palisot de Beauvois, 1805): nouvelle espèce pour la Corse et la faune de France (Odonata, Anisoptera: Libellulidae). *Martinia* 5: 23–4.

Rödel, M.-O., 1991. Erstnachweis von *Ophiogomphus cecilia* (Fourcroy, 1785) für Griechenland (Odonata: Gomphidae). *Mitt. int. ent. Ver.* 16: 93–6.

Ruddele, J., 1998. *Leucorrhinia albifrons* (Burm.) in coastal W. France. *Notul. odonatol.* 5: 11.

Sahlén, G., 1994. Tundratrollsländan *Somatochlora sahlbergi* funnen i nordligaste Sverige. *Ent. Tidskr.* 115: 137–42.

Salamanca-Ocaña, J. C., Canovillegas, F. & Ferreras-Romero, M., 2001. Contribución al conocimiento de la distribución ibérica actual de *Onychogomphus costae* Sélys, 1885 (Odonata: Gomphidae). *Boln. Asoc. esp. Ent.* 25: 187–8.

Salzi, S., 1993. Segnalazioni faunistiche italiane. 219. *Cordulia aenea* (Linneo, 1758) (Odonata Corduliidae). *Boll. Soc. ent. ital.* 125: 71.

Schwaller, T., 1989. Beobachtungen an einer vorübergehenden Population von *Lestes barbarus* (Fabricius) bei Derendingen, Bezirk Wasseramt, Kanton Solothurn, Schweiz (Odonata: Lestidae). *Opusc. zool. flumin.* 38: 1.

Stauder, A., 1991. Water fauna of a Madeiran stream with notes on the zoogeography of the Macaronesian Islands. *Bolm Mus. munic. Funchal* 43: 243–99.

Sternberg, K., 1996. Colour, colour change, colour patterns and 'cuticular windows' as light traps; their thermoregulatoric and ecological significance in some *Aeshna* species (Odonata: Aeshnidae). *Zool. Anz.* 235: 77–88.

Stobbe, H., 1990. Ein Beitrag zur Kenntnis der Verbreitung von *Calopteryx splendens* und *Calopteryx virgo* in Griechenland./ Bemerkungen zur Gattung *Calopteryx* in Griechenland. *Naturk. Rundbr.* 4: 2–4/ 5–19.

Terzani, F., 1987. Odonati dell'Italia meridionale i nuovi dati (Insecta: Odonata). *Redia* **70**: 229–43.

Terzani, F., 1991. Segnalazioni faunistiche italiane. 165. *Trithemis annulata* (Palisot de Beauvois, 1805) (Odonata: Libellulidae). *Boll. Soc. ent. ital.* **123**: 67–8.

Toth, S., 1985. Libellen und ihre Biotope im Bakony-Gebirge. *Folia Mus.Hist.-nat. Bakonyensis* 2(1983): 45–54.

Trockur, B. & Mauersberger, R., 2000. Vergleichende ökologische Untersuchungen an *Epitheca bimaculata* Charpentier, 1825 im Saarland und in der Uckermark (Odonata: Corduliidae). *Beitr. Ent.* **50**: 487–518.

Tromp, J. & Wasscher, M., 2000. Once in a century: the second record of *Sympetrum meridionale* in the Netherlands. *Brachytron* 4: 25–7.

Utzeri, C., 1989. Segnalazioni faunistiche italiane: 132. *Gomphus vulgatissimus. Boll. Soc. ent. ital.* **121**: 73.

Valtonen, P., 1995. Litteähukankorennon (*Libellula depressa* L.) uusekspansio Suomessa. *Diamina* 4: 4–11.

Wakeham-Dawson, A., Benton. T. & Barnham, V., 1999. Butterflies and dragonflies in northern Greece. *Entomologist's Rec. J. Var.* **111**: 121–8.

Wildermuth, H., 1999. *Somatochlora alpestris* (Sélys, 1840) in den Schweizer Alpen: eine Verbreitungs- und Habitatanalyse (Anisoptera: Corduliidae). *Odonatologica* **28**: 399–416.

Line Drawings and Maps

All line-drawings in the text have been prepared by the author, but certain figures are based in part on the published work of the authors which is gratefully acknowledged here: Aguesse (1968), figs 392–394; Cannings & Cannings (1985), fig. 465; Conci & Nielsen (1956), figs 68, 374, 375, 396–403, 410, 412, 413, 427, 428, 432, 438, 443, 446–457, 461, 483–486, 488, 489; Corbet (1957a), fig. 491; Dumont (1977b), fig. 351; Franke (1979), fig. 421; Gardner (1954b, 1955a), figs 317–320, 332, 373, 379, 384–391, 395, 409, 414–420, 424, 426, 429–431, 433, 434, 436, 439–442, 462–464, 470–482, 487, 501, 502; Lieftinck (1966), figs 250, 254; Marinov (2001), fig. 510; Robert (1958), figs 6, 83, 9a–e, 380, 382, 402–404, 423, 435, 459, 467, 468; Schmidt, Er. (1936b), figs 452–457; Theischinger (1979), figs 242, 243, 253; Waterston (1976), figs 240, 241, 251, 252; Zahner (1959), figs 392–394.

The distribution maps have all been prepared by the author using an outline of the region specially drawn for this work by Richard Geiger.

Notes and Addenda

Notes and Addenda

COLOUR PLATES

The following plates are from water-colour paintings prepared from mounted specimens. Damselflies (Plates 1–13) are all depicted ×2.16 life-size, the dragonflies (Plates 14–30) ×1.08 life-size. An enlarged image of each specimen (×4 for Zygoptera, ×2 for Anisoptera) was projected on a screen and traced. In this way, accuracy of shape and proportions could be achieved. Any apparent differences between species, not mentioned in the text, such as shape of the head or proportions of the legs, result from slightly differing angles of view. Colours are intended to represent those of the living insect. Rapid post-mortem changes in colour made it necessary to use, to achieve a degree of verisimilitude, colour transparencies or colour photographs of live insects. In the small number of cases where these were not available, paintings were completed from notes and sketches made in the field. Whenever possible, the finished paintings have been checked against living specimens. The data pertaining to the illustrated specimens are those for the insect which was used to prepare the initial drawing. The paintings for Plate 30 were prepared from digital photographs of the specimens, enlarged to twice natural size and traced.

Plate 1: Calopterygidae

Figs 1,2, x 2.16

1   *Calopteryx virgo* (Linnaeus) ♂
    Sigoules, Dordogne, France. 1.viii.1977, D. R. Askew
    (*Page* 54)

2   *Calopteryx virgo* (Linnaeus) ♀
    Ste Foy la Grande, Dordogne, France. 4.viii.1977,
    D. R. Askew (*Page* 54)

1

♂

2

♀

Plate 2: Calopterygidae

Figs 3–5, x 2.16

3 *Calopteryx splendens splendens* (Harris) ♂
   Vibraye, Sarthe, France. 13.vii.1977, R. R. Askew
   (*Page* 55)
4 *Calopteryx splendens xanthostoma* (Charpentier) ♂ right front
   wing
   Lalinde, Dordogne, France. 14.viii.1978, R. R. Askew
   (*Page* 55)
5 *Calopteryx splendens splendens* (Harris) ♀
   Whitchurch, Shropshire, England. vi.1973, R. R. Askew
   (*Page* 55)

Plate 2: Calopterygidae (x 2.16)

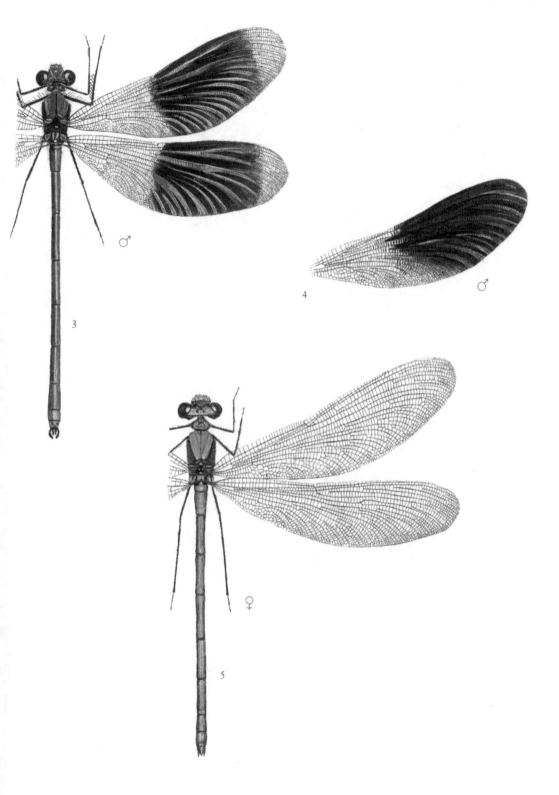

Plate 3: Calopterygidae

Figs 6–8, x 2.16

6 *Calopteryx haemorrhoidalis* (Vander Linden) ♀
   Lalinde, Dordogne, France. 14.viii.1978, R. R. Askew
   (*Page* 56)
7 *Calopteryx haemorrhoidalis* (Vander Linden) ♂
   Data as above (*Page* 56)
8 *Calopteryx haemorrhoidalis* (Vander Linden) ♂ apex of
   abdomen in ventral view
   Data as above (*Page* 56)

Plate 3: Calopterygidae (x 2.16)

Plate 4: Euphaeidae and Lestidae

Figs 9–12, x 2.16

9 *Epallage fatime* (Charpentier) ♂
Anatolia, Turkey. McLachlan coll., British Museum
(Natural History) (*Page* 57)

10 *Epallage fatime* (Charpentier) ♂ immature in dorsolateral
view with only right hindwing shown
Palestine. 15.v.1931, A. Grünnberg, British Museum
(Natural History) (*Page* 57)

11 *Lestes viridis* (Vander Linden) ♀
Sigoules, Dordogne, France. viii.1977, R. R. Askew
(*Page* 61)

12 *Lestes viridis* (Vander Linden) ♂ abdomen
Duras, Lot et Garonne, France. viii.1977, D. R. Askew
(*Page* 61)

Plate 4: Euphaeidae and Lestidae (x 2.16)

Plate 5: Lestidae

Figs 13–15, x 2.16

13 *Lestes virens* (Charpentier) ♂
    Tournon, Indre, France. 31.vii.1979, R. R. Askew
    (*Page* 62)

14 *Lestes barbarus* (Fabricius) ♂
    Estigarde, Landes, France. vii.1977, D. R. Askew
    (*Page* 61)

15 *Lestes macrostigma* (Eversmann) ♂
    Albania. McLachlan coll., British Museum (Natural History)
    (*Page* 62)

Plate 5: Lestidae (x 2.16)

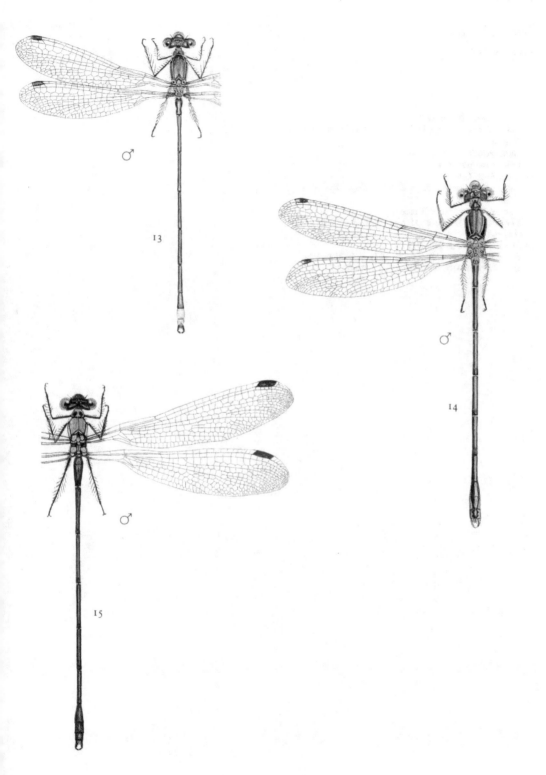

13

♂

14

♂

15

♂

Plate 6: Lestidae

Figs 16–20, x 2.16

16 *Lestes sponsa* (Hansemann) ♀
   Abbots Moss, Cheshire, England. 10.vii.1979, R. R. Askew
   (*Page* 63)
17 *Lestes sponsa* (Hansemann) ♂
   Data as above (*Page* 63)
18 *Lestes dryas* Kirby ♂
   Lapeyrade, Landes, France. 22.viii.1979, R. R. Askew
   (*Page* 64)
19 *Sympecma fusca* (Vander Linden) ♂
   Tournon, Indre, France. 31.vii.1979, R. R. Askew
   (*Page* 65)
20 *Sympecma annulata* (Sélys) ♂
   Ankeveen, Netherlands. 21.viii.1951, C. E. Longfield,
   British Museum (Natural History) (*Page* 66)

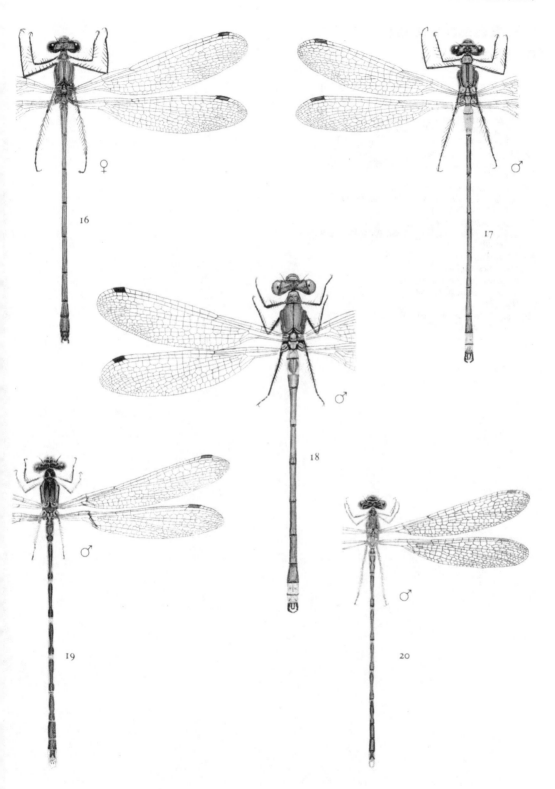

♀

16

♂

17

♂

18

♂

19

♂

20

Plate 7: Platycnemididae

Figs 21–27, x 2.16

21 *Platycnemis pennipes* (Pallas) immature ♀
abdomen Duras, Lot et Garonne, France. viii.1977,
R. R. Askew (*Page* 68)

22 *Platycnemis pennipes* (Pallas) ♂
Sigoules, Dordogne, France. viii.1977, R. R. Askew
(*Page* 68)

23 *Platycnemis pennipes* (Pallas) ♀
Ste Foy la Grande, Dordogne, France. viii.1977,
R. R. Askew (*Page* 68)

24 *Platycnemis latipes* Rambur ♂
La Force, Dordogne, France. viii.1978, R. R. Askew
(*Page* 69)

25 *Platycnemis latipes* Rambur ♀
Data as above (*Page* 69)

26 *Platycnemis acutipennis* Sélys ♂
La Force, Dordogne, France. 31.v.1978, R. R. Askew
(*Page* 70)

27 *Platycnemis acutipennis* Sélys ♀
Data as above (*Page* 70)

Plate 7: Platycnemididae (x 2.16)

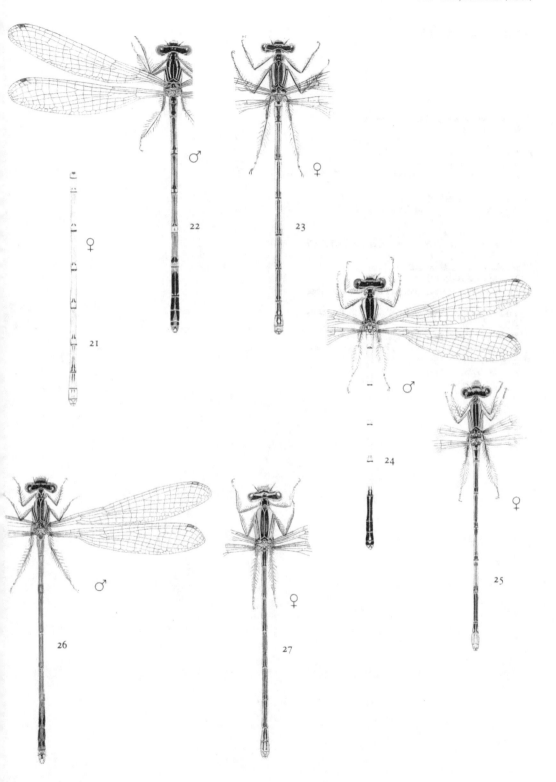

Plate 8: Coenagrionidae

Figs 28–35, x 2.16

28 *Pyrrhosoma nymphula* (Sulzer) ♀ form *melanotum*
(heterochrome)
Anglesey, Wales. 12.vi.1978, R. R. Askew (*Page* 72)

29 *Pyrrhosoma nymphula* (Sulzer) ♀ form *fulvipes*
(homeochrome/andromorph)
Dinnet, Aberdeenshire, Scotland. 12.vi.1977, R. R. Askew
(*Page* 72)

30 *Pyrrhosoma nymphula* (Sulzer) ♂
Witherslack, Westmorland, England. 3.vi.1977,
R. R. Askew (*Page* 72)

31 *Erythromma najas* (Hansemann) ♀
Woodchester Park, Gloucestershire, England. vii.1979,
R. R. Askew (*Page* 73)

32 *Erythromma najas* (Hansemann) ♂
Data as above (*Page* 73)

33 *Erythromma najas* (Hansemann) ♂ in lateral view (small
form)
Fraisse, Dordogne, France. 18.viii.1978, D. R. Askew
(*Page* 73)

34 *Erythromma viridulum* (Charpentier) ♀
Camargue, France. 13.viii.1981, R. R. Askew (*Page* 74)

35 *Erythromma viridulum* (Charpentier) ♂
Data as above (*Page* 74)

Plate 8: Coenagrionidae (x 2.16)

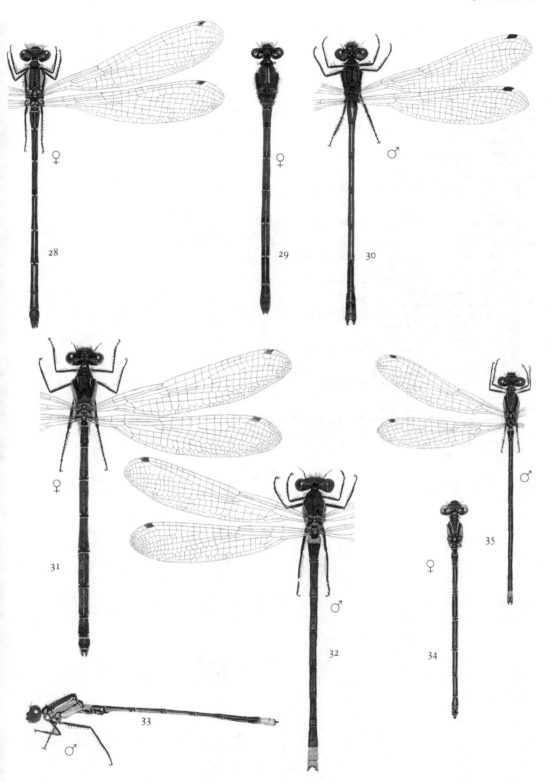

28 ♀

29 ♀

30 ♂

31 ♀

32 ♂

33 ♂

34 ♀

35 ♂

Plate 9: Coenagrionidae

Figs 36–46, x 2.16

36 *Coenagrion mercuriale* (Charpentier) ♀
Rochebeaucourt, Dordogne, France. 30.v.1981,
R. R. Askew (*Page* 80)

37 *Coenagrion mercuriale* (Charpentier) ♂
Data as above (*Page* 80)

38 *Coenagrion scitulum* (Rambur) ♂
Benfleet, Essex, England. 24.vi.1951, A. E. Gardner,
P. S. Corbet coll. (*Page* 80)

39 *Coenagrion scitulum* (Rambur) ♀
Bologna, Italy. 15.v.1933, C. Nielsen, Royal Museum of
Scotland (*Page* 80)

40 *Coenagrion caerulescens* (Fonscolombe) ♀
Spain. Morton coll., Royal Museum of Scotland (*Page* 81)

41 *Coenagrion caerulescens caesarum* Schmidt ♀
Nonza, Corsica. 23–30.vii.1929, K. J. Morton, Royal
Museum of Scotland (*Page* 81)

42 *Coenagrion caerulescens* (Fonscolombe) ♂
Digne, Provence, France. 1–15.vii.1914, K. J. Morton,
Royal Museum of Scotland (*Page* 81)

43 *Coenagrion hastulatum* (Charpentier) ♂
Jostedalen, Indre Sogn, Norway. vii.1979, Ross Andrew
(*Page* 82)

44 *Coenagrion hastulatum* (Charpentier) ♀
Saeterstoen, Norway. Morton coll., Royal Museum of
Scotland (*Page* 82)

45 *Coenagrion lunulatum* (Charpentier) ♂
Overijssel, Netherlands. 8.vi.1925, M. A. Lieftinck, Royal
Museum of Scotland (*Page* 82)

46 *Coenagrion lunulatum* (Charpentier) ♀
Data as above (*Page* 82)

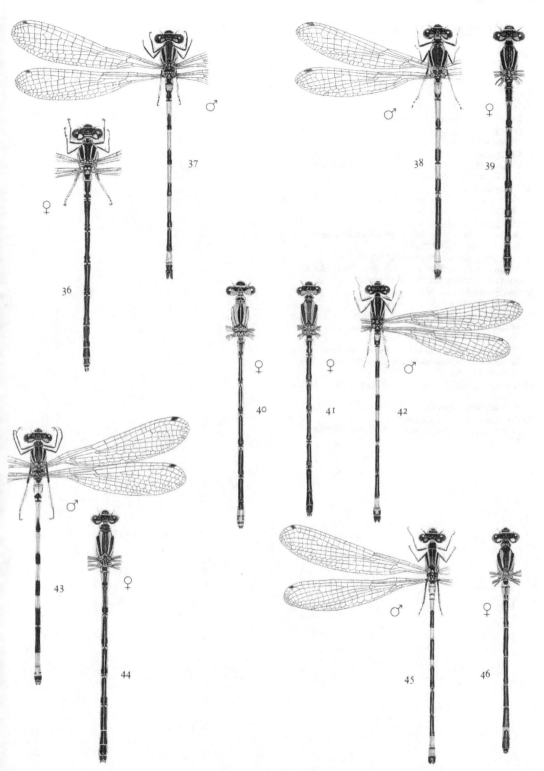

Plate 10: Coenagrionidae

Figs 47–56, x 2.16

47 *Coenagrion armatum* (Charpentier) ♀
Soborg, Mors, Denmark. 12.v.1906, Morton coll., Royal
Museum of Scotland (*Page* 83)

48 *Coenagrion armatum* (Charpentier) ♂
Stalham, Norfolk, England. 2.vi.1951, A. E. Gardner,
P. S. Corbet coll. (*Page* 83)

49 *Coenagrion ornatum* (Sélys) ♀
Tapolca, Balaton, Hungary. 5.vi.1928, Erich Schmidt, Royal
Museum of Scotland (*Page* 84)

50 *Coenagrion ornatum* (Sélys) ♂
Ankara, Turkey. 26.vi.1962, K. M. Guichard, Royal
Museum of Scotland (*Page* 84)

51 *Coenagrion johanssoni* (Wallengren) ♀ abdomen
Jääski, Karelia, Finland. 5.vii.1927, K. J. Valle, Morton
coll., Royal Museum of Scotland (*Page* 85)

52 *Coenagrion johanssoni* (Wallengren) ♂
Data as above (*Page* 85)

53 *Coenagrion hylas* (Trybôm) ♂ in lateral view,
Siberia. 20.vi.1909, A. Bartenef, Morton coll.,
Royal Museum of Scotland (*Page* 85)

54 *Coenagrion puella* (Linnaeus) ♀ heterochrome form
Woodchester Park, Gloucestershire, England. 4.vii.1977,
R. R. Askew (*Page* 86)

55 *Coenagrion puella* (Linnaeus) ♀ abdomen of homeochrome
form
Data as above (*Page* 86)

56 *Coenagrion puella* (Linnaeus) ♂
Monestier, Dordogne, France. vi.1978, R. R. Askew
(*Page* 86)

Plate 10: Coenagrionidae (x 2.16)

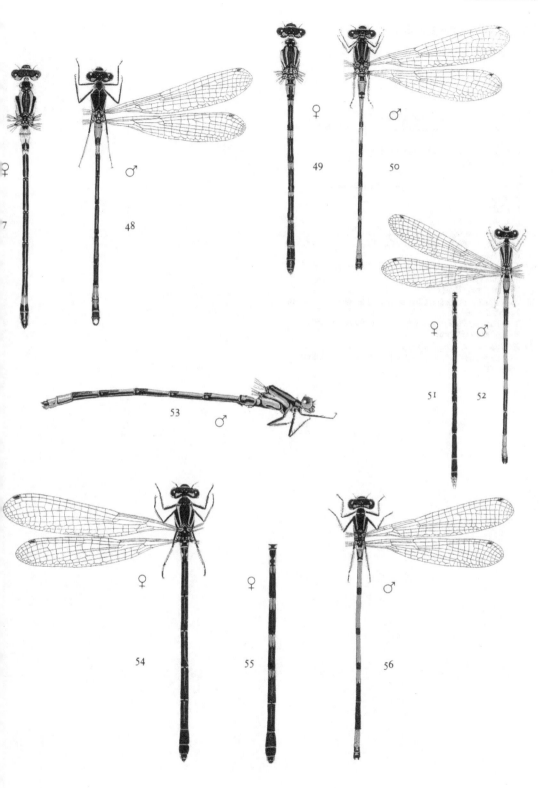

47 ♀

48 ♂

49 ♀

50 ♂

51 ♀

52 ♂

53 ♂

54 ♀

55 ♀

56 ♂

Plate 11: Coenagrionidae

Figs 57–64, x 2.16

57 *Coenagrion pulchellum* (Vander Linden) ♂
Beckermet, Cumberland, England. 17.vii.1979, R. R. Askew
(*Page* 87)

58 *Coenagrion pulchellum* (Vander Linden) ♀
Data as above (*Page* 87)

59 *Cercion lindeni* (Sélys) ♀
St Georges-Blancaneix, Dordogne, France. viii.1978,
R. R. Askew (*Page* 88)

60 *Cercion lindeni* (Sélys) ♂
Data as above (*Page* 88)

61 *Enallagma cyathigerum* (Charpentier) ♂
Woodchester Park, Gloucestershire, England. vii.1978,
R. R. Askew (*Page* 89)

62 *Enallagma cyathigerum* (Charpentier) ♀ homeochrome
(andromorph) form
Data as above (*Page* 89)

63 *Enallagma cyathigerum* (Charpentier) ♀ homeochrome form
in lateral view
Woodchester Park, Gloucestershire, England. vii.1978,
R. R. Askew (*Page* 89)

64 *Enallagma cyathigerum* (Charpentier) ♀ heterochrome form
Abbots Moss, Cheshire, England. vi.1979, R. R. Askew
(*Page* 89)

Plate 11: Coenagrionidae (x 2.16)

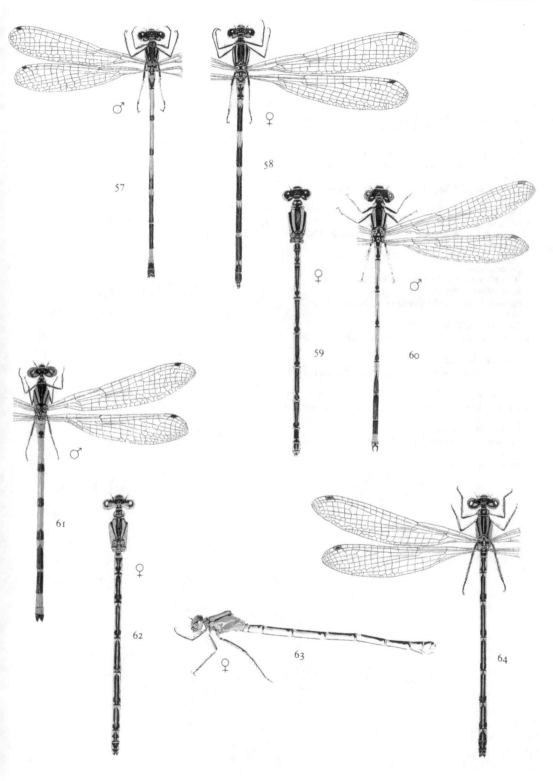

57 ♂

58 ♀

59 ♀

60 ♂

61 ♂

62 ♀

63 ♀

64

Plate 12: Coenagrionidae

Figs 65–74, x 2.16

65 *Ischnura pumilio* (Charpentier) ♂
Monestier, Dordogne, France. 18.vii.1976, D. R. Askew
(*Page* 92)

66 *Ischnura pumilio* (Charpentier) ♀ homeochrome form
Faux, Dordogne, France. 14.viii.1979, R. R. Askew
(*Page* 92)

67 *Ischnura pumilio* (Charpentier) ♀ form *aurantiaca*
(heterochrome)
Data as above (*Page* 92)

68 *Ischnura elegans* (Vander Linden) ♂
Delamere, Cheshire, England. 20.viii.1976, R. R. Askew
(*Page* 92)

69 *Ischnura elegans* (Vander Linden) ♀ form *rufescens*
(heterochrome)
Woodchester Park, Gloucestershire, England. vii.1978,
R. R. Askew (*Page* 92)

70 *Ischnura elegans* (Vander Linden) ♀ form *typica* in lateral
view
Data as above (*Page* 92)

71 *Ischnura elegans* (Vander Linden) ♀ form *violacea* in lateral
view
Data as above (*Page* 92)

72 *Ischnura elegans* (Vander Linden) ♀ form *infuscans* in lateral
view
Data as above (*Page* 92)

73 *Ischnura elegans* (Vander Linden) ♀ form *infuscans-obsoleta* in
lateral view
Data as above (*Page* 92)

74 *Ischnura elegans* (Vander Linden) ♀ form *rufescens* in lateral
view
Data as above (*Page* 92)

Plate 12: Coenagrionidae (x 2.16)

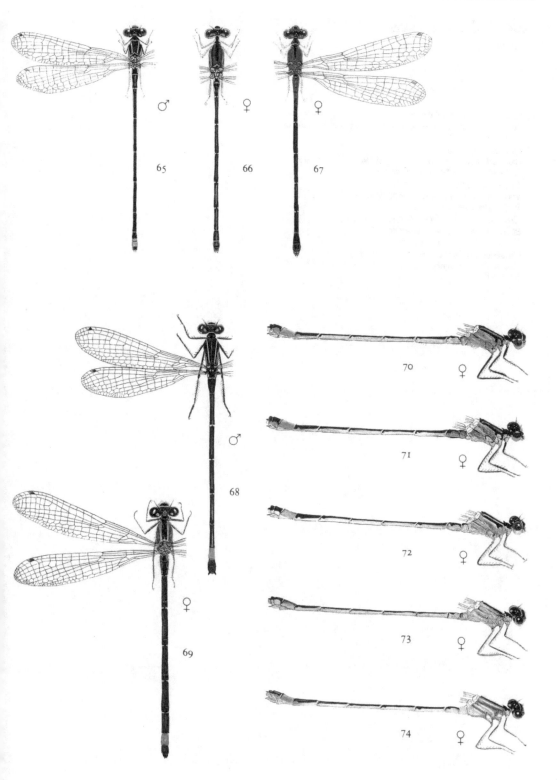

Plate 13: Coenagrionidae

Figs 75–79, x 2.16

75 *Ischnura graellsi* (Rambur) ♂
Sobradiel, Spain. 3.vii.1904, Morton coll., Royal Museum of
Scotland (*Page* 94)

76 *Nehalennia speciosa* (Charpentier) ♂
Canton Zurich, Switzerland. 3–10.vii.1904, K. J. Morton,
Royal Museum of Scotland (*Page* 95)

77 *Ceriagrion tenellum* (Villers) ♂
St Georges-Blancaneix, Dordogne, France. viii.1977,
R. R. Askew (*Page* 97)

78 *Ceriagrion tenellum* (Villers) ♀ form *melanogastrum*
St Georges-Blancaneix, Dordogne, France.
12.viii.1979, R. R. Askew (*Page* 97)

79 *Ceriagrion tenellum* (Villers) ♀ typical form
Monestier, Dordogne, France. viii.1977, R. R. Askew
(*Page* 97)

Plate 13: Coenagrionidae (x 2.16)

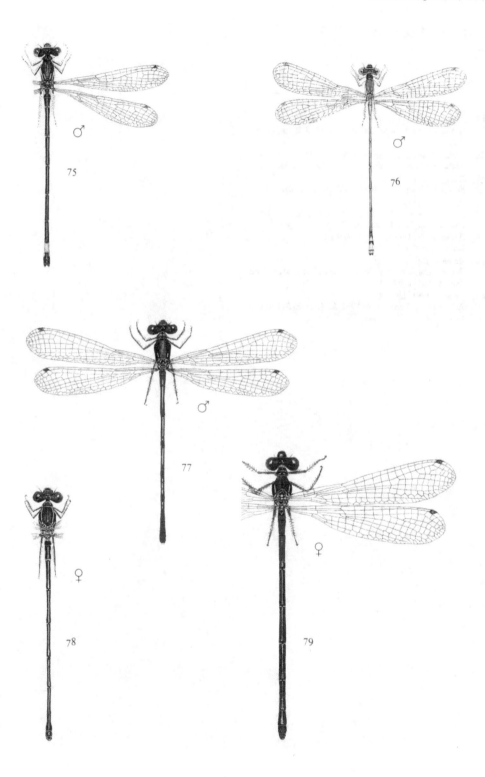

Plate 14: Aeshnidae

Figs 80–86, x 1.08

80 *Aeshna caerulea* (Ström) ♂
Jøstedalen, Indre Sogn, Norway. vii.1979, Ross Andrew
(*Page* 103)

81 *Aeshna juncea* (Linnaeus) ♀ in lateral view
Arosa, Switzerland. viii.1981, R. R. Askew (*Page* 104)

82 *Aeshna juncea* (Linnaeus) ♂
Swansea, Wales. 16.ix.1978, R. R. Askew (*Page* 104)

83 *Aeshna subarctica* Walker ♂
Lüneburg Heide, Germany. 2.ix.1928, Axel Rosenbohm,
Morton coll., Royal Museum of Scotland (*Page* 105)

84 *Aeshna subarctica* Walker ♀ in lateral view
Data as above (*Page* 105)

85 *Aeshna crenata* Hagen ♂
South Finland. K. J. Valle, Morton coll., Royal Museum
of Scotland (*Page* 106)

86 *Aeshna serrata* Hagen ♂
Abo, Turku, Finland. 9.ix.1931, K. J. Valle, Morton coll.,
Royal Museum of Scotland (*Page* 107)

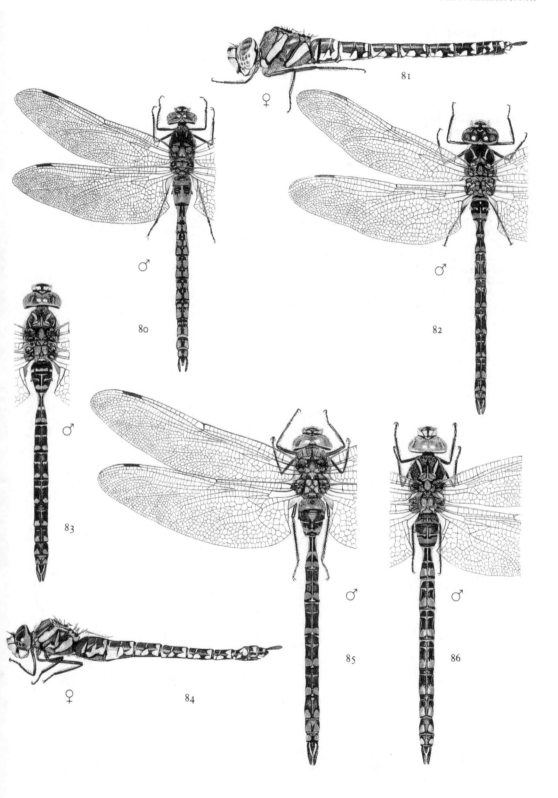

Plate 14: Aeshnidae (x 1.08)

81 ♀

80 ♂

82 ♂

83 ♂

84 ♀

85 ♂

86 ♂

Plate 15: Aeshnidae

Figs 87–92, x 1.08

87   *Aeshna mixta* Latreille ♂
     Monestier, Dordogne, France. 11.viii.1978, R. R. Askew
     (*Page* 107)

88   *Aeshna mixta* Latreille ♀ in lateral view
     Tulcea, Romania. 31.vii.1983, R. R. Askew (*Page* 107)

89   *Aeshna affinis* Vander Linden ♂
     Duras, Lot et Garonne, France. 11.viii.1977, D. R. Askew
     (*Page* 108)

90   *Aeshna affinis* Vander Linden ♀ in lateral view
     Monestier, Dordogne, France. 9.viii.1977, D. R. Askew
     (*Page* 108)

91   *Aeshna cyanea* (Müller) ♀
     Swansea, Wales. 17.ix.1978, R. R. Askew (*Page* 108)

92   *Aeshna cyanea* (Müller) ♂
     Data as above (*Page* 108)

Plate 15: Aeshnidae (x 1.08)

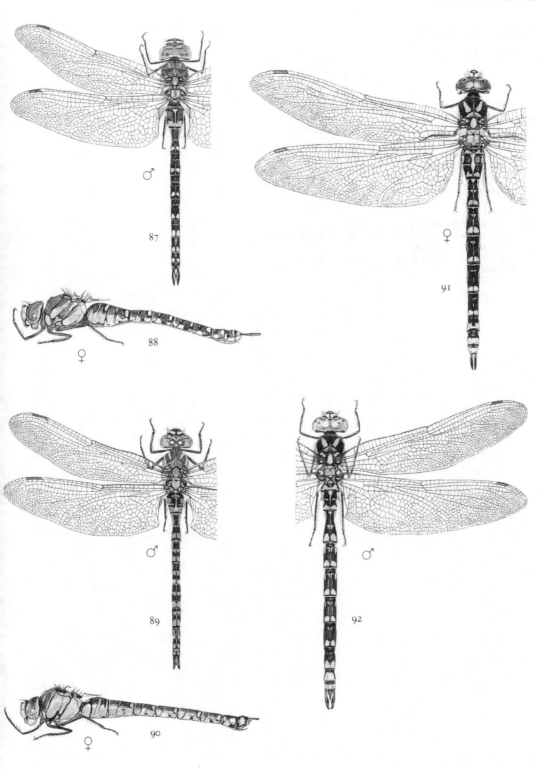

♂

87

♀

88

♀

91

♂

89

♂

92

♀

90

Plate 16: Aeshnidae

Figs 93–97, x 1.08

93 *Aeshna viridis* Eversmann ♂
    USSR. 5.ix.1902, Morton coll., Royal Museum of Scotland
    (*Page* 110)
94 *Aeshna viridis* Eversmann ♀ in lateral view
    Kaliningrad (Königsberg), USSR. 11.ix.1902, Morton
    coll., Royal Museum of Scotland (*Page* 110)
95 *Aeshna grandis* (Linnaeus) ♂
    Hawes Water, Westmorland. ix.1979, M. A. Kirby
    (*Page* 111)
96 *Aeshna grandis* (Linnaeus) ♀ in lateral view
    St Moritz, Switzerland. viii.1981, R. R. Askew (*Page* 111)
97 *Aeshna isosceles* (Müller) ♂
    Canton Zurich, Switzerland. 3–10.vii.1904, K. J. Morton,
    Royal Museum of Scotland (*Page* 111)

Plate 16: Aeshnidae (x 1.08)

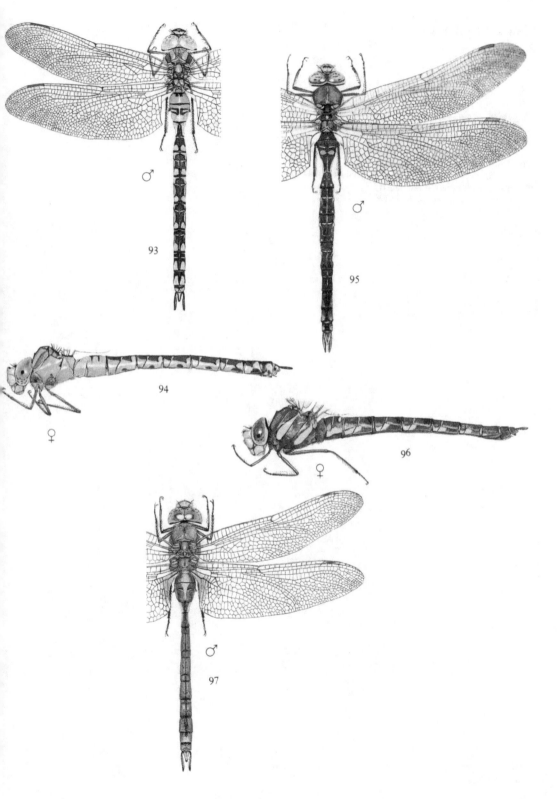

♂

93

♂

95

♀

94

♀

96

♂

97

Plate 17: Aeshnidae

Figs 98–101, x 1.08

98 *Anax imperator* Leach ♂
    Cillibion, Gower, Wales. vii.1965, R. R. Askew (*Page* 114)

99 *Anax imperator* Leach ♀
    Santa Cruz, Madeira. 16.iv.1981, D. R. Askew (*Page* 114)

100 *Anax parthenope* (Sélys) ♂
    Camargue, France. 14.viii.1981, R. R. Askew (*Page* 115)

101 *Hemianax ephippiger* (Burmeister) ♂
    Kardamena, Kos, Greece. 13.iv.1982, R. R. Askew
    (*Page* 116)

Plate 17: Aeshnidae (x 1.08)

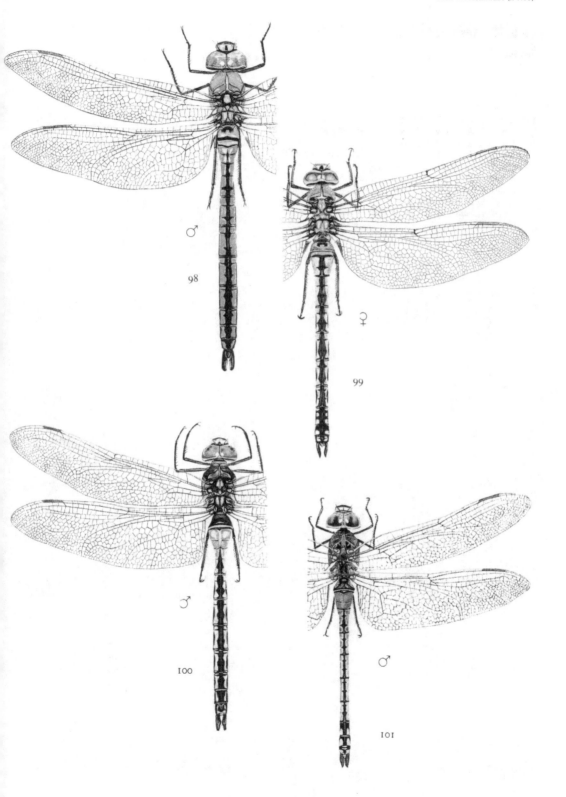

98

♂

99

♀

100

♂

101

♂

Plate 18: Aeshnidae

Figs 102–105, x 1.08

102 *Brachytron pratense* (Müller) ♂
Anglesey, Wales. 6.vi.1980, R. R. Askew (*Page* 117)

103 *Brachytron pratense* (Müller) ♀
Sine loco. 11.vi.1934, Royal Museum of Scotland (*Page* 117)

104 *Caliaeschna microstigma* (Schneider) ♂
Ankara, Turkey. 26.vi.1962, K. M. Guichard, Royal
Museum of Scotland (*Page* 118)

105 *Boyeria irene* (Fonscolombe) ♂
La Force, Dordogne, France. 11.viii.1978, R. R. Askew
(*Page* 119)

Plate 18: Aeshnidae (x 1.08)

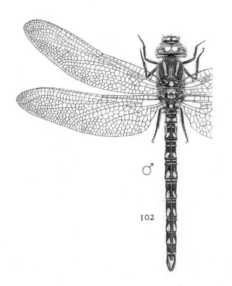

102 ♂

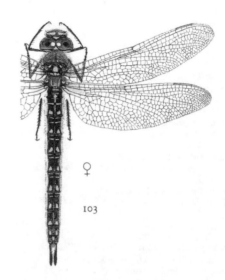

103 ♀

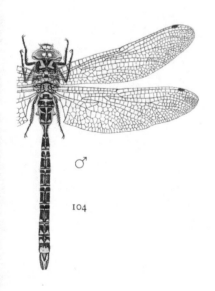

104 ♂

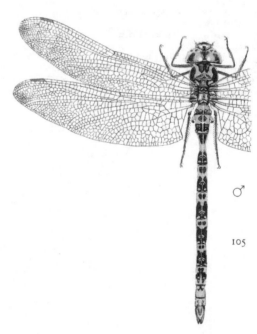

105 ♂

Plate 19: Gomphidae

Figs 106–117, x 1.08

106 *Gomphus flavipes* (Charpentier) ♂
Arles, Bouches du Rhône, France. 2.vii.1911,
K. J. Morton, Royal Museum of Scotland
(*Page* 123)

107 *Gomphus flavipes* (Charpentier) ♀ in lateral view
Vienna, Austria. Morton coll. Royal Museum of Scotland
(*Page* 123)

108 *Gomphus vulgatissimus* (Linnaeus) ♀
Loches, Indre et Loire, France. 25.v.1980, R. R. Askew
(*Page* 125)

109 *Gomphus vulgatissimus* (Linnaeus) ♂
La Force, Dordogne, France. vi.1978, R. R. Askew
(*Page* 125)

110 *Gomphus simillimus* Sélys ♂
Arles, Bouches du Rhône, France. 2–10.vii.1911,
K. J. Morton, Royal Museum of Scotland (*Page* 125)

111 *Gomphus simillimus* Sélys ♀ in lateral view
La Bugue, Dordogne, France. 26.vi.1931, K. J. Morton,
Royal Museum of Scotland (*Page* 125)

112 *Gomphus schneideri* Sélys ♀ in lateral view
Mauri, Kocaeli, Turkey. 27.v.1921, P. P. Graves, British
Museum (Natural History) (*Page* 126)

113 *Gomphus schneideri* Sélys ♂
Diakoflo, Peloponnese, Greece. 26.v.1939, F. C. Fraser
bequest, British Museum (Natural History) (*Page* 126)

114 *Gomphus pulchellus* Sélys ♂
St Georges Blancaneix, Dordogne, France. vi.1978,
R. R. Askew (*Page* 127)

115 *Gomphus pulchellus* Sélys ♀ in lateral view
Data as above (*Page* 127)

116 *Gomphus graslini* Rambur ♀ in lateral view
Le Blanc, Indre, France. Morton coll., Royal Museum of
Scotland (*Page* 127)

117 *Gomphus graslini* Rambur ♂
Lalinde, Dordogne, France. 14.viii.1978, R. R. Askew
(*Page* 127)

Plate 19: Gomphidae (x 1.08)

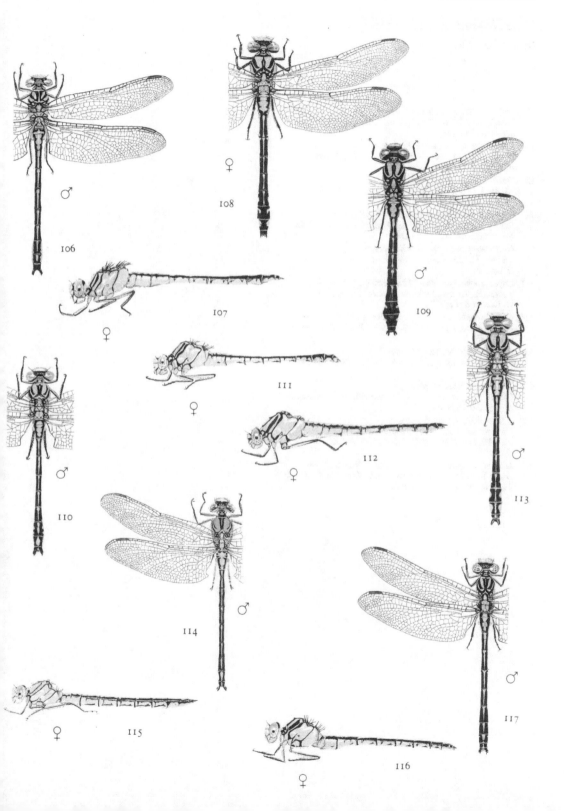

106 ♂

108 ♀

107 ♀

109 ♂

111 ♀

110 ♂

112 ♀

113 ♂

114 ♂

115 ♀

116 ♀

117 ♂

Plate 20: Gomphidae

Figs 118–128, x 1.08

118 *Paragomphus genei* (Sélys) ♂
Taiz, Yemen. 29.iv.1971, D. Davidson, Royal Museum of
Scotland (*Page* 128)

119 *Paragomphus genei* (Sélys) ♀ in lateral view
Wadi Sahama, Yemen. 16.ix.1962, G. Popov, Royal
Museum of Scotland (*Page* 128)

120 *Ophiogomphus cecilia* (Fourcroy) ♂
Nieder-Lausitz, East Germany. Morton coll. Royal
Museum of Scotland (*Page* 130)

121 *Ophiogomphus cecilia* (Fourcroy) ♀ in lateral view
Reuver, Netherlands. 30.vi.1924, M. A. Lieftinck, Royal
Museum of Scotland (*Page* 130)

122 *Onychogomphus forcipatus* (Linnaeus) ♂
Lalinde, Dordogne, France. 14.viii.1978, R. R. Askew
(*Page* 132)

123 *Onychogomphus forcipatus* (Linnaeus) ♀ in lateral view
Tursac, R. Vézere, Dordogne, France. 17.viii.1979,
D. R. Askew (*Page* 132)

124 *Onychogomphus uncatus* (Charpentier) ♀ in lateral view
La Force, Dordogne, France. 29.viii.1981, R. R. Askew
(*Page* 134)

125 *Onychogomphus uncatus* (Charpentier) ♂
La Force, Dordogne, France. 4.viii.1978, D. R. Askew
(*Page* 134)

126 *Onychogomphus costae* Sélys ♂
Sobradiel, Spain. 1.vii.1904, Morton coll., Royal Museum
of Scotland (*Page* 134)

127 *Lindenia tetraphylla* (Vander Linden) ♂
Hammar Lake, R. Euphrates, Iraq. 18.v.1918,
P. A. Buxton, Morton coll., Royal Museum of Scotland
(*Page* 135)

128 *Lindenia tetraphylla* (Vander Linden) ♀ in lateral view
Basra, Iraq. 5.vi.1926, Brevitt Taylor, Morton coll., Royal
Museum of Scotland (*Page* 135)

Plate 20: Gomphidae (x 1.08)

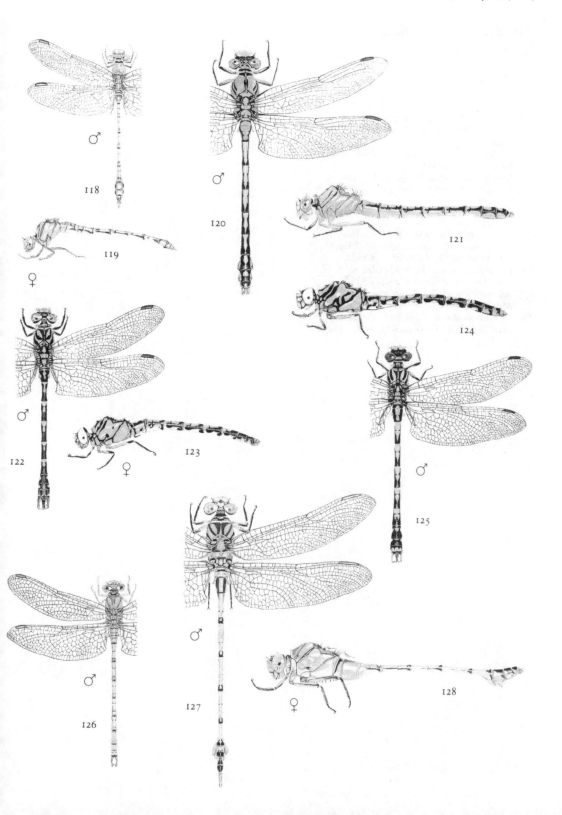

118 ♂

119 ♀

120 ♂

121

122 ♂

123 ♀

124

125 ♂

126 ♂

127 ♂

128 ♀

Plate 21: Cordulegastridae

Figs 129–136, x 1.08

129 *Cordulegaster boltoni* (Donovan) ♂
Monestier, Dordogne, France. 23.vii.1977, D. R. Askew
(*Page* 138)

130 *Cordulegaster boltoni* (Donovan) ♀ in lateral view
Data as above (*Page* 138)

131 *Cordulegaster bidentata* Sélys ♀ in lateral view
Gèdre, Hautes-Pyrenees, France. 18.vii.1977, R. R. Askew
(*Page* 143)

132 *Cordulegaster boltoni immaculifrons* Sélys & Hagen ♂ abdomen
After Fraser (1929; Basses Alpes) (*Page* 138)

133 *Cordulegaster boltoni algirica* Morton ♂ abdomen
After Fraser (1929) from the type (*Page* 138)

134 *Cordulegaster princeps* Morton ♂ abdomen
After Fraser (1929) from the type (*Page* 139)

135 *Cordulegaster picta* Sélys ♂ abdomen
After Waterston (1976; as *charpentieri*) (*Page* 139)

136 *Cordulegaster heros* Theischinger ♂ abdomen
After Theischinger (1979) from the type (*Page* 140)

Plate 21: Cordulegastridae (x 1.08)

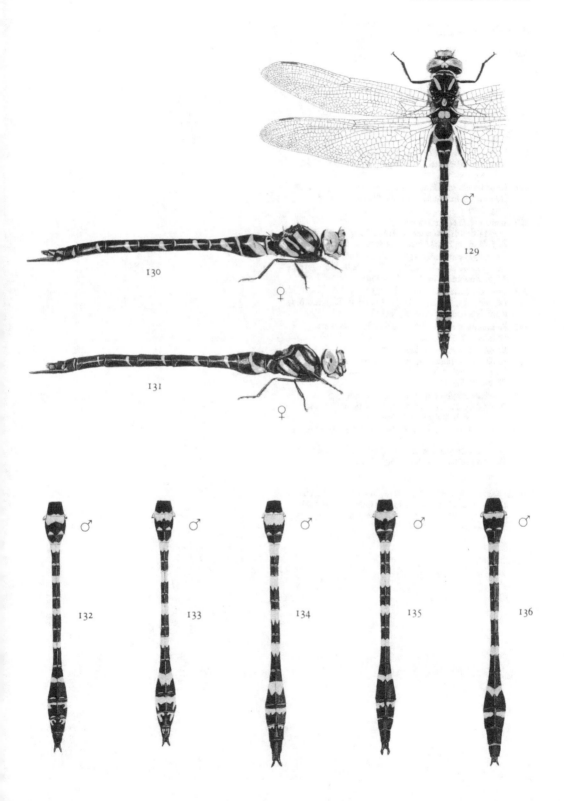

♂

129

130

♀

131

♀

132 ♂

133 ♂

134 ♂

135 ♂

136 ♂

Plate 22: Corduliidae

Figs 137–146, x 1.08

137 *Cordulia aenea* (Linnaeus) ♂
St Georges Blancaneix, Dordogne, France. 30.v.1979,
R. R. Askew (*Page* 145)

138 *Somatochlora metallica* (Vander Linden) ♂
St Georges Blancaneix, Dordogne, France. 18.viii.1978,
R. R. Askew (*Page* 149)

139 *Somatochlora metallica* (Vander Linden) ♀ in lateral view
St Moritz, Switzerland. 23.viii.1981, R. R. Askew
(*Page* 149)

140 *Somatochlora alpestris* (Sélys) ♂
San Martino di Castrozza, Trentino, Italy. 31.vii.1928,
K. J. Morton, Royal Museum of Scotland (*Page* 150)

141 *Somatochlora alpestris* (Sélys) ♀ in lateral view
Virtasalmi, Finland. 8.vii.1931, K. J. Valle, Morton coll.,
Royal Museum of Scotland (*Page* 150)

142 *Somatochlora arctica* (Zetterstedt) ♂
Lenzerheide, Switzerland. 11–17.vii.1904, K. J. Morton,
Royal Museum of Scotland (*Page* 151)

143 *Somatochlora arctica* (Zetterstedt) ♀ in lateral view
Data as above (*Page* 151)

144 *Somatochlora sahlbergi* Trybôm ♂
Parklimo, Petsamo, Lapponica Fennica (now USSR).
30.vii.1929, K. J. Valle, Morton coll., Royal Museum of
Scotland (*Page* 151)

145 *Somatochlora flavomaculata* (Vander Linden) ♂
Bourg d'Oisans, Isère, France. 13.vii.1925, K. J. Morton,
Royal Museum of Scotland (*Page* 152)

146 *Somatochlora flavomaculata* (Vander Linden) ♀ in lateral
view
Oerlikon, Zurich, Switzerland. 7.vii.1887, Morton coll.,
Royal Museum of Scotland (*Page* 152)

NOTE: *Somatochlora borisi* Marinov ♂, in dorsal view, from
Diavolorema, Thracia, north-east Greece, 25.v.2001,
is illustrated on the front cover. Colour photograph by
B. Grebe (see also *Page* 216)

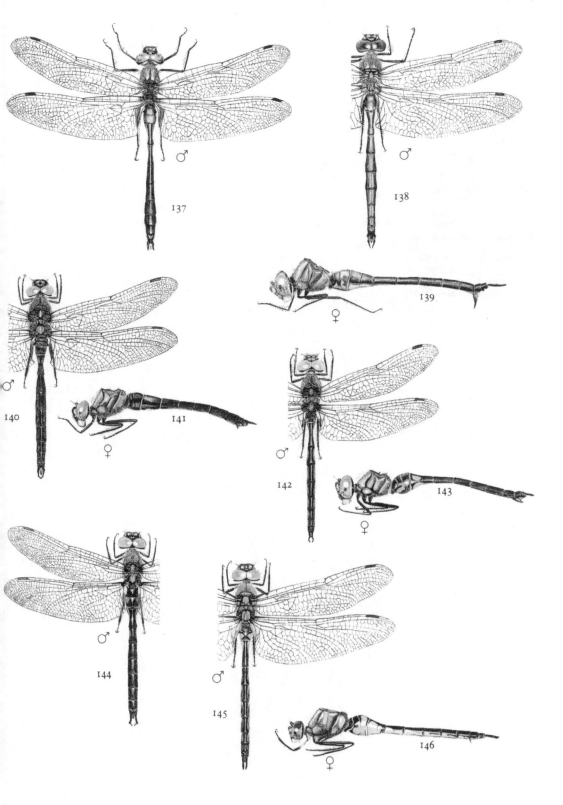

Plate 22: Corduliidae (x 1.08)

137

138

139

140

141

142

143

144

145

146

Plate 23: Corduliidae

Figs 147–152, x 1.08

147 *Epitheca bimaculata* (Charpentier) ♀ in lateral view
     Sine loco. Morton coll., Royal Museum of Scotland
     (*Page* 153)

148 *Epitheca bimaculata* (Charpentier) ♂
     Janōw, Poland. 31.v.1910, Morton coll., Royal Museum of
     Scotland (*Page* 153)

149 *Oxygastra curtisi* (Dale) ♂
     Albas, Lot, France. 3.viii.1979, D. R. Askew (*Page* 154)

150 *Oxygastra curtisi* (Dale) ♀
     Luzech, Lot, France. 3.viii.1979, R. R. Askew (*Page* 154)

151 *Macromia splendens* (Pictet) ♀ in lateral view
     Cahors, Lot, France. 7.vii.1923, K. J. Morton, Royal
     Museum of Scotland (*Page* 155)

152 *Macromia splendens* (Pictet) ♂
     Cahors, Lot, France. 30.vi.1924, K. J. Morton, Royal
     Museum of Scotland (*Page* 155)

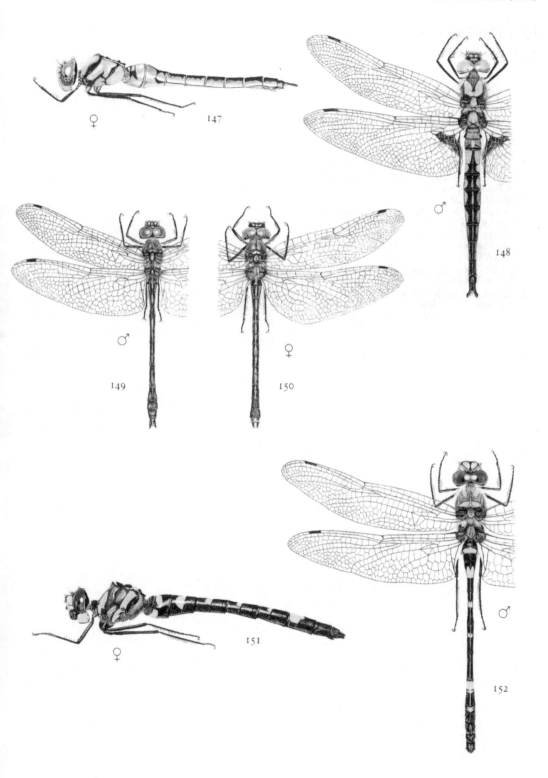

Plate 23: Corduliidae (x 1.08)

♀
147

♂
148

♂
149

♀
150

♀
151

♂
152

Plate 24: Libellulidae

Figs 153–158, x 1.08

153 *Libellula quadrimaculata* Linnaeus ♂ and left wings of form
   praenubila
   Whixall Moss, Shropshire, England. vi.1975, R. R. Askew
   (*Page* 158)
154 *Libellula fulva* Müller ♂
   St Georges Blancaneix, Dordogne, France. 17.viii.1978,
   D. R. Askew (*Page* 159)
155 *Libellula fulva* Müller ♀
   Hurn, New Forest, Hampshire, England. 15.vii.1951,
   P. S. Corbet (*Page* 159)
156 *Libellula depressa* Linnaeus ♂
   Woodchester Park, Gloucestershire, England. vii.1976,
   R. R. Askew (*Page* 160)
157 *Libellula depressa* Linnaeus ♀ abdomen of old specimen
   Data as above (*Page* 160)
158 *Libellula depressa* Linnaeus ♀
   Data as above (*Page* 160)

Plate 24: Libellulidae (x 1.08)

153

154

156

155

157

158

Plate 25: Libellulidae

Figs 159–166, x 1.08

159 *Orthetrum trinacria* (Sélys) ♂
Sakaraha, Madagascar. 20.iii.1968, K. M. Guichard, Royal
Museum of Scotland (*Page* 161)

160 *Orthetrum trinacria* (Sélys) ♀ in lateral view
Data as above (*Page* 161)

161 *Orthetrum chrysostigma* (Burmeister) ♂
Muscat, Oman. 30.iii.1977, F. J. Walker, Royal Museum of
Scotland (*Page* 163)

162 *Orthetrum chrysostigma* (Burmeister) ♀ in lateral view
Kebili, Tunisia. 16.iv.1984, R. R. Askew (*Page* 163)

163 *Orthetrum cancellatum* (Linnaeus) ♀
Duras, Lot et Garonne, France. 29.vii.1976, D. R. Askew
(*Page* 163)

164 *Orthetrum cancellatum* (Linnaeus) ♂
Luzech, Lot, France. 3.viii.1979, R. R. Askew (*Page* 163)

165 *Orthetrum albistylum* (Sélys) ♂
St Georges Blancaneix, Dordogne, France. 18.viii.1978,
R. R. Askew (*Page* 164)

166 *Orthetrum albistylum* (Sélys) ♀
Data as above (*Page* 164)

Plate 25: Libellulidae (x 1.08)

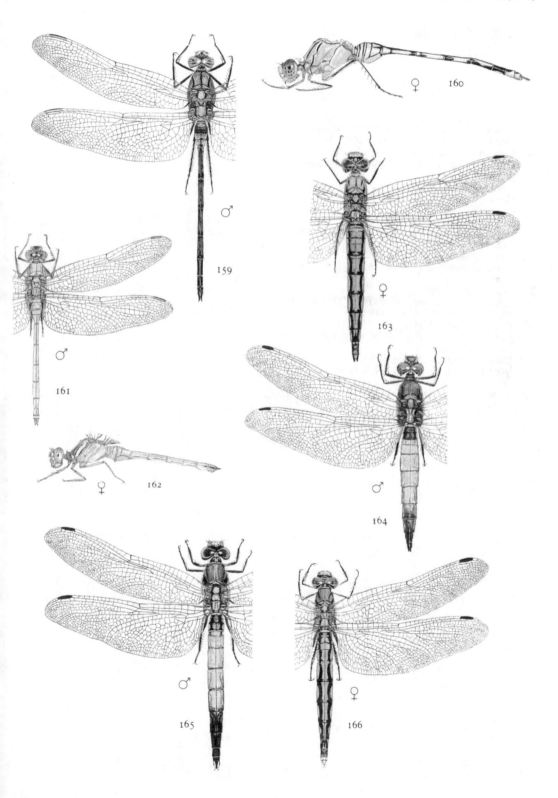

Plate 26: Libellulidae

Figs 167–175, x 1.08

167 *Orthetrum nitidinerve* (Sélys) ♂
Téniet-al-Haâd, Algeria. 17.vi.1904, M. E. Fountaine,
Morton coll., Royal Museum of Scotland (*Page* 165)

168 *Orthetrum brunneum* (Fonscolombe) ♂
Duras, Lot et Garonne, France. 2.viii.1977, R. R. Askew
(*Page* 166)

169 *Orthetrum brunneum* (Fonscolombe) ♀ in lateral view
Duras, Lot et Garonne, France. viii.1979, D. R. Askew
(*Page* 166)

170 *Orthetrum coerulescens* (Fabricius) ♂
St Georges Blancaneix, Dordogne, France. 18.viii.1978,
R. R. Askew (*Page* 167)

171 *Orthetrum coerulescens* (Fabricius) ♀
Duras, Lot et Garonne, France. 21.vii.1977, D. R. Askew
(*Page* 167)

172 *Orthetrum ramburi* (Sélys) ♂
Samothráki, Thráki, Greece. 16.viii.1962, K. M. Guichard,
Royal Museum of Scotland (*Page* 168)

173 *Diplacodes lefebvrei* (Rambur) ♂
Muscat, Oman. 30.iii.1977, F. J. Walker, Royal Museum of
Scotland (*Page* 169)

174 *Brachythemis leucosticta* (Burmeister) ♀
Lake George, Uganda. 19.i.1970, N. Morgan, Royal
Museum of Scotland (*Page* 170)

175 *Brachythemis leucosticta* (Burmeister) ♂
Data as above (*Page* 170)

Plate 26: Libellulidae (x 1.08)

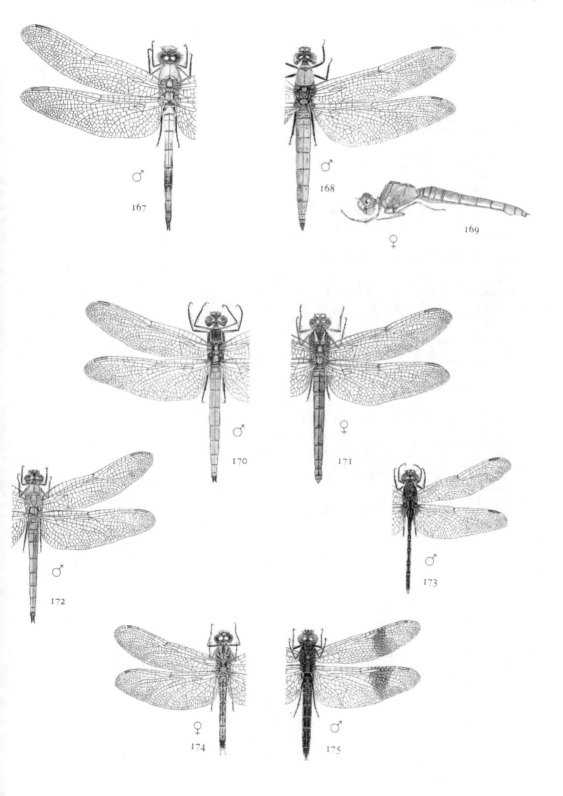

Plate 27: Libellulidae

Figs 176–188, x 1.08

176 *Crocothemis erythraea* (Brullé) ♂
Lapeyrade, Landes, France. 11.viii.1978, R. R. Askew
(*Page* 171)

177 *Crocothemis erythraea* (Brullé) ♀
St Georges Blancaneix, Dordogne, France. 18.viii.1978,
R. R. Askew (*Page* 171)

178 *Sympetrum vulgatum* (Linnaeus) ♂
Leende, Netherlands. 26.viii.1951, P. S. Corbet (*Page* 177)

179 *Sympetrum vulgatum* (Linnaeus) ♀ in lateral view
Irkutsk, Siberia, USSR. 23.viii.1980, R. R. Askew
(*Page* 177)

180 *Sympetrum striolatum* (Charpentier) ♀ in lateral view
Lapeyrade, Landes, France. 6.viii.1978, R. R. Askew
(*Page* 176)

181 *Sympetrum striolatum* (Charpentier) ♂
Swansea, Wales. ix.1977, R. R. Askew (*Page* 176)

182 *Sympetrum striolatum* (Charpentier) ♀
Oxwych, Gower, Wales. 20.ix.1977, R. R. Askew
(*Page* 176)

183 *Sympetrum nigrescens* Lucas ♀ in lateral view
Hachon Lochan, Tongue, Sutherland, Scotland. viii.1978,
A. Whitelaw (*Page* 177)

184 *Sympetrum meridionale* (Sélys) ♀ in lateral view
Tournon, Indre et Loire, France. 31.vii.1979, R. R. Askew
(*Page* 178)

185 *Sympetrum fonscolombei* (Sélys) ♂
Sigoules, Dordogne, France. 21.vii.1977, D. R. Askew
(*Page* 180)

186 *Sympetrum fonscolombei* (Sélys) ♀ in lateral view
Paros, Cyclades, Greece. ix.1980, A. Quayle (*Page* 180)

187 *Sympetrum flaveolum* (Linnaeus) ♀ in lateral view
St Nicholas, Lake Baikal, Siberia, USSR. 24.viii.1980,
R. R. Askew (*Page* 180)

188 *Sympetrum flaveolum* (Linnaeus) ♂
Wisley, Surrey, England. 9.ix.1955, A. E. Gardner, Corbet
coll. (*Page* 180)

Plate 27: Libellulidae (x 1.08)

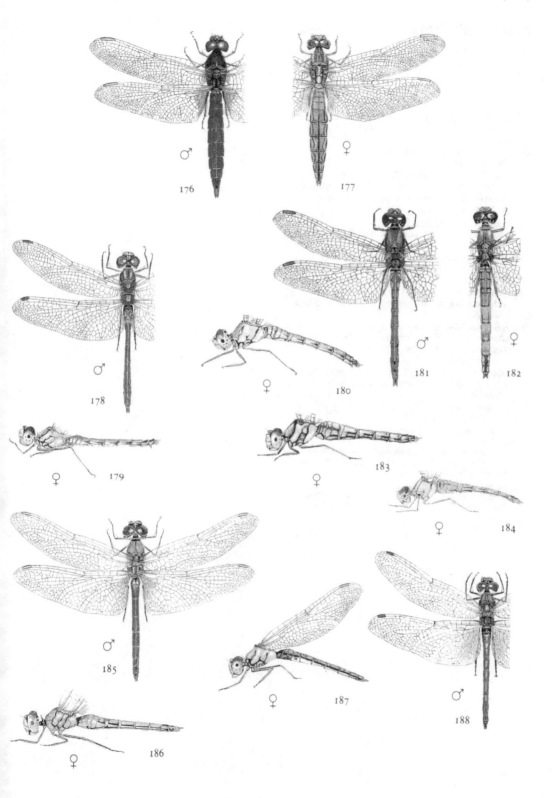

176 ♂

177 ♀

178 ♂

179 ♀

180 ♀

181 ♂

182 ♀

183 ♀

184 ♀

185 ♂

186 ♀

187 ♀

188 ♂

Plate 28: Libellulidae

Figs 189–200, x 1.08

189 *Sympetrum sanguineum* (Müller) ♀ in lateral view
Lapeyrade, Landes, France. 6.viii.1978, R. R. Askew
(*Page* 181)

190 *Sympetrum sanguineum* (Müller) ♂
Data as above (*Page* 181)

191 *Sympetrum depressiusculum* (Sélys) ♂
Leende, Netherlands. 26.viii.1951, P. S. Corbet (*Page* 182)

192 *Sympetrum depressiusculum* (Sélys) ♀ in lateral view
Data as above (*Page* 182)

193 *Sympetrum danae* (Sulzer) ♀
Chat Moss, Manchester, England. ix.1979, R. R. Askew
(*Page* 182)

194 *Sympetrum danae* (Sulzer) ♀ in lateral view
Acharacle, Argyllshire, Scotland. viii.1961, R. R. Askew
(*Page* 182)

195 *Sympetrum danae* (Sulzer) ♂
Chat Moss, Manchester, England. ix.1979, R. R. Askew
(*Page* 182)

196 *Sympetrum pedemontanum* (Allioni) ♂
Gravedona, Lake Como, Italy. 21.viii.1925, K. J. Morton,
Royal Museum of Scotland (*Page* 183)

197 *Leucorrhinia albifrons* (Burmeister) ♀
'Katzens' (= Katzenbuckel, Odenwald, West Germany?).
2.vi.1884, Morton coll., Royal Museum of Scotland
(*Page* 186)

198 *Leucorrhinia albifrons* (Burmeister) ♂
Data as above (*Page* 186)

199 *Leucorrhinia caudalis* (Charpentier) ♀ (form with apical wing
marks)
Weissen See, Märkische Schweiz, Berlin, East Germany.
7.vi.1931, Erich Schmidt, Morton coll., Royal Museum of
Scotland (*Page* 186)

200 *Leucorrhinia caudalis* (Charpentier) ♂
Sortavala, Finland (now USSR). 1.vii.1932, L. Tiensun,
Morton coll., Royal Museum of Scotland (*Page* 186)

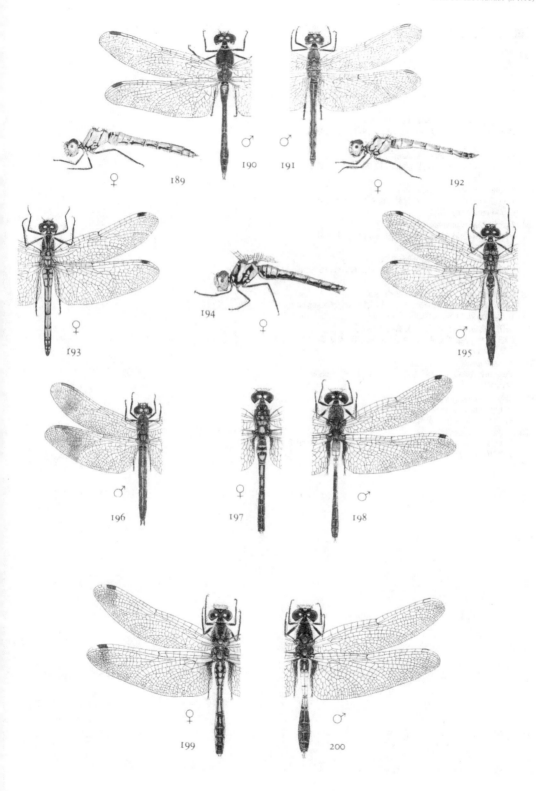

Plate 28: Libellulidae (x 1.08)

Plate 29: Libellulidae

Figs 201–210, x 1.08

201 *Leucorrhinia dubia* (Vander Linden) ♂
Delamere, Cheshire, England. 10.vii.1977, R. R. Askew
(*Page* 187)

202 *Leucorrhinia dubia* (Vander Linden) ♀
Abbots Moss, Cheshire, England. 12.vii.1979, R. R. Askew
(*Page* 187)

203 *Leucorrhinia rubicunda* (Linnaeus) ♂
Locality illegible. 26.v.1910, Morton coll., Royal Museum
of Scotland (*Page* 188)

204 *Leucorrhinia rubicunda* (Linnaeus) ♀
Braunschweig, West Germany. Morton coll., Royal
Museum of Scotland (*Page* 188)

205 *Leucorrhinia pectoralis* (Charpentier) ♀
Janów, Poland. 13.v.1906, Morton coll., Royal Museum of
Scotland (*Page* 189)

206 *Leucorrhinia pectoralis* (Charpentier) ♂
Hamburg, West Germany. 9.vii.1896, F. Ris, Morton coll.,
Royal Museum of Scotland (*Page* 189)

207 *Trithemis annulata* (Palisot de Beauvois) ♂
Immouzer, Agadir, Morocco. 2.vi.1983, R. R. Askew
(*Page* 189)

208 *Pantala flavescens* (Fabricius) ♂
Dagomys, Georgia, USSR. 23.vii.1982, R. R. Askew
(*Page* 191)

209 *Selysiothemis nigra* (Vander Linden) ♂
Al Qurnah, R. Tigris, Iraq. 17.v.1910, P. A. Buxton,
Morton coll., Royal Museum of Scotland (*Page* 193)

210 *Zygonyx torridus* (Kirby) ♂
Taiz, Yemen. 29.iv.1971, D. Davidson, Royal Museum of
Scotland (*Page* 191)

Plate 29 Libellulidae (x 1.08)

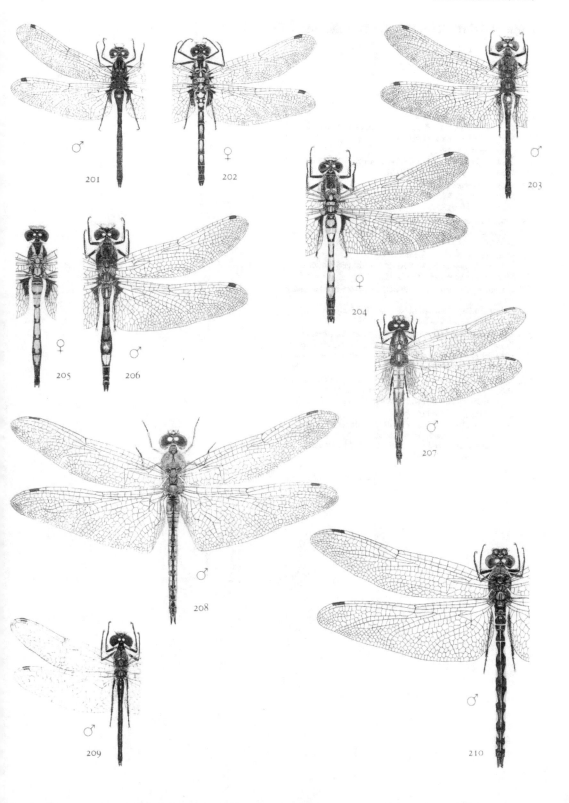

Plate 30 (Supplement): Aeshnidae and Libellulidae

Figs 211–219, x 1.08

211 *Anax immaculifrons* Rambur ♂
No locality stated. 13.xi.1920, Morton coll., National
Museums of Scotland (*Page* 216)

212 *Anax junius* (Drury) ♂
Bluffton, Indiana, USA. 17.v.1903, Morton coll., National
Museums of Scotland (*Page* 214)

213 *Orthetrum sabina* (Drury) ♂
Muscat, Oman. 1.iv.1977, F. J. Walker, National Museums
of Scotland (*Page* 217)

214 *Orthetrum sabina* (Drury) ♀ in lateral view
Oman. 18.x.1978, F. J. Walker, National Museums of
Scotland (*Page* 217)

215 *Orthetrum taeniolatum* (Schneider) ♂
Akrounda, Cyprus. 26.x.1930, ex Staudinger, National
Museums of Scotland (*Page* 217)

216 *Orthetrum taeniolatum* (Schneider) immature ♂ in lateral
view
Sikkim, India, 4300 feet, no date given. Lindgren leg.,
National Museums of Scotland (*Page* 217)

217 *Sympetrum sinaiticum tarraconensis* Jödicke immature ♂
Gava, Barcelona, Spain. viii.1906, Museo Nacional de
Ciencias Naturales, Madrid, loaned courtesy of A. Compte-
Sart (*Page* 213)

218 *Sympetrum sinaiticum tarraconensis* Jödicke ♂ in lateral view
Ribesalbes, Spain. 9.x.1996, drawn partly after a
photograph by Jens Kählert of a living insect (*Page* 213)

219 *Trithemis festiva* (Rambur) ♂
Akrounda, Cyprus. 27.x.1930, ex Staudinger, National
Museums of Scotland; drawn partly after a photograph by
Antoine van der Heijden of a living insect, Dalyan area,
Turkey, 25.v.2000 (*Page* 217)

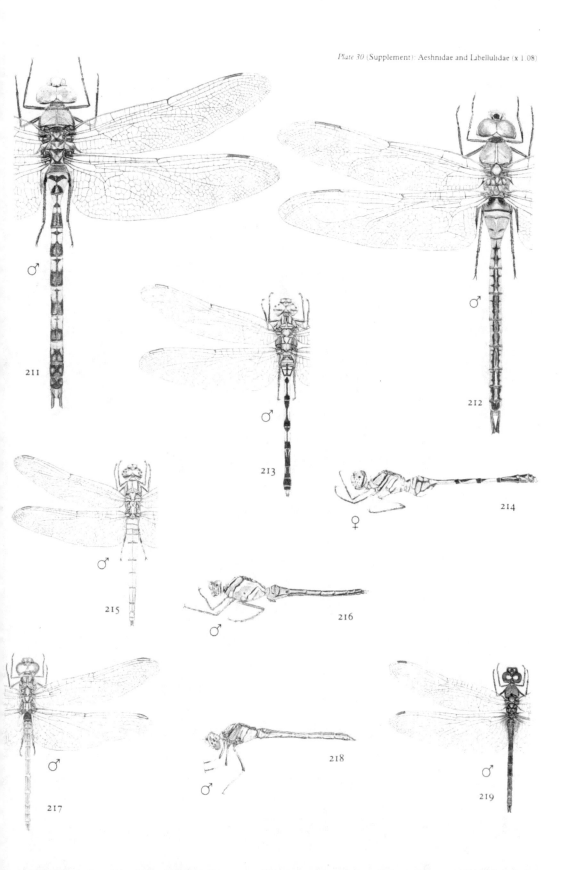

Plate 30 (Supplement): Aeshnidae and Libellulidae (x 1.08)

♂

211

♂

213

♂

215

♀

216

♂

212

214

♂

217

218

♂

219

♂

303

Printed in the United States
by Baker & Taylor Publisher Services